Springer Transactions in Civil and Environmental Engineering

Springer Transactions in Civil and Environmental Engineering (STICEE) publishes the latest developments in Civil and Environmental Engineering. The intent is to cover all the main branches of Civil and Environmental Engineering, both theoretical and applied, including, but not limited to: Structural Mechanics, Steel Structures, Concrete Structures, Reinforced Cement Concrete, Civil Engineering Materials, Soil Mechanics, Ground Improvement, Geotechnical Engineering, Foundation Engineering, Earthquake Engineering, Structural Health and Monitoring, Water Resources Engineering, Engineering Hydrology, Solid Waste Engineering, Environmental Engineering, Wastewater Management, Transportation Engineering, Sustainable Civil Infrastructure, Fluid Mechanics, Pavement Engineering, Soil Dynamics, Rock Mechanics, Timber Engineering, Hazardous Waste Disposal Instrumentation and Monitoring, Construction Management, Civil Engineering Construction, Surveying and GIS Strength of Materials (Mechanics of Materials), Environmental Geotechnics, Concrete Engineering, Timber Structures.

Within the scopes of the series are monographs, professional books, graduate and undergraduate textbooks, edited volumes and handbooks devoted to the above subject areas.

More information about this series at https://link.springer.com/bookseries/13593

B. V. Venkatarama Reddy

Compressed Earth Block & Rammed Earth Structures

B. V. Venkatarama Reddy
Department of Civil Engineering & Centre
for Sustainable Technologies
Indian Institute of Science
Bangalore, India

ISSN 2363-7633 ISSN 2363-7641 (electronic)
Springer Transactions in Civil and Environmental Engineering
ISBN 978-981-16-7876-9 ISBN 978-981-16-7877-6 (eBook)
https://doi.org/10.1007/978-981-16-7877-6

This Springer imprint is published by the registered company Springer Nature Singapore Pte Ltd.
The registered company address is: 152 Beach Road, #21-01/04 Gateway East, Singapore 189721, Singapore

Dedicated to my parents, wife Annapurna
and the kids Dileep and Jayanth

Foreword

Building with earth (also referred to variously as clay, mud and (sub)soil construction) takes many forms, with a history as long as human settlement, and a continued use to this day longer and more extensive than most other construction materials. Traditional forms of earth building include adobe (sun-dried mud bricks), cob (layers of mud and straw stacked in-situ in layers), rammed earth (layers of soil compacted inside formwork), as well as decorative internal plasters and masonry mortars.

Low mechanical resistance and a susceptibility to water deterioration of traditional earth building materials are two of the reasons why over the past 50+ years stabilised earth construction techniques have grown in popularity across the world, including parts of North America, Australia, across Africa, as well as India and other Asian countries. Stabilised earthen materials have proven able to provide safe, affordable and sustainable buildings.

Over the past four decades, Prof. Venkatarama Reddy has established himself as the world's foremost engineering expert on stabilised earth construction. Through research and development, education, training and stakeholder engagement, as well as building standard development, Prof. Venkatarama Reddy has played a leading role in the growth and reputation of stabilised earth construction in India and beyond.

This book, focusing two of the most popular forms of modern stabilised earth construction—compressed earth blocks and rammed earth—brings together, for the very first time, unequalled and collected research and professional experience for the benefit of a wide readership. Students and practitioners of engineering and architecture, as well as builders and developers, but also accessible to others interested in exploring alternative sustainable forms of construction, will benefit from its content. Across 17 chapters, the book provides comprehensive coverage from fundamental materials to material performance characteristics, manufacture and construction technologies, structural design, as well as considering life cycle impacts. The latest technologies and developments, such as geopolymer stabilisation, are also covered. Enjoy!

June 2021

Prof. Pete Walker
Director, BRE Centre for Innovative
Construction Materials
University of Bath
Bath, UK

Preface

The word "earth" means the planet on which we live and also the substance of the land surface: the soil—a layer of earth. The words "soil" and "earth" are commonly used phrases in the context of earthen structures. When the humans attempted to shed the nomadic life, need arose for the construction of the shelters. The construction techniques were evolved for utilising the locally available soil or the earth for the dwellings. There are footprints of earthen structures in all the main cradles of the civilisations. The earthen materials have nearly zero carbon footprint and complete recyclability with zero environmental costs. With the advent of the modern construction materials and the construction techniques after the Industrial Revolution, the developments as well as use of earthen constructions took a back seat. The modern building materials are energy intensive and consume unsustainably extracted mined raw materials from the planet earth threatening the sustainability of the living habitats on the planet. The soil or the earth provides scope for devising low carbon and sustainable options to build the modern earthen structures.

The traditional earth constructions suffer from strength loss on moisture absorption and durability against adverse environmental actions. Such problems can be mitigated through the process of stabilisation using inorganic binders and industrial by-products. The stabilised earth roads and pavements were popular in the early part of the twentieth century. The lime stabilised rammed earth can be seen in the centuries old famous Chinese tulou houses and rammed earth used in Alcazaba Cadima in Granada, Spain (eighth century AD).

Since the last 5–6 decades, immense technological developments can be noticed on the stabilised earth construction. Research and innovation into earthen construction materials and technologies for earthen structures has grown significantly in recent years fostering the development of relevant codes of practices dealing with diverse functional performances of the buildings, viz. materials, durability, climatic-response, indoor air quality, acoustics, aesthetics and environment. The stabilised earth building technologies are being commercially exploited across the globe, and there is a considerable stock of modern stabilised earth buildings in many countries. There is a great deal of interest among many professionals, organisations and individuals in utilising the stabilised earth technology for the buildings.

The stabilised Compressed Earth Block (CEB) and rammed earth form the core of discussions in the book. The topics covered are mainly supported by scientific data generated over four decades of rigorous R&D, helped by the doctoral research programmes, sponsored research and the experience gained over technology dissemination activities leading to a large number of CEB and rammed earth buildings and structures. The structural design aspects presented on stabilised CEB masonry and rammed earth walls shed more light on the basic design principles for the structural elements using the new materials and illustrated through design examples. The novel materials using industrial by-products and new binders (geopolymers), promotion of circular economy demonstrated through recycling of natural clay minerals from the stabilised earth products and the concept of sustainable materials are elaborated in greater detail. The book will serve as a very useful document for researchers, teachers and students of engineering and architecture, professionals, builders and the individuals. The author will be grateful if the discerning readers point out any shortcomings and mistakes in the book.

Bengaluru, India
June 2021

B. V. Venkatarama Reddy

Acknowledgements

The scientific contributions from many of my research students culminated into compilation of this book. It is difficult to find students willing to work on research topics pertaining to earth construction, recycled materials and energy in buildings. I was lucky to be associated with such research scholars: Mr. Ajay Gupta, Mr. Richardson Lal, Prof. P. Prasanna Kumar, Prof. M. S. Latha, Prof. M. Muttharam, Ms. Anitha M., Dr. K. I. Praseeda, Dr. S. N. Ullas, Dr. K. Gourav, Dr. N. C. Balaji, Dr. Lepakshi Raju, Dr. R. K. Preethi, Dr. V. Vibha, Mr. Ch. V. Uday Vyas and Dr. R. Sri Bhanupratap Rathod. The knowledge I have accumulated over the last four decades, on earth construction, energy and sustainable materials, would not have been possible but for the efforts of these research scholars. I wish to acknowledge my beloved students' efforts for allowing me to use the information from their theses and the published work.

My colleague Dr. K. S. Nanjunda Rao co-supervised few students, and I used to bank upon him for the advice on numerical analysis work. I wish to place on record his intellectual support during the last three decades. I sincerely thank Prof. Monto Mani for the assistance extended in reviewing the chapter on sustainability of construction materials. During the course of the book compilation, I used to encounter gaps in the data represented in many illustrations and these gaps were filled by the experimental research work carried out by research associates Dr. H. N. Abhilash, Mr. V. Nikhil, Mr. M. Nikhilash, Mr. Vishwas Raj and Mr. Vadhiraj. Their efforts are sincerely acknowledged. I appreciate and thank Dr. S. N. Ullas and Mr. Vishwas Raj for the help extended in preparing several sketches and drawings. My mentor Prof. K. S. Jagadish was passionate about research into earthen materials and advised me to pursue research in the exciting area of earthen materials and structures. I am indebted to him for the valuable advise which took me into the exciting world of mechanics of earthen materials, where I spent major part of my over four decades of academic life. Last but certainly not the least, acknowledgement must go to my wife, Annapurna, who relieved me from domestic chores and bore the brunt of my long hours of absence at home, almost every day including weekends, where I used to spend time in the laboratory.

The sponsored research grants were essential to pursue R&D into earthen materials and the dissemination of the ideas to demonstrate the earth building technologies. These activities were supported by the research grants from the Department of Science and Technology (Government of India), Karnataka State Council for Science and Technology, University Grants Commission, Indo-French collaborative Projects (CEFIPRA), UK-India Project (UKIERI) initiative, H. T. Parekh Foundation under CSR initiative of HDFC limited and the Indian Institute of Science, Bangalore. I greatly acknowledge this financial assistance in fostering research and developing, and dissemination of many alternative building technologies.

B. V. Venkatarama Reddy

About This Book

This is the first ever comprehensive seminal book on stabilised earth construction. The book is a result of highly focused R&D and dissemination work on the stabilised compressed earth blocks and the rammed earth over a period of four decades at the Department of Civil Engineering and the Centre for Sustainable Technologies, Indian Institute of Science, Bangalore, India. The book provides insights into fundamental aspects of soils and soil stabilisation, static soil compaction, principles of designing the machine for the compressed earth blocks, in-depth analysis on the characteristics of stabilised Compressed Earth Blocks (CEBs), earth mortars, the CEB masonry, geopolymer stabilised CEB and the concepts on utilising non-organic solid wastes for the CEB. The characteristics of stabilised rammed earth and behaviour of stabilised rammed earth under compression, tension and shear are dealt in greater detail. The structural design of the stabilised CEB masonry and the rammed earth buildings with design examples is a unique feature of the book. The later part of the book deals with recycling, embodied energy and embodied carbon of stabilised CEB and the rammed earth; the book ends with a thought-provoking aspect of sustainability of construction materials and green buildings. The book will serve as a useful document for the researchers, design professionals, students and teachers of engineering and architecture, professionals and builders, and the individuals.

Contents

About the Author

B. V. Venkatarama Reddy was a Professor at the Department of Civil Engineering and Chairman, Centre for Sustainable Technologies, Indian Institute of Science (IISc), India. Professor Reddy's research interests include structural masonry, mechanics of materials, energy in buildings, green buildings, low carbon construction materials, geopolymers, recycling solid wastes, circular economy and sustainable construction. Professor Reddy has over four decades of R&D experience into earthen materials and earthen structures, and mentored many research scholars. His translational R&D was of direct relevance to the society and the industry, and practiced by many professionals resulting in over millions of tonnes of carbon savings. Apart from publishing over 100 papers, jointly authored a book on alternative building materials and technologies, he has edited several books on masonry and materials. Professor Reddy has served as consultant for several innovative projects on low-carbon and alternative building technologies, and served as a member of several technical committees in Bureau of Indian Standards and other state and central government agencies. He was DAAD Visiting professor at Bauhaus University Germany and Visiting Professor at University of Bath, UK.

Part I
Introduction to Earth Construction and Soil Stabilisation

Chapter 1
Earthen Materials and Earthen Structures

1.1 History of Earthen Structures

Earthen structures are spread across the world, and still a large population is living in earthen dwellings. History of building with earth or soil is not documented systematically. There are many attempts to document and display the examples of earthen structures in the past. As we dig more into the history of construction, we encounter more and more examples of earthen structures. When the humans first decided to shed their nomadic and hunter-gatherer lifestyle and settle down at one place, need arose for the buildings and other structures. Locally available soil, stones and biomass were the obvious choice for the construction materials. There are evidences for earthen buildings and structures in all the main cradles of the civilisations (Fig. 1.1): the Indus Valley Civilisation, civilisation of Mesopotamia and Babylonia, the civilisations of Egypt, China, Mexico and Peru. Why our ancestors used soil for the building construction? There are many arguments highlighting the points in favour of soil/earth-based constructions: the advantages of cost, easy and ready availability of local soil, and lack of access to transport systems and energy to process the materials.

Construction of dwellings and other structures by our ancestors commenced soon after the shedding of highly nomadic life maybe 10,000 years ago. The early civilisations dependent on agriculture for their livelihood started dwelling in river valleys, e.g. Nile river valley, Indus Valley, etc. Generally, these river valleys are rich in transported fertile soils and may lack in the availability of stones and also, difficulty in breaking stones to smaller sizes. In such situations, soil becomes the material of choice for the construction.

Turkey's central Anatolian city of Konya, Çatalhöyük, is among the oldest settlements of the Neolithic Age, dating back to 7400–5200 BC, where earthen houses can be seen (Fig. 1.2). The unbaked earthen structures in Mohenjo-daro city in the Indus Valley date back to 3rd millennium BC (CRAterre 2012). The stupa mound and several other major structures show the earthen structures implemented on a massive

B. V. V. Reddy, *Compressed Earth Block & Rammed Earth Structures*,
Springer Transactions in Civil and Environmental Engineering,
https://doi.org/10.1007/978-981-16-7877-6_1

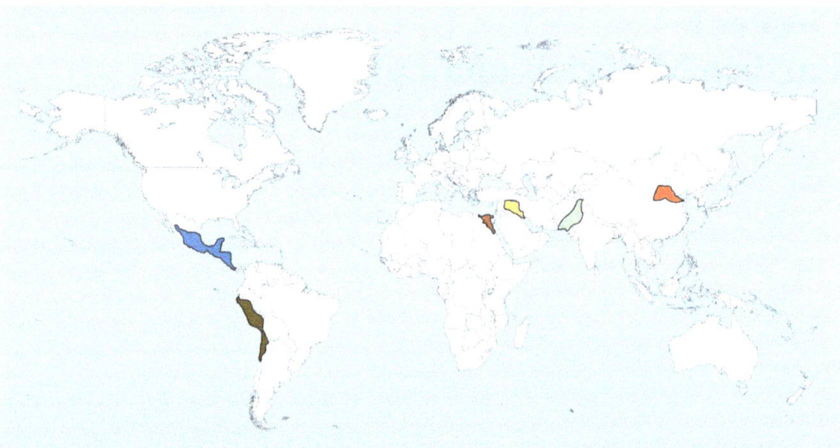

Fig. 1.1 Main cradles of civilisations (Indus Valley, Mesopotamia and Babylonia, civilisations of Egypt, China, Mexico and Peru) (prepared by H. N. Abhilash)

Fig. 1.2 Archaeological excavation site, Anatolia Turkey (7400–5200 BC) (*Source* https://www.flickr.com/photos/catalhoyuk/albums/72157647113315030)

scale (Fig. 1.3). The Bam Citadel (Fig. 1.4) located in the city of Bam in Iran is the largest adobe brick building in the world. There is no precise archaeological dating of the buildings of the Citadel of Bam. The first human settlements in this area are dated back to fourth–sixth century BC. The Citadel construction activities can be

Fig. 1.3 Stupa mound and other structures, 3rd millennium BC (*Source* https://whc.unesco.org/en/list/138/gallery/)

Fig. 1.4 Ancient Bam citadel and surroundings (575–300 BC), (Courtesy: Prof. Ing. Aleš Kocourek)

traced back to twelfth century AD and lasting for several centuries (CRAterre 2012, https://en.wikipedia.org). One surviving section of the Great Wall of China, in the Shandong Province, is made of rammed earth and is estimated to be 2,500 years old. The Great Wall of China, near Jiayuguan in Gansu Province, was built using the unfired clay, the locally available material. The techniques of rammed earth as well as adobe blocks have been adopted (Anger and Fontaine 2009) (Fig. 1.5). Very thick adobe block masonry vaults at the Ramesseum, Luxor, Egypt, were built about 3,500 years ago (Association la Voûte Nubienne 2015). These are called Nubian vaults (Fig. 1.6).

The Ait Ben Haddou is the best-known example of earthen architecture of southern Morocco (Fig. 1.7). These are mainly rammed earth buildings, with some parts of the structures built with adobe block masonry. The earth constructions in these regions date back to eleventh century AD to the present (CRAterre 2012). The Shibam

Fig. 1.5 Section of the Great Wall of China built around third century BC (*Source* © Gwydion Williams Flickr, Creative Commons, CRAterre)

Fig. 1.6 Adobe block vaults at the Ramesseum, Luxor, Egypt, built ~3,500 years ago (*Source* © Thierry Joffroy, CRAterre)

Fig. 1.7 Traditional earth houses in Ait Benhaddou Qsar, southern Morocco (picture by Maureen from Buffalo, USA—Ait Benhaddou Qsar, CC BY 2.0, https://commons.wikimedia.org/w/index.php?curid=73697816)

city in Yemen has impressive tower houses up to seven-storey height (~30 m high) nicknamed "the Manhattan of the Desert" (Fig. 1.8). This is a sixteenth-century walled city and is the pioneering example on the principle of vertical construction. These structures were built entirely of adobe block masonry (CRAterre 2012). In

Fig. 1.8 Tall structures with load bearing adobe block masonry in Shibam city, Yemen, built in sixteenth century AD (picture by Jaysegul Tastaban, © CRAterre)

1968, Riyadh in Saudi Arabia had 46% of residential buildings constructed with earth (Mubarak 1999). Earth has been used throughout central Arabia Najd, in both the sand desert areas and the fertile valleys, much of the interior of Yemen and Oman, extending northwards into Iraq and the Syrian Desert (King 1998). Culturally, Arab population is attached and familiar in living in different forms of earth shelters. The great religious leaders used mud brick (libin) for the residences and mosque in Medina in AD 622 (Mortada 2016; King 1998).

Germany has examples of earthen buildings since 4500 BC. Traditional earth construction techniques such as cob, wattle and daub, adobe block masonry and rammed earth were practised in Germany (Schroeder 2016). The Weilburg a. d. Lahn town in Germany has 200 rammed earth buildings built during eighteenth–nineteenth century (Schick 1987). Figure 1.9 shows the famous six-storey load bearing rammed earth wall building built in 1826. This is a remarkable structure designed and built during the period when the robust structural design methods were non-existent. The building wall thickness varies from 750 mm on the ground floor to 300 mm at the top floor. The rammed earth construction thrived in France. France has more than half a million rammed earth houses. Figure 1.10 shows a traditional rammed earth building in the Rhone Valley. Francois Cointeraux's publications (Cointeraux 1791) on rammed earth in Lyon in 1791 paved the way for the rammed earth technique spreading across Europe and USA.

Fig. 1.9 Six-storey unstabilised rammed earth building, Weilburg a. d. Lahn, Germany (built in 1826)

Fig. 1.10 Rammed earth building, Rhone Valley, France

Jiyao and Weitung's (1990) compilation mentions the recorded history of earth construction since the 2100–1600 BC. There is a mention of 30,000 earth buildings, dating mostly from the Ming (1368–1644 AD) and Qing (1644–1911 AD) dynasties, in the Fujian Province (https://www.eartharchitecture.org). The Hakka rammed earth dwellings called Tulou (Fig. 1.11) were built with thick walls incorporating escape passages and enclosed in rectangular or circular form. These structures were mainly designed for defence instead of climatic conditions. They were 3–4 storeys high and can accommodate more than 500 people. The oldest Tulou was built in 1308 AD, and the construction of such structures continued till twentieth century (Jaquin 2012). The rammed earth walls of some of the Hakka buildings have considerable quantity of lime, indicating that there is some evidence of lime stabilisation in rammed earth structures (Liang et al. 2011; Stanislawski 2011). The thickness of the rammed earth walls in the Hakka buildings was in the range of 1.5–2.0 m (Liang et al. 2013). The Great Mosque of Djenné (Fig. 1.12) located in Djenné, Mali, dates back to 1907. The walls are with adobe block masonry and provided with earth plaster that gives the building its smooth and sculpted look. The building walls are decorated with bundles of Rodier palm sticks, projecting about 600 mm from the surface.

Use of earth building techniques such as rammed earth, adobe block masonry and stone with earth mortar can be found in Tibetan Plateau and Himalayan region. Many examples of rammed earth constructions can be seen in Ladakh in India, Nepal and Bhutan. Figure 1.13 shows 200 year-old four-storey residential structure in Bhutan.

Fig. 1.11 Fujian Tulou houses (picture by Basile Cloquet © CRAterre)

Fig. 1.12 Great Mosque of Djenné, Mali (*Source* © Thierry Joffroy, CRAterre)

Fig. 1.13 Four-storey rammed earth house in Bhutan, wall thickness at the bottom: 1.35 m (built in eighteenth century, picture in 2009)

Even now, the unstabilised rammed earth construction is in vogue for building the houses in Bhutan. Apart from the adobe masonry earthen structures in the Harappa and Mohenjo-daro civilisations (3000 BC), all the major forms of earth construction techniques (cob, adobe, wattle and daub, rammed earth and laterite block) can be seen, spreading across the Indian subcontinent. The 1971 and 2001 India census shows 61 and 74 million earthen dwellings in India, respectively. This represents 49 and 30% of the housing stock in 1971 and 2001, respectively (http://www.censusindia.net). Rammed earth constructions are mainly restricted to the Himalayan region and the state of Rajasthan. Predominantly, wattle and daub constructions are found in north-eastern states, laterite block masonry in the west coast, adobe and cob structures in the rest of the Indian states. Figure 1.14 shows ruins of a palace structure built during sixteenth–seventeenth century AD, with cob amidst heaps of granite rocks on a hilltop, and Fig. 1.15 shows a masonry with special shape adobe blocks (hand made without using any mould) in the palace adjacent area.

Spain has many historical rammed earth structures. The World Heritage Site of Alhambra Palace was constructed using rammed earth around 1238 AD. The Royal Palace of Cordoba in the city of Cordoba is another notable example of rammed earth construction in Spain. Royal rammed earth (1 lime: 3 earth) walls constructed in eighth century can be found in Alcazaba Cadima in Granada (Fig. 1.16) (Arango Gonzalez 1999).

Fig. 1.14 Ruins of a cob wall palace, Chitradurga Fort, Karnataka state, India (built during sixteenth–seventeenth century AD, cob walls, picture in 2018)

Fig. 1.15 Special shape adobe block (hand made without using any mould) masonry, Nayaka rulers' palace, Chitradurga Fort, Karnataka, India (built in seventeenth century AD, picture in 2018)

Fig. 1.16 Royal rammed earth in Alcazaba Cadima, Granada, Spain (*Source* https://commons.wikimedia.org/wiki/File:Alcazabas_de_Granada.JPG)

Apart from the massive, monumental structures such as Pyramid of Sun (at Teotihuacan 200 AD), native Americans in Mexico and southern states used earth as building material. McHenry (1989) provides brief history of earth construction and a pictorial overview of adobe and rammed earth structures in USA and the world. The Pueblo people-built adobe structures spread in the current New Mexico state (thirteenth–fourteenth century AD) (https://en.wikipedia.org/wiki/Taos_Pueblo). Figure 1.17 shows multi-tier and multi-storey adobe Pueblo de Taos in New Mexico state. The California, Arizona, New Mexico and Colorado states witnessed many old adobe and rammed earth buildings built during 1800–1900 AD. Figure 1.18 shows an adobe structure in Sonoma, California, built in early 1900.

Fig. 1.17 Pueblo de Taos, New Mexico, USA, adobe buildings built in thirteenth–fourteenth century (picture by John Mackenzie Burke—Own work, CC BY-SA)

Fig. 1.18 Adobe block masonry quarters, Sonoma, California, USA, in the early 1900s (picture: in 2015)

1.2 Earthen Structures in Twentieth–Twenty-First Century

There is a noticeable surge of interest in the construction of earthen buildings since the 1950s. Large numbers of new earthen buildings have been built since the last 6–7 decades. Many R&D professionals and several universities are involved across the world in the development and the standardisation of earthen materials and techniques. The basic stimulus for this development is the desire to reduce environmental impact of constructions while exploring more sustainable building methods using natural and low embodied carbon materials. The R&D efforts can be seen in (a) developing methods and materials for the repair and conservation of the existing earthen structures, (b) refining the formwork, standardising mix proportions and construction process for the rammed earth and (c) emergence of the compressed earth block technology. Large number of publications on rammed earth and the compressed earth block technology, in the recent past, is a testimony to the rigorous R&D work pursued across the world. Also, some attempts can be seen in the learning and the standardisation of cob construction process/technique.

The development and dissemination of the modern earthen structures can be grouped under two broad categories: (a) the unstabilised earth structures and (b) the stabilised earth structures. The natural soil/earth has clay minerals, which control the characteristics of the natural soil, mainly causing swelling and shrinkage due to the moisture movement. The natural soil characteristics should be modified mainly to minimise the swelling and shrinkage of the earthen building product. The first simple step in achieving this is by blending the natural soil with the inert materials such as sand, gravel and aggregates such that the clay content of the mix comes to an optimum level. This method of modifying the soil characteristics is termed as mechanical stabilisation. The mechanical stabilisation is common to both the unstabilised and the stabilised earth construction techniques. In the case of stabilised earth construction techniques, in addition to the mechanical stabilisation inorganic stabilisers such as cement and lime are used. The stabilised earth products possess higher strength (especially in the wet condition) and better durability characteristics.

1.2.1 Unstabilised Earthen Structures in Twentieth–Twenty-First Century

Many professionals, builders and research groups are involved in promoting the unstabilised earth structures since the last 6–7 decades. There are efforts in repairing, retrofitting and conservation of the old earthen structures. Easton (1996) provides a summary of developments and details of unstabilised rammed earth constructions during 1920–1970 in USA. The Department of Agriculture published in 1926 a Farmer's Bulletin No. 1500 on rammed earth walls for buildings giving technical information and construction methods. The literature shows more than 100 articles on rammed earth in the journals and the magazines during 1926–1950 (Patty

1936). The academic research work on optimum soil grading in controlling erosion of unstabilised rammed earth was pursued during 1935–1945 (Patty 1936; Patty and Minium 1945). There are many mechanised commercial adobe block production industries in the New Mexico state of USA (Wilson 2008). Exploiting the thermal and humidity control features of the adobe block masonry, 400–600 potato cellars (barns) were built during 1900–1960 using the adobe masonry in the San Luis Valley, USA (Wilson 2012). Hassan Fathy with his two Nubian masons built Dar Al-Islam village community mosque in New Mexico (in 1980), using adobe bricks and without using formwork for the adobe masonry vaults and domes (Abdel-Moniem 2001).

Germany witnessed adobe and rammed earth construction in twentieth century. In order to meet the severe shortage of housing post-world wars, the government in former East Germany imposed rules to build the houses using natural and locally available materials. Construction of buildings using adobe masonry, rammed earth and timber frame filled with earthen materials can be noticed after 1945 (Fig. 1.19). An estimate shows 18,000 rammed earth buildings were built in the year 1948 (Röhlen and Ziegert 2011). Figure 1.20 shows a reconstructed rammed earth building (in 2008) in Xanten Archaeological Park using a traditional historic shuttering. The building shows burnt clay brick lining at crucial points such as eves drops, where there is a possibility of flashing water from the eve projections. The Chapel of Reconciliation

Fig. 1.19 Multi-family apartment building using rammed earth in Muechel near Merseburg, Germany, built in the 1950s (picture *Source* https://images.app.goo.gl/FxmJSH5zCc7YgbE66)

Fig. 1.20 Reconstructed rammed earth building (in 2008) in Xanten Archaeological Park, Germany

in Berlin Mitte (Germany) rebuilt in 1999 is a classic example of the unstabilised rammed earth walls utilising the local materials. Figure 1.21 shows an unstabilised curved rammed earth wall for a hospital structure in Germany.

The UK has many earthen structures built using unstabilised earth and construction techniques such as cob and rammed earth after the 1970s. Several individuals, professionals, organisations and universities are responsible for the development, construction, dissemination and education on earthen materials and structures. Some of the unstabilised rammed earth structures include the Eden project in Cornwall (1999), the main building of Mount Pleasant Ecological Park (2003), Pines Calyx Conference Centre, Kent (2005), Atrium at the Rivergreen Centre near Durham (2005), WISE lecture theatre, Centre for Alternative Technology Wales (Hall and Swaney 2012), and few of the other cob wall and rammed earth structures are shown in Figs. 1.22, 1.23, 1.24 and 1.25. France has a huge stock of earthen structures. R&D into earth construction, restoration and construction of unstabilised earth buildings persists in France. The unstabilised rammed earth constructions are prominently seen in France (Figs. 1.26 and 1.27). Terra Europae (2011), a compilation on the earthen architecture in the European Union, gives a good overview of the earthen structures in the Europe. Otto Kapfinger and Marko Sauer (2015) edited monograph compiles Martin Rauch's rammed earth projects executed during 1998–2014. The rammed earth projects spread across Europe. Some of the prominent rammed earth building works include Swiss Ornithological Institute at Sempach (2013–2014, Fig. 1.28), Ricola Kräuter Zentrum at Laufen, Switzerland (2012–2013, Fig. 1.29), Mezzana

Fig. 1.21 Curved unstabilised rammed earth wall, Central Hospital Suhl, Germany

Fig. 1.22 Unstabilised circular rammed earth wall, WISE lecture theatre, Centre for Alternative Technology, Wales, UK (*Source* https://cat.org.uk/info-resources/free-information-service/building/wise-building/)

Fig. 1.23 Unstabilised rammed earth external wall, Eden Project Visitor Centre, Cornwall, UK (picture in 2009)

Agricultural College at Coldrerio, Switzerland (2010–2012), Cinema Sil Plaz at Ilanz/Glion, Switzerland (2009–2010), House of Rauch at Schlins, Austria (2005–2008), Sihlhölzli Sports Complex at Zurich (2001–2002), Chapel of Reconciliation at Berlin (1999–2000) and many other residential buildings.

The China has a huge stock of earthen buildings still being occupied. There is a keen interest among the researchers, architects and engineers in developing the methods to build traditional earthen buildings, especially rammed earth. The number of new earthen buildings built in the last few decades is limited (Hu and Liu 2015). Millions of active earthen buildings exist in India. Figure 1.30 shows the status of earthen dwellings in India. The census data shows 65–75 million earthen dwellings in India during the period 1971–2011. The bulk (~90%) of these dwellings are in the rural areas. Though the number of earthen dwellings remains constant, the percentage

Fig. 1.24 Cob wall buildings, Devon, UK (built by: Kevin McCabe, picture in 2009)

Fig. 1.25 Cob bus shelter, Cornwall, UK (built by: Jackie Abey and Jill Smallcombe, picture: in 2009)

of these buildings with reference to the total housing stock decreased from 51 to 22% during the period 1971–2011. The housing stock in India has increased from 123 to 305 million during 1971–2011. The construction of unstabilised earthen structures is stagnant since the last couple of decades, except for a few enthusiastic professionals attempting to build such structures and to a very limited extent constructions in the

Fig. 1.26 A factory built in 1870 at Saint-Siméon-de-Bressieux, in south-eastern France, unstabilised rammed earth walls, currently rehabilitated into a residential building complex, inset: exposed rammed earth (picture by H. N. Abhilash in 2014)

rural areas. Figure 1.31 shows a cluster of cob buildings in a village situated in the heavy precipitation region of the Indian west coast.

Centuries old residential, educational and commercial earthen buildings built were repaired, restored and brought back to use in Yazd region of Iran (Farahza et al. 2012). Eike Klinge et al. (2016) narrate the repaired and reconstructed rammed earth works of three important projects: Celestial Stairs, Golden Spiral and City of Orion in Morocco during 1985–2003. Africa has a great tradition in the construction of earthen buildings, and bulk of the rural African population is still inhabited in earthen dwellings. Hassan Fathy devoted his entire lifetime in mastering and practising earthen architecture in the twentieth century. Fathy advocated the use of local materials, training masons to solve housing problems of the poor and the peasants, and promoted the appropriate and sustainable construction practices. Fathy's repeated attempts to build mud brick (sun-dried brick) masonry vaults without formwork for the buildings of Royal Society of Agriculture Farm in Bahtim village near Cairo in 1941 failed. Fathy discovered masons in Nubia (a village west of Aswan, Upper Egypt) building the masonry vaults without using the formwork and respected the vernacular vocabulary and the building techniques of the Nubian villages. The Nubian masons helped in completing the mud brick masonry vaults in the agriculture farm buildings. The famous Fathy's project was planning and construction of

Fig. 1.27 New unstabilised rammed earth housing complex at Saint-Antoine-l'Abbaye, south-eastern France, built in 2014 (picture by Dr. H. N. Abhilash)

Fig. 1.28 Visitor's centre, Swiss Ornithological Institute, Sempach (2013–2014), prefabricated unstabilised rammed earth panels (courtesy: Lehm Ton Erde Baukunst GmbH, Österreich/Austria, www.lehmtonerde.at/alexander-jaquemet)

Fig. 1.29 Ricola Kräuter Zentrum, Laufen, Switzerland (2012–2013), external wall envelope with unstabilised prefabricated rammed earth (courtesy: Lehm Ton Erde Baukunst GmbH, Österreich/Austria, www.lehmtonerde.at/ Markus Bühler)

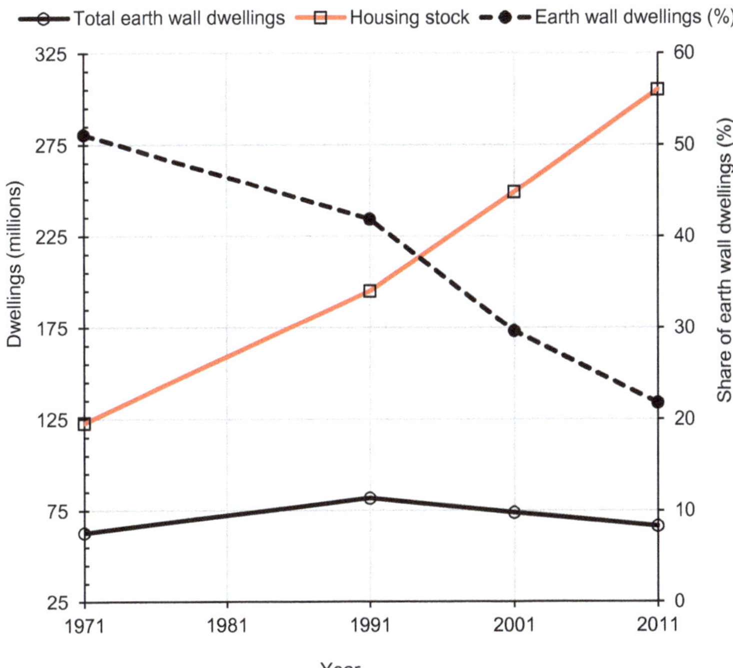

Fig. 1.30 Housing and earthen housing statistics for India

Fig. 1.31 Cluster of cob buildings in Kumta village, west coast of India, where annual precipitation >3000 mm

New Gourna village to rehabilitate 7000 peasants living in the old Gourna village. This project was a large-scale demonstration opportunity for building mud brick masonry vaults and domes during 1945–1948 (Fig. 1.32). Figure 1.33 shows few buildings of New Bariz village, Kharga Oasis, Egypt, executed during 1964–1967. Fathy's many remarkable projects across many continents and his contributions to skill building, education and appropriate technology gave the earthen architecture a great fillip (Abdel-moniem 2001; Fathy 1973). Earth building enthusiasts associated

Fig. 1.32 New Gourna village (1945–48) (*Source* https://whc.unesco.org/en/activities/637/)

Fig. 1.33 New Baris public buildings with the museum in the foreground and the market vaults in the background, Egypt (1964–1967) (*Source* picture in 2010 by Viola Bertini https://cdn.wallpaper.com/main/styles/responsive_1460w_scale/s3/p227_-viola-bertini-2_0.jpg?itok=gCe9wL2F)

with Earth Building Association of New Zealand (EBANZ) have promoted adobe and cob buildings in New Zealand in the last 2–3 decades. New Zealand has brought out three standard codes on earth building (NZS 4297 1998; NZS 4298 1998; NZS 4299 1998).

1.2.2 Stabilised Earth Structures in Twentieth–Twenty-First Century

The natural soil possesses certain distinct characteristics such as swelling, shrinkage, strength, cohesion and internal friction, mainly controlled by the type and quantity of the clay minerals present in the soil. The earthen products such as bricks, blocks, cob and rammed earth derived by using natural soils are expected to perform certain structural functions in a structure. The earthen structural element should possess adequate strength and should be durable. The unstabilised earthen structures lose strength upon saturation and suffer damage when exposed to wet weather conditions. The swelling and shrinkage characteristics of unstabilised earthen products are controlled by mixing inert materials such as sand/aggregates and sometimes natural plant-based fibres. Mixing of the soil and the sand/gravel or mixing of two different soils results in a product which has different characteristics than that of the original soil. This kind of soil constituent's manipulations is termed as mechanical stabilisation. In addition to the mechanical stabilisation, chemical admixtures (such as cement and lime) or biological admixtures are added, and the earthen product is densified through the compaction process. More detailed discussions on soil stabilisation are dealt in the subsequent chapters. A review of the developments in the cement or lime stabilised earthen structures is as follows.

The lime (basically calcium hydroxide) and the lime–pozzolana mixtures have been used in modifying the soil characteristics used for the manufacture of earthen products. Some of the rammed earth Hakka (Tulou) building walls show considerable quantity of lime, indicating the evidence for lime stabilisation in rammed earth in Hakka structures (Daniel Stanislawski 2011; Liang et al. 2013). Lime stabilised rammed earth called royal rammed earth was used in the constructions of Alcazaba Cadima in Granada in eighth century AD (Arango Gonzalez 1999). The walls surrounding Horyuji Temple in Japan have been built using lime stabilised rammed earth during 600–750 AD. Another example of lime stabilised rammed earth wall built in 1610 for Sanjusangen-Do Temple is in Kyoto, Japan (Hall et al. 2012). The stabilised expansive soil deposits exist in many regions of India. These soils have been produced by adopting a stabilisation process involving ageing of the expansive black cotton soil–wood ash-organic matter (leaves and grass) mixture for several decades under ambient climatic conditions (Rao et al. 2000). Such soils have been extensively used for the earth construction in India since many centuries and have been explored for the cement/lime stabilised compressed earth blocks since 1990 (Reddy et al. 2003). The stabilised soil has been used successfully for the construction of roads in different parts of the world since 1935 (Lambe 1962).

The development of the CINVA-Ram press in 1956 in Columbia and the publication of a number of monographs on earth construction (Dept. of HUD 1955; Bulletin 5 1952; Fitzmaurice 1958) encouraged the use of stabilised earth in the buildings. The Portland cement stabilised compressed earth block and rammed earth structures are being built across the globe since the last 6–7 decades. A number of cement stabilised compressed earth block buildings using CINVA-Ram press came up in Columbia, Chile, Venezuela, Bolivia and Brazil (UN Report 1964). Application of cement stabilised rammed earth construction for the buildings can be seen from the 1940s in Arizona and New Mexico states in the USA. Easton started his earth builder journey in 1976 in promoting cement stabilised rammed earth (CSRE) buildings especially in California (Easton 1996). Many CSRE buildings have been built by Easton in California. Easton established an industry to produce hollow compressed earth block (CEB) stabilised with cement and built buildings in the California state using cement stabilised hollow CEBs. There are many individuals and groups involved in the stabilised earth constructions in USA responsible for many CSRE and stabilised

CEB buildings in the states of California, Arizona, Colorado and New Mexico (Easton 1996). Figures 1.34 and 1.35 show CSRE and hollow CEB buildings in California. The North America (Canada) has a limited stock of CSRE and insulated CSRE buildings constructed since the 1990s (Krayenhoff 2012).

Fig. 1.34 Reinforced and cement stabilised rammed earth house, California, USA (built by David Easton, picture in 2015)

Fig. 1.35 Cement stabilised hollow CEB masonry house, California, USA (built by David Easton, picture in 2015)

UK has limited number of cement stabilised rammed earth buildings built since the 1990s in Devon, Norfolk and Derbyshire. France has many stabilised earth buildings. The stabilised earth construction received a fresh impetus after the formation of CRAterre (Centre for the Research and Application of Earth) in Grenoble, France, in 1979. CRAterre was involved in stabilised earth constructions in France, Algeria, Mauritania, Morocco, Ivory Coast, Upper Volta and many other places since 1980. CRAterre initiated a major demonstration of earth housing project, Domaine de la Terre near the village of L'Isle-d' Abeau in France. Sixty-five buildings demonstrating the potential of the cement stabilised compressed earth blocks and the cement stabilised rammed earth were constructed during 1982–1985 (Figs. 1.36, 1.37 and 1.38). Some of them are three-storey load bearing structures. In collaboration with Ecole Nationale des Travaux Publics de 'Etat (ENTPE), many individuals/construction firms have built stabilised rammed earth structures in France'.

In the last 3–4 decades, rammed earth constructions flourished in Australia. Vast majority of the rammed earth buildings in Australia are cement stabilised rammed earth. Since the last 4–5 decades, cement stabilised rammed earth (CSRE) has been used increasingly in Australia, with few thousands of CSRE buildings built. In some regions of Western Australia such as Margret River, >20% of the new home constructions are built with CSRE. Some of the contemporary CSRE buildings include Port Philip Estate Winery, Red Hill in Victoria, TarraWarra Museum of Modern Art,

Fig. 1.36 Cement stabilised compressed earth block load bearing building, Domaine de la Terre at Isle d' Abeau, France (picture: in 2006)

Fig. 1.37 Cement stabilised rammed earth residential building, Domaine de la Terre at Isle d' Abeau, France (picture in 2006)

Fig. 1.38 Three-storey cement stabilised rammed earth building, Domaine de la Terre at Isle d' Abeau, France (picture: in 2006)

Healesville in Victoria, Lauriston Science and Resource Centre in Melbourne, 3–4 storey Kooralbyn Hotel in Queensland, St Francis Xavier College Berwick Victoria, Charles Sturt University Thurgoona in NSW and many other commercial, institutional and residential buildings. Figures 1.39, 1.40 and 1.41 show some of the CSRE buildings in Australia. The Australian earth building professionals follow the bulletin 5 guidelines and EBAA (2005) document for the design of earthen buildings. The "Earthsong" is an innovative urban cohousing in Auckland, New Zealand, built during 2002–2007, using natural materials and the cement stabilised rammed earth (Fig. 1.42).

The number of stabilised rammed earth buildings in China is limited (Hu and Liu 2015). The first cement stabilised rammed earth in China was Just Grapes wine store in Shanghai built in 2006 (Wallis 2012). The buildings for the river house eco-retreat in Zhangjiagang were the first stabilised insulated rammed earth buildings. Southeast Asian countries such as Nepal, Sri Lanka and Bhutan have new modern rammed earth structures. The Bayalpata Hospital Complex in Nepal is a marvellous example of cement stabilised rammed earth buildings (Sharon Davis Design). Five medical buildings, an administrative block, 10 houses and a dormitory form the campus on a hilltop in the Seti River Valley (Fig. 1.43). Cement stabilised CEB and rammed earth buildings were built since the mid-2000s, in many rehabilitation projects affected by 2004 Tsunami in Indian Ocean. Later R&D work continued in Sri Lankan universities on cement stabilised CEB and rammed earth, resulting in several buildings. The prince Ahmad Bin Salman Mosque in Riyadh, Saudi Arabia, was constructed using cement stabilised CEB during 2008–2010 (Bader 2012).

Zami and Mohammad (2012) provide a brief narration of stabilised earth construction in Africa. Many international agencies and individuals assisted in the construction of cement stabilised rammed earth and compressed earth block buildings in many African countries. Assisted by Scientific and Industrial Research Development Council (SIRDC) of Zimbabwe, many cement stabilised rammed earth and CEB

Fig. 1.39 Cement stabilised rammed earth building for St. Thomas Catholic Church, Margret River, Western Australia (picture in 2015)

Fig. 1.40 Cement stabilised rammed earth building, Margret River, Western Australia (picture: in 2015)

Fig. 1.41 Cement stabilised rammed earth residential building, Sydney suburbs, Australia (picture in 2005)

buildings were constructed since the late 1990s. The cement stabilised CEBs have been successfully used for social housing in Sudan (Adam and Agib 2001). El Haj Yousif School in Sudan (Fig. 1.44) and Gando Primary School in Burkina Faso are the attempts to popularise the cement stabilised CEB concepts. South Africa has cement stabilised CEB buildings and interesting vaults using cement stabilised thin CEBs (Fig. 1.45). Hydraform in Johannesburg in South Africa promoted cement stabilised

Fig. 1.42 Earthsong—a residential community complex in Auckland, New Zealand, built using cement stabilised rammed earth (picture in 2005)

Fig. 1.43 Cement stabilised rammed earth (6% cement) buildings at Bayalpata Hospital Medical Complex in Achham, Nepal (photograph by Elizabeth Felicella, courtesy: Sharon Davis Design)

Fig. 1.44 El Haj Yousif School in Sudan, built using cement stabilised CEB (*Source* Adam and Agib 2001)

Fig. 1.45 Cement stabilised CEB tile vault at Mapungubwe in South Africa (https://www.notechmagazine.com/2009/12/timbrel-vaulting-in-south-africa-by-peter-rich-architects.html)

CEB and is responsible for many CEB buildings in Africa and other countries since 1990.

1.2.2.1 Stabilised Earth Constructions in India

The stabilised soils were explored for the construction of roads/pavements in India since 1935 (Swaminathan et al. 1983). The soil–cement was explored for the construction of houses for the workers in the big dam sites (Mitra 1951). The earliest attempt to build houses using soil–cement, in a larger scale in India, was in 1948, to house the influx of people when the country got divided. Four thousand cement stabilised rammed earth houses were built in the states of Punjab and Haryana. The project was backed by three years of earlier research at Lahore and Karnal (Verma et al. 1950). Figure 1.46 shows the picture of one of these rammed earth houses taken in 1981. The soil characteristics used in the rammed earth constructions were controlled, and 2.5% cement was used. The houses were rendered with a layer of cement plaster. The next attempt to use soil–cement in housing was in Bangalore,

Fig. 1.46 Rammed earth house built in 1948 in Karnal, Haryana state, India (picture in 1981 Subhas C Basu)

Fig. 1.47 Handmade cement stabilised earth block house built in 1949, Bangalore city, India (picture in 1985)

India, in 1949. About 260 houses were built using hand rammed earth blocks with 4.0–5.0% cement. The mix composition was adjusted by adding some gravel to the soil (Madhavan and Narasinga Rao 1949). The exposed blockwork masonry walls were given a coat of lime wash. A few of these houses are still in use (Fig. 1.47). These houses had performed well without much maintenance, in the weather conditions of Bangalore city.

The next phase of stabilised earth construction for the buildings was using machine pressed blocks. In the early 1970s, a cement stabilised CEB residential building was built using CEBs produced from CINVA-Ram press in the Bangalore city. The R&D on CEBs commenced at the Indian Institute of Science (IISc), Bangalore, since 1976, through the construction of cement stabilised CEB building in the IISc campus (Fig. 1.48). A South African origin machine (Elson Block Master) was manufactured in India in Gujarat state in the 1970s. Initially, this machine was used to construct CEB buildings in Bangalore and the surrounding regions (Fig. 1.49). The IISc developed manually operated and semi-automatic machines for the production of CEBs since 1980. These machines are now commercially produced and marketed. The construction of cement stabilised CEB buildings continued steadily during two decades since 1985. Both the residential and commercial buildings were built in this period. Figures 1.50, 1.51 and 1.52 illustrate some of these structures. Many individuals, NGOs and other organisations constructed cement stabilised CEB buildings in this two-decade period.

Fig. 1.48 Biogas laboratory, Indian Institute of Science, Bangalore, built with 5% cement stabilised CEB made from CINVA-Ram press in 1976

Fig. 1.49 Library for St. Peter's Seminary, Bangalore city, built in 1976 using 5% cement CEB manufactured using Elson Block Master

Fig. 1.50 Load bearing 230 mm-thick cement stabilised CEB masonry building, KR Puram, Bangalore, India

Fig. 1.51 Centre for Sustainable Technology Office Complex, Indian Institute of Science, Bangalore, built in 1985–86 using 6% cement CEB (designed by Architect Subhas C Basu)

Fig. 1.52 Canteen building, Tamil Nadu, India, built in 1993 using cement stabilised CEB's (designed by Architect Krishna)

The R&D on alternative building materials and technologies, especially on CEB and rammed earth, was rigorously pursued at IISc Bangalore. The IISc work on CEB and rammed earth resulted in large number of journal publications, monographs, books, Indian standard codes, more than a dozen PhDs, many national and international conferences and collaborations. IISc pursued capacity building activities in facilitating CEB and rammed earth technology delivery mechanisms across the country. Architects, engineers, building professionals and academicians were trained through lectures and hands-on training workshops continuously since 1990. The capacity building activities resulted in a large number of professionals commercially exploiting the CEB and rammed earth technology for housing and other structures. Parallelly, few other private entities such as Mrinmayee, Auroville Earth Institute, Biome Solutions, Development Alternatives and GoodEarth are promoting the construction of stabilised CEB and rammed earth structures in India. The spread of the CEB/rammed earth technology across India is depicted in Fig. 1.53. The CEB technology has spread to the neighbouring Asian countries as well. An approximate estimate shows over 150,000 CEB and rammed earth structures in India. Figures 1.54, 1.55, 1.56 and 1.57 show some of the recent buildings using cement stabilised CEB and rammed earth. GoodEarth Eco Homes (https://goodearth.org.in) is one of the large-scale projects adopting cement stabilised CEB technology. Four hundred load bearing masonry villa houses of 1–3 storey height were built in this project. About 6 million cement stabilised CEBs were made at the construction site using manual machines (Reddy et al. 2016). Figures 1.58 and 1.59 show some of the houses in the GoodEarth project.

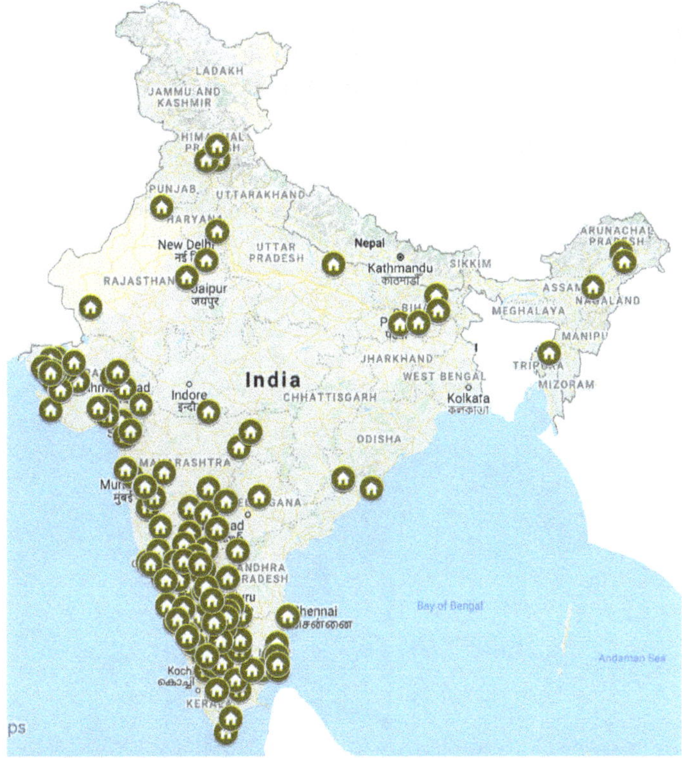

Fig. 1.53 Spread of stabilised earth buildings in India (prepared by Dr. H. N. Abhilash)

Fig. 1.54 Multi-storey load bearing cement stabilised CEB masonry and rammed earth dormitory building, Indian Institute of Science, Challakere Campus, India, built in 2019

Fig. 1.55 C-BELT Complex, Indian Institute of Science, Challakere Campus, India, built in 2019, cement stabilised CEB masonry walls, vaults, domes and jack-arch roof

Fig. 1.56 Residence of Mr. R. C. Manjunath, Bangalore, India, built in 2019 using cement stabilised CEB load bearing masonry (picture by S. N. Ullas)

Fig. 1.57 Residence of the author, Bangalore, India, built in 2019 using cement stabilised CEB load bearing masonry and rammed earth (only the parking space uses reinforced concrete frame with CEB masonry infill), energy-positive building with solar power and rainwater harvesting

Fig. 1.58 An individual cement stabilised CEB masonry wall house at GoodEarth Eco Homes, Bangalore, India

Fig. 1.59 A cluster of cement stabilised CEB masonry houses at GoodEarth Eco Homes, Bangalore, India

1.3 Prefabricated Rammed Earth Construction

The in-situ rammed earth constructions are widely practised, and such construction practices are economical especially when the labour costs are low. Since the last 2–3 decades, there are attempts towards the off-site fabrication of rammed earth wall elements assembling them into a building. Prefabrication potentially allows for better quality control in the factory production, can minimise the construction time and is convenient to carry out rammed earth works in congested places. Rammed earth prefabricated wall elements are heavy, demand heavy specialised equipment for lifting and placing, and also add to the transportation and handling costs. Such operations might increase the cost of prefabricated rammed earth. Prefabrication concepts have been explored for both the stabilised and unstabilised rammed earth constructions.

Otto Kapfinger and Marko Sauer (2015) and Rauch (2007) provide details of some of the prefabricated unstabilised rammed earth buildings and the projects completed using prefabricated rammed earth since 1997. The Kräuter Zentrum in Laufen and the Swiss Ornithological Institutes Visitor Centre in Sempach are the most recent ones using prefabricated rammed earth panels (Figs. 1.28 and 1.29). Figure 1.60 shows the positioning of the prefabricated rammed earth panels. There are few prefabricated rammed earth structures in France (Hall and Swaney 2012). M/s. Rammed Earth Works Group has built prefabricated stabilised rammed earth buildings in USA.

Fig. 1.60 Unstabilised prefabricated rammed earth panels transported and assembled at site (courtesy: Lehm Ton Erde Baukunst GmbH, Österreich/Austria, www.lehmtonerde.at)

1.4 Codes and Norms on Stabilised CEB and Stabilised Rammed Earth

The engineered construction demands guidelines and specifications for the manufacture of the materials, construction of the structural elements, the thermal comfort and the durability of the built structures. Even though earth construction exists since the dawn of civilisation, there is lack of universally accepted standardisation on the material production and the construction methods as compared to the standards available on conventional materials such as concrete, masonry and steel. Since the last 6–7 decades, there are attempts to develop standard codes and normative documents for the modern earth construction, facilitating the professionals in the design and the construction of the modern earth buildings.

The buildings consist of different structural elements (foundations, walls, roofs/floors, etc.). Table 1.1 gives the details of the earth construction techniques for different building components and the status of codes or standards available. One of the major problems in developing codes/standards is the common terminology

Table 1.1 Earth construction techniques for buildings

Building component	Types of construction techniques	Status of codes or standards
Foundation	- Rammed earth - Earth mortars	Few codes exist
Masonry walls	- Adobe - Compressed earth - Extruded earth - Cut earth (sod and laterite)	Codes exist for some of the techniques
Monolithic walls	- Cob - Rammed earth - Poured earth - Flowable earth mix concrete	Codes exist for some of the techniques
Infill walls	- Wattle and daub - Bagged earth	No codes
Floors/roofs	- Earth panels - Precast elements - In-situ techniques	No codes
Plastering and renderings/finishes	- Earth mortars - Plastering/renderings - Clay panels	Few codes exist

for different types of earth construction techniques and the earthen materials. There are regional specific terms for some of the modern earth construction techniques and the materials. The regional definitions may have to be avoided to suit with the common terminology for the global usage. The common definitions or terminology for earthen materials and earthen structures highlighted in Table 1.1 are as follows.

Earth: Earth is the basic material required for the manufacture of earthen building materials. Another commonly used term for the earth is soil, which is a result of weathering process of rocks over a period of millions of years. Earth or soil is a granular material consisting of inert crystalline silica particles and clay minerals. The silica particles are called silt, sand and gravel (depending upon the grain size of the particle). The swell–shrink characteristics of the soil or the earth and the strength are mainly controlled by the type and quantity of the clay minerals. There are different regional terminologies for the earth or the soil.

Earth mortar (EM): The EM is basically composed of earth and sand. Sometimes, fibres (generally natural fibres) are mixed in order to control the cracks due to shrinkage. The earth mortars find applications in plasters or renders as well as bed joints in the earth block masonry. The earth mortars used for rendering and plastering can contain colouring agents such as natural soils or pigments.

Stabilised earth mortar (SEM): The SEM consists of earth, sand and inorganic stabilisers such as cement or lime. Such mortars are used for the construction of masonry using stabilised earth bricks or blocks. The stabilisers and the granular composition of the mix are dictated by the strength and the workability characteristics of the mortar.

Earth block (EB): The EB is a masonry unit. There are varieties of EBs, such as adobe brick or block, compressed earth block (CEB), extruded earth brick (EEB) and cut earth block (sod and laterite). Adobe and CEB can be further classified as the stabilised adobe and the stabilised compressed earth block. The stabilised EBs have inorganic binders in addition to the earth-based granular mix.

Cob (Cb): For the cob wall, processed earth mixture in a plastic state is piled up with some tamping to form a monolithic wall. Frequently, organic fibres such as straw are mixed while processing the earth mix. The cob wall construction does not require any formwork.

Rammed earth (RE): RE is a monolithic construction involving compaction of partially saturated soil or the earth aggregate mixture in layers inside a rigid formwork. A layered texture resembling a sedimentary rock is ascertained for the wall. When inorganic stabilisers (such as cement or lime) are added to the earthen mixture, it is designated as stabilised rammed earth (SRE).

Wattle and daub: For wattle and daub, processed earth of plastic consistency is applied on either face of a skeleton structure and finished manually to get even surfaces for the wall. Generally, the skeleton structure is made up of unshaped wooden members and wooden sticks. This skeleton is the wattle, which is daubed from both the sides with the processed earth material.

Poured earth (PE): For PE, processed earth in a slurry consistency is poured into a formwork and the formwork is stripped after the earth ascertains stiffness. Generally, it is stabilised with an inorganic binder. Alternatively, the mix can fill bags that are piled, while the earth/soil is in plastic state to form walls (bagged earth).

Flowable earth mix (FEM) concrete: This is nearly similar to poured earth but for the inclusion of coarse aggregates. Also, this is termed as mud concrete. The FEM concrete is poured into a formwork and can be vibrated like a conventional cement concrete. The stiffness of the formwork used is much less than the stiffness of formwork used in the rammed earth construction.

Earthen or clay panels: These are thin precast panels (25–30 mm thick) finding application in the partitioning of spaces and cladding the inner surfaces of the walls. The composition can be diverse. One of the possibilities is basically the processed earth mixed with the fibres.

Soil or earth consists of inert particles such as gravel, sand and silt, and the active clay minerals. The characteristics of the soil are primarily controlled by the type and quantity of the clay mineral. The standard codes attempt to specify the limits on the sand and clay content for the production of stabilised compressed and rammed earth products. The range of values for the soil composition and Atterberg's limits specified for some of the standards on stabilised earth construction are given in Table 1.2. The range of values specified in the existing codes on stabilised CEB and rammed earth vary widely. The specifications and the information on the earth construction in these standards differ widely. There is a need for developing comprehensive codes on each type of earth construction technology where a range of values is specified for the soil or the earth properties, which are universally acceptable.

Table 1.2 Earth or soil composition and other properties specified in different standards/codes on stabilised rammed earth and stabilised CEB

Composition parameter	Stabilised			
	Rammed earth		CEB	
	Range	Ref	Range	Ref
Clay content (%) Sand (0.075–0.425 mm) Gravel (%)	– >35 –	IS 2110 (1980)	2–30 >60 0–10	IS 1725 (2013), PCH-2-87 (1988), SLS 1382-Part 1 (2009)
Liquid limit (%) Plasticity index (%)	≤27.0 8.5–10.5	IS 2110 (1980)	≤45 ≤12	NBR 8491 (2012), NBR 8492 (2012), NBR 10833 (2012), IS 1725 (2013)
pH	–	–	6–8.5	IS 1725 (2013), SLS 1382-Part 1 (2009)

Schroeder (2012) and Reddy et al. (2022) provide a review of current standards available on earth construction. It appears there are a greater number of attempts in developing the codes/standards on stabilised earth (especially stabilised CEB and stabilised rammed earth). Developing comprehensive global standards on the earthen products will be more useful for better promotion of the earth construction. Moreover, it is an absolute necessity to develop new codes and standards in order to encourage the designers to build with earth, as well as to convince the regulatory bodies, which are sometimes reluctant to use earthen products (MacDougall 2008). There is a need for international laboratory standards on testing the earthen materials and the earthen building products. The new comprehensive global standards should address the following generic items:

1. Earth or soil selection, optimum composition/grading
2. Moulds and machinery for earth construction
3. Production or manufacturing techniques
4. Testing and quality control
5. Structural design guidance including earthquake resistance design
6. Construction methodology and construction procedure
7. Thermal performance, hygroscopicity and moisture buffering
8. Durability, maintenance and limitations
9. Common glossary on earthen products.

A list of standard codes and normative documents is provided in the bibliography.

1.5 Energy, Emissions, Environment and Earthen Structures

Currently, there is a great concern about the climate change and global warming. The buildings and the related habitat are energy and material resources guzzlers.

The current practices of construction and maintaining the buildings consume energy and mined material resources, and generate wastes (solid, liquid and gaseous). The buildings consume nearly 50% of the total energy produced for their construction and maintenance. The modern construction practices heavily depend on mined material resources. Maintaining the comfortable living conditions in the modern buildings is associated with greater demand on energy resources. The energy, material and buildings nexus in the current scenario ultimately ends up in unsustainable habitats. The solutions based on traditional and modern earth construction practices, and the related built environment provide a ray of hope for sustainable building habitats. The whole concept of revival of the earth construction techniques stems from the arguments such as low specific energy consumption and low carbon emission during the production, easily recyclable material and healthy built environment. Energy, carbon emissions and sustainability of construction materials and buildings are discussed in Chap. 17.

1.6 Scope and Structure

There are arguments for and against stabilised earth construction mainly from the consideration of the carbon emissions. The stabilised earth products have low embodied carbon when compared with the embodied carbon of many conventional materials. Such products possess strength in wet condition and are highly durable. The stabilised earth products using stabilisers such as Portland cement possess residual clay minerals almost equal to the clay content of the original soils used in the product (Reddy and Latha 2014). The science and technology of stabilised earth especially for the construction of the buildings and other structures has progressed considerably in the last 5–6 decades. There is a great deal of interest among many professionals in utilising stabilised earth technology for the buildings. The book is focused on stabilised earth construction mainly used for the structural components of the buildings.

The book comprises 17 chapters under four major divisions, dealing with different topics pertaining to the stabilised earth construction. Apart from the introduction to the earthen structures, Part I covers fundamental aspects of soils and soil stabilisation. Part II encompasses the scientific principles of static soil compaction, basic principles of designing a machine for the compressed earth blocks, analysis of block production, characteristics of stabilised compressed earth blocks (CEBs), mortars and their masonry, geopolymer stabilised CEB and the concepts on utilising non-organic solid wastes for CEB. Strength and stress–strain behaviour of rammed earth and structural design aspects of rammed earth walls are dealt under Part III. Recycling of stabilised earth, energy and carbon emissions in the stabilised earth products, and an overview on sustainability of construction materials and green buildings are highlighted under Part IV.

The topics covered are mainly supported by scientific data generated over four decades of rigorous R&D work helped by the doctoral research programmes, sponsored research and the experience gained over technology dissemination activities leading to a large number of CEB and rammed earth buildings and structures. The structural design aspects presented on the stabilised CEB masonry and the rammed earth walls shed more light on the basic design principles for the structural elements using new materials and illustrated through design examples. The novel materials using industrial by-products and new binders (geopolymers), promotion of circular economy demonstrated through recycling of natural clay minerals from the stabilised earth products and the concept of sustainable materials are elaborated in greater detail. The book serves as a useful document for the researchers, design professionals, teachers of engineering and architecture, construction professionals, students and the individuals.

References

Adam EA, Agib ARA (2001) Compressed stabilised earth block manufacture in Sudan. Printed by Graphoprint for UNESCO, Paris France

Anger R and Fontaine L (2009) Bâtir en Terre. Belin, Paris, France. ISBN 2701152046

Arango Gonzalez JR (1999) Uniaxial deformation-stress strain behaviour of the rammed earth of the Alcazaba Cadima. Mater Struct 32:70–74

Association la Voûte Nubienne (2015) 7 rue Jean Jaurès—34190 Ganges, February, France. www.lavoutenubienne.org

Bader A (2012) Prince Ahmad Bin Salman Mosque, Riyadh, Saudi Arabia. In: Proc. LEHM 2012, 6th international conference on building with earth, Dachverband Lehm e.V., 5–7 October, Weimar, Germany. pp 23–27. ISBN 978-3-00-039649-6

Bulletin 5 (Middleton GF 1952) Revised by Schneider LM (1987) Earth-wall construction, Fourth Edition. CSIRO Division of Building, Construction and Engineering, North Ryde, NSW 2113, Australia

Cointeraux F (1791) Traite des constructions rurales et e leur disposition, Paris, Franca

CRAterre ENSAG WHEAP Programme (2012) World heritage inventory of earthen architecture. ISBN 978-2-906901-70-4

Dept. of HUD (1955) Earth for homes, prepared by the department of housing and urban development. Division of International Affairs, Washington DC, 20410

Easton D (1996) The rammed earth house. Chelsea Green Publishing Company, USA. ISBN 0-930031-79-2

EBAA (2005) Building with earth bricks and rammed earth in Australia. Earth Building Association of Australia (ebaa.org.au). ISBN 0-9756036-0-4

El-shorbagy AM (2001) The architecture of Hassan Fathy: between Western and non-Western perspectives. PhD thesis, University of Canterbury, New Zealand

Farahza N, Golchin M, Safi S, Khajehrezaei I (2012) New uses for earthen architecture in Yazd—a critique. In: Proc. LEHM 2012, 6th international conference on building with earth, Dachverband Lehm e.V., 5–7 October. Weimar, Germany, pp 242–251, ISBN 978-3-00-039649-6

Fathy H (1973) Architecture for the poor. The University of Chicago Press, USA. ISBN-10-0226239160

Fitzmaurice RF (1958) Manual on stabilised soil construction for housing. UN Technical Assistance Programme, New York

Hall MR, Swaney W (2012) Soil stabilisation and earth construction materials, properties and techniques. In: Hall MR, Lindsay R, Krayenhoff M (eds) Modern earth buildings: materials, engineering, construction and applications. Woodhead Publishing Ltd., UK, pp 650–687. ISBN 978-0-85709-026-3

Hall MR, Najim KB, Keikhaei Dehdezi P (2012) Soil stabilisation and earth construction materials, properties and techniques. In: Hall MR, Lindsay R, Krayenhoff M (eds) Modern earth buildings: materials, engineering, construction and applications. Woodhead Publishing Ltd., UK, pp 222–255. ISBN 978-0-85709-026-3
https://en.wikipedia.org/wiki/Arg-e_Bam
https://en.wikipedia.org/wiki/Taos_Pueblo
https://goodearth.org.in
https://www.eartharchitecture.org
Hu R, Liu J (2015) Rescuing a sustainable heritage: prospects for traditional rammed earth housing in China today and tomorrow. In: Ciancino D, Beckett C (eds) Rammed earth construction—cutting edge research on traditional and modern rammed earth. CRC Press/Balkema, London UK. ISBN 978-1-138-02770-1
IS 1725 (2013) Stabilized soil blocks used in general building construction—specification (2nd revision). Bureau of Indian Standards, New Delhi, India
IS 2110 (1980) (reaffirmed 1998) Code of practice for in situ construction of walls in buildings with soil-cement (first revision). Bureau of Indian Standards, New Delhi, India
Jaquin P (2012) History of earth building techniques. In: Hall MR, Lindsay R, Krayenhoff M (eds) Modern earth buildings: materials, engineering, construction and applications. Woodhead Publishing Ltd., Cambridge, UK, pp 307–323
Jiyao H, Weitung J (1990) Earth-culture-architecture: the protection and development of rammed earth and adobe architecture in China. In: 6th international conference on the conservation of earthen architecture: adobe 90 preprints: Las Cruces, New Mexico, USA, pp 72–76
King G (1998) The traditional architecture of Saudi Arabia. I. B. Tauris & Co Ltd, Victoria House, Bloomsbury Square, London
Klinge ER, Voth H, Brockmann H, Klinge A, Ziegert C (2016) Celestial Stairs, Golden Spiral and City of Orion, Maintaining the works of Hannsjörg Voth in the Plaine de Marha, Morocco. In: Proc. LEHM 2016, 7th international conference on building with earth, Dachverband Lehm e.V., 12–14 November. Weimar, Germany
Krayenhoff M (2012) North American modern earth construction. In: Hall MR, Lindsay R, Krayenhoff M (eds) Modern earth buildings: materials, engineering, construction and applications. Woodhead Publishing Ltd., UK, pp 561–608
Lambe TW (1962) Soil stabilisation Chapter-4. Foundation Engineering, Edited by Leonards
Liang R, Hota G, Lei Y, Li Y, Stanislawski D, Jiang Y (2013) Non-destructive evaluation of historic Hakka rammed earth structures. Sustainability 5:298–315. https://doi.org/10.3390/su5010298
Liang R, Stanislawski D, Hota G (2011) Structural responses of Hakka rammed earth buildings under earthquake loads. In: International symposium on innovation and sustainability of structures in civil engineering, 22–26 October. Xiamen University, Xiamen, China
MacDougall C (2008) Natural building materials in mainstream construction: lessons from the U. K. J Green Build 3:1–14
Madhavan R, Narasingha Rao CN (1949) Report on labour housing scheme. City Improvement Trust Board, Bangalore, India
McHenry PG (1989) Adobe and rammed earth buildings-design and construction. The University of Arizona Press, Tucson, USA
Mitra JN (1951) Suitability of soil for stabilised soil houses for Rangawan dam colony. Indian Concr J 15:234–238
Mortada H (2016) Sustainable desert traditional architecture of the central region of Saudi Arabia. Sustain Dev 24:383–393. https://doi.org/10.1002/sd.1634
Mubarak (1999) Cultural adaptation to housing needs: a case study, Riyadh, Saudi Arabia. In: IAHS conference proceedings, San Francisco, USA
NBR 10833 (2012) Fabricação de tijolo e bloco de solo-cimento com utilização de prensa manual ou hidráulica—procedimento (Production of earth-cement bricks and blocks with manual or hydraulic press—procedure). ABNT, Rio de Janeiro

NBR 8491 (2012) Tijolo de solo-cimento—Requisitos. Rio de Janeiro (Earth-cement bricks—requirements), ABNT
NBR 8492 (2012a) Tijolo de solo-cimento—análise dimensional, determinação da resistência à compressão e da absorção de água—método de ensaio (Earth-cement bricks—dimensions, compressive strength and water adsorption—test procedures). ABNT, Rio de Janeiro
NZS 4297 (1998) Engineering design of earth buildings. Standards New Zealand, Wellington, New Zealand
NZS 4298 (1998) Materials and workmanship for earth buildings. Standards New Zealand, Wellington, New Zealand
NZS 4299 (1998) Earth buildings not requiring specific design. Standards New Zealand, Wellington, New Zealand
Kapfinger O, Sauer M (2015) Martin Rauch refined earth construction and design with rammed earth. Eberl Print GmbH, Immenstadt, Germany
Patty RL (1936) Clay soil unfavourable for rammed earth walls, Bulletin 298. South Dakota State College Agricultural Experiment Station, South Dakota State College, Brookings, South Dakota, USA
Patty RL, Minium LW (1945) Rammed earth walls for farm buildings, Bulletin 277. Agricultural Engineering Department, Agricultural Experiment Station, South Dakota State College, Brookings, South Dakota, USA
PCH-2-87 (1988) State building committee of the Republic of Kyrgyzstan/Gosstroi of Kyrgyzstan: Возведение малоэтжных зданий и сооружений из грунтоцементобетона PCH-2-87 (Building of low storied houses with stabilized rammed earth), Republic Building Norms RBN-2-87, Frunse (Bischkek) Republic of Kyrgyzstan
Rao SM, Venkatarama Reddy BV, Muttharam M (2000) Engineering behaviour of wood-ash-modified soils. Ground Improv (Thomas Telford) 4:137–140
Rauch M (2007) Earth house with European Standard: a review of the project 'rammed earth house' by the company Lehm Ton Erde Baukunst GmbH (Soil clay earth building art Ltd.) in Schlins, Austria. In: Venkatarama Reddy BV, Mani M (eds) Proc. International symposium on earthen structures. Interline Publishing, pp 47–52
Reddy BVV, Latha MS (2014) Retrieving clay minerals from stabilised soil compacts. Appl Clay Sci 101:362–368
Reddy BVV, Rao SM, Arun Kumar MK (2003) Characteristics of stabilised mud blocks using ash-modified soils. Indian Concr J 77(2):903–911
Reddy BVV, Morel JC, Faria P, Fontana P, Oliveira DV, Serclerat I, Walker P, Maillard P (2022) Codes and standards on earth construction, Ch 7 In: Fabbri A, Morel JC, Aubert JE, Bui QB, Gallipoli D, Reddy BVV (Eds) Testing and characterisation of earth-based building materials and elements, State of the art report of the RILEM Technical Committee 274-TCE RILEM, Springer publication, pp 243–259. ISBN 978-3-030-83297-1
Reddy BVV, Iype N, George S (2016) Large-scale dissemination of stabilised soil blocks—a case study of GoodEarth eco-homes. In: Proceedings of the LEHM-2016, 7th international conferences on building with earth, Dachverband Lehm e.V., 7–11 November. Weimar, Germany
Röhlen U, Ziegert C (2011) Earth building practice, planning-design-building. Beuth Verlag GmbH Berlin, Germany
Schick W (1987) Der Pise-Bau Zu Weilburg an der Lahn. Burgerintiative, Alt Weilburg, Germany
Schroeder H (2012) Modern earth building codes, standards and normative development. In: Hall MR, Lindsay R, Krayenhoff M (eds) Modern earth buildings: materials, engineering, construction and applications. Woodhead Publishing Ltd., UK, pp 688–711. ISBN 978-0-85709-026-3
Schroeder H (2016) Sustainable building with earth. Springer International Publishing, Switzerland
Sharon Davis Design. https://sharondavisdesign.com/project/bayalpata-community-hospital-nepal/
SLS 1382 Part 1 (2009) Specification for compressed stabilized earth blocks, part 1: requirements. Sri Lanka Standards Institution, Colombo, Sri Lanka

Stanislawski D (2011) Structural responses and finite element modelling of Hakka Tulou rammed earth structures. MSc thesis, Department of Civil and Environmental Engineering Morgantown, West Virginia, USA
Swaminathan CG, Lal NB, Wason OP (1983) Relevance of soil stabilisation techniques in low cost road construction. J Indian Roads Congr 44:613–648
Terra Europae (2011) Earthen architecture in the European Union. Progetti Seperi Sentieri. ISBN 978-884672957-6
U.N. Report (1964) Soil–cement—its use in building. Dept. of Economic and Social Affairs, United Nations, New York, USA
Verma PL, Mehra SR (1950) Use of soil–cement in house construction in the Punjab. Indian Concr J 91–96
Wallis RK (2012) Modern rammed earth construction in China. In: Hall MR, Lindsay R, Krayenhoff M (eds) Modern earth buildings: materials, engineering, construction and applications. Woodhead Publishing Ltd., UK, pp 688–711. ISBN 978-0-85709-026-3
Wilson Q (2008) Manufacturing sun-cured adobe bricks in the USA. In: Proc. LEHM 2008, 5th international conference on building with earth, Dachverband Lehm e.V., 9–12 October. Koblenz, Germany. ISBN 978-3-00-025956-2
Wilson Q (2012) The adobe potato cellars (barns) of the San Luis Valley, Colorado, USA. In: Proc. LEHM 2012, 6th international conference on building with earth, Dachverband Lehm e.V., 5–7 October. Weimar, Germany. ISBN 978-3-00-039649-6
Zami MS, Mohammad B (2012) Adoption of contemporary earth construction in Africa to alleviate the urban housing crisis. In: Proc. LEHM 2012, 6th international conference on building with earth, Dachverband Lehm e.V., 5–7 October. Weimar, Germany, pp 41–49. ISBN 978-3-00-039649-6

(a) Books, Monographs and Papers

Chayet A, Jest C, Sanday J (1990) Earth used for building in the Himalayas, the Karakoram, and central Asia-recent research and future trends. In: Proc. 6th international conference on the conservation of earthen architecture: Adobe 90 preprints: Las Cruces. New Mexico, USA, pp 29–34
Dethier J (1982) Down to earth. Thames and Hudson, London, UK
Hall MR, Lindsay R, Krayenhoff M (2012b) Modern earth buildings, materials engineering, construction and applications. Woodhead Publishing, Cambridge, UK
HB 195 (2002) Peter Walker and Standards Australia, The Australian earth building handbook
Houben H, Guillaud H (2003) Earth construction: a comprehensive guide. Intermediate Technology Publications/CRAterra-EAG, London, UK
Jagadish KS (2009) Building with stabilized mud. I K International Publishing House, New Delhi, India
Keable J (1996) Rammed earth structures: a code of practice. Intermediate Technology Publications, London, UK
King BPE (1996) Buildings of earth and straw: structural design for rammed earth and straw-bale architecture. Ecological Design Press, Sausalito, California, USA
Minke G (2000) Earth construction handbook—the building material earth in modern architecture. WIT Press, Southampton, UK
Morton T (2008) Earth masonry—design and construction guidelines. HIS BRE Press, Bracknell, Berkshire, UK. ISBN 978-1-86081-978-0
Norton J (1997) Building with earth—a handbook, 2nd edn. Intermediate Technology Publications, London, UK

Lehmbau Regeln (2009) Dachverband Lehm e.V. (Hrsg.): Lehmbau Regeln—Begriffe, Baustoffe, Bauteile, 3, überarbeitete Aufl (Rules of Earth Construction—Terms and Definitions, Building Materials, Building Elements, 3rd revised edition). Vieweg+Teubner/GWV Fachverlage, Wiesbaden, Germany
Spence RJS, Cook DJ (1983) Building materials in developing countries. Wiley, Brisbane, Australia
Walker P, Keable R, Martin J, Maniatidis V (2005) Rammed earth: design and construction guidelines. BRE Bookshop, UK, Watford

(b) Standards and Codes

ACP-EU/CRAterre-BASIN Center for the Development of Industry: Compressed Earth Blocks, Series Technologies. Nr. 5 Production equipment (1996); Nr. 11: Standards (1998); Nr. 16 Testing procedures (1998). Brussels (1996–1998)
ARS 670 (1996) Standard for terminology. Centre for Development of Industry (CDI), African Regional Organisation for Standardisation ARSO, Nairobi
ARS 671 (1996) Standard for definition, classification and designation of compressed earth blocks. Centre for Development of Industry (CDI), African Regional Organisation for Standardisation ARSO, Nairobi
ARS 672 (1996) Standard for definition, classification and designation of earth mortars. Centre for Development of Industry (CDI), African Regional Organisation for Standardisation ARSO, Nairobi
ARS 673 (1996) Standard for definition, classification and designation of compressed earth block masonry. Centre for Development of Industry (CDI), African Regional Organisation for Standardisation ARSO, Nairobi
ARS 674 (1996) Technical specifications for ordinary compressed earth blocks. Centre for Development of Industry (CDI), African Regional Organisation for Standardisation ARSO, Nairobi
ARS 675 (1996) Technical specifications for facing compressed earth blocks. Centre for Development of Industry (CDI), African Regional Organisation for Standardisation ARSO, Nairobi
ARS 676 (1996) Technical specifications for ordinary mortars. Centre for Development of Industry (CDI), African Regional Organisation for Standardisation ARSO, Nairobi
ARS 677 (1996) Technical specifications for facing mortars. Centre for Development of Industry (CDI), African Regional Organisation for Standardisation ARSO, Nairobi
ARS 678 (1996) Technical specifications for ordinary compressed earth block masonry. Centre for Development of Industry (CDI), African Regional Organisation for Standardisation ARSO, Nairobi
ARS 679 (1996) Technical specifications for facing compressed earth block masonry. Centre for Development of Industry (CDI), African Regional Organisation for Standardisation ARSO, Nairobi
ARS 680 (1996) Code of practice for the production of compressed earth block. Centre for Development of Industry (CDI), African Regional Organisation for Standardisation ARSO, Nairobi
ARS 681 (1996) Code of practice for the preparation of earth mortars. Centre for Development of Industry (CDI), African Regional Organisation for Standardisation ARSO, Nairobi
ARS 682 (1996) Code of practice for the assembly of compressed earth block masonry. Centre for Development of Industry (CDI), African Regional Organisation for Standardisation ARSO, Nairobi
ARS 683 (1996) Standard for classification of material identification tests and mechanical tests. Centre for Development of Industry (CDI), African Regional Organisation for Standardisation ARSO, Nairobi

ASTM E2392/E2392-10 (2010) Standard guide for design of earthen wall building systems. ASTM International, West Conshohocken

CID-GCB-NMBC-14.7.4 (2006) Regulation and Licensing Dept., Construction Industries Div., General Constr. Bureau: 2006, New Mexico Earthen Building Materials Code. Santa Fe

DIN 18945-08 (2013) Lehmsteine—Begriffe, Anforderungen, Prüfverfahren (Earth blocks—terms and definitions, requirements, test methods)

DIN 18946-08 (2013) Lehmmauermörtel—Begriffe, Anforderungen, Prüfverfahren (Earth masonry mortar—terms and definitions, requirements, test methods)

DIN 18947-08 (2013) Lehmputzmörtel—Begriffe, Anforderungen, Prüfverfahren (Earth plasters—terms and definitions, requirements, test methods)

IS 13827 (1993) Improving earthquake resistance of earthen buildings—guidelines. Indian Standard, New Delhi

IS 17165 (2020). Manufacture of stabilized soil blocks—guidelines. Bureau of Indian Standards, New Delhi, India

IS 4332 Part IV (1967) Methods of tests for stabilised soils. Wetting and drying, and freezing and thawing tests for compacted soil–cement mixtures. Bureau of Indian Standards, New Delhi, India

IS 4332 Part V (1967) Methods of tests for stabilised soils. Determination of unconfined compressive strength of stabilised soils. Bureau of Indian Standards, New Delhi, India

KS02-1070 (1993) Specifications for stabilized soil blocks. Kenya Bureau of Standards, Nairobi, Kenya

Legge 24 Diciembre n. 378 (2003) Disposizioni per la tutela e la valorizzazione dell'architettura rurale. Gazzetta Ufficiale 13 (2004)

LNEC (1953) The use of earth as a building material (in Portuguese). CIT 9, Laboratório Nacional de Engenharia Civil, Lisbon, Portugal

National Building Code (2006) 1st edn. Federal Republic of Nigeria: LexisNexis Butterworths, Cape Town, South Africa

NBC 203 (1994) Nepal National Building code, Nepal, Guidelines for earthquake resistant building construction: low strength masonry

NBC 204 (1994) Nepal National Building code, Nepal, Guidelines for earthquake resistant building construction: earthen buildings (EB)

NBR 10833 (2012) Fabricação de tijolo e bloco de solo-cimento com utilização de prensa manual ou hidráulica—Procedimento (Production of earth-cement bricks and blocks with manual or hydraulic press—procedure). Associação Brasileira de Normas Técnicas, Rio de Janeiro

NBR 10834 (2012) Bloco de solo-cimento sem função estrutural—Requisitos (Soil–cement block without structural function—Requirements). Associação Brasileira de Normas Técnicas (ABNT), Rio de Janeiro

NBR 10836 (2013) Bloco de solo-cimento sem função estrutural—Análise dimensional, determinação da resistência à compressão e da absorção de água (Soil–cement block without structural function—dimensional analysis, compressive strength determination and water absorption—test method). Associação Brasileira de Normas Técnicas, Río de Janeiro

NBR 12023 (2012) Solo-cimento. Ensaio de compactação (Soil–cement. Compaction test method). Associação Brasileira de Normas Técnicas, Rio de Janeiro

NBR 12024 (2012) Solo-cimento. Moldagem e cura de corpos de prova cilíndricos, Procedimenyo (Soil–cement. Molding and curing of cylindrical specimens. Procedure). Associação Brasileira de Normas Técnicas, Rio de Janeiro

NBR 12025 (2012) Solo-cimento. Ensaio de compressão simples de corpos de prova cilíndricos. Método de ensaio (Soil–cement. Simple compression test of cylindrical specimens. Method of test). Associação Brasileira de Normas Técnicas, Rio de Janeiro

NBR 12253 (2012) Soil–cement—mixture for use in pavement layer—procedure. Associação Brasileira de Normas Técnicas, Rio de Janeiro

NBR 13553 (2012) Materiais para emprego em parede monolítica de solo-cimento sem função estrutural. Requisitos (Soil–cement materials for monolithic walls of soil–cement without structural function. Requirements). Associação Brasileira de Normas Técnicas, Rio de Janeiro

NBR 13554 (2012) Solo-cimento. Ensaio de durabilidade por molhagem e secagem. Método de ensaio (Soil–cement. Durability test by wetting and drying. Test method). Associação Brasileira de Normas Técnicas, Rio de Janeiro

NBR 13555 (2012) Solo-cimento. Determinação da absorção de água. Método de ensaio (Soil–cement. Determination of water absorption. Test method). Associação Brasileira de Normas Técnicas, Rio de Janeiro

NBR 16096 (2012) Solo-cimento. Determinação do grau de pulverização. Método de ensaio (Soil–cement. Determination of pulverization rate. Test method). Associação Brasileira de Normas Técnicas, Rio de Janeiro

NBR 8491 (2012) Tijolo de solo-cimento—Requisitos, Rio de Janeiro (Earth-cement bricks—requirements). Associação Brasileira de Normas Técnicas, Rio de Janeiro

NBR 8492 (2012b) Tijolo de solo-cimento—Análise dimensional, determinação da resistência à compressão e da absorção de água—Método de ensaio (Earth-cement bricks—dimensions, compressive strength and water adsorption—test procedures). Associação Brasileira de Normas Técnicas, Rio de Janeiro

NBR11798 DE 08 (2012) Materiais para base de solo-cimento—Requisitos. Associação Brasileira de Normas Técnicas, Río de Janeiro

Nch 3332 (2013) Estructuras—Intervención de construcciones patrimoniales de tierra cruda—Requisitos del proyecto structural (Structural design—retrofitting of historic earth buildings—requirements for the structural design planning). Instituto Nacional de Normalización, Santiago de Chile

NF XP P13-901 (2001) Blocs de terre comprimée pour murs et cloisons. Definitions, spécifications, méthodes d'essai, conditions de reception. AFNOR, Paris

NIS 369 (1997) Standard for stabilized earth bricks. Standards Organisation of Nigeria, Lagos, Nigeria

NMX-C-508-ONNCCE (2015) Industria de la construcción—Bloques de tierra comprimida estabilizados con cal—Especificaciones y métodos de ensayo (Construction industry—earth blocks—specifications and test methods)

NT 21.33 (1996) Blocs de terre comprimée ordinaires—Spécifications techniques (Common compressed earth blocks—Technical specifications). Institut National de la Normalisation et de la Proprieté. Industrielle (INNOPRI), Tunis

NT 21.35 (1996) Blocs de terre comprimée. Définition, classification et désignation (Compressed earth blocks. Definition, classification and designation). INNOPRI, Tunis

NTC 5324 (2004) Bloques de suelo cemento para muros y divisones. Definiciones. Especificaciones. Métodos de Ensayo. Condiciones de entrega (Earth-cement blocks for structural and partition walls. Definitions. Specifications. Test methods. Delivery conditions). ICONTEC, Bogotá

NTE E. 080 (2000) Adobe. Reglamento Nacional de Construcciones (Adobe. National Building Standards of Peru. SENCICO, Technical Building Standard, Lima, Peru

NTP 331.201 (1978) Elementos de suelo sin cocer: adobe estabilizado con asfalto para muros: Requisitos (Non-fired earth elements: bitumen stabilized adobe for walls. Requirements). Instituto Nacional de Defensa de la Competencia y de la Protección de la Propiedad Intelectual (INDECOPI), Lima, Peru

NTP 331.202 (1978) Elementos de suelos sin cocer: adobe estabilizado con asfalto para muros: Métodos de ensayo (Non-fired earth elements: bitumen stabilized adobe for walls. Test methods). Instituto Nacional de Defensa de la Competencia y de la Protección de la Propiedad Intelectual (INDECOPI), Lima, Peru

NTP 331.203 (1978) Elementos de suelos sin cocer: adobe estabilizado con asfalto para muros: Muestra y recepción (Non fired earth elements: bitumen stabilized adobe for walls. Samples and acceptance). Instituto Nacional de Defensa de la Competencia y de la Protección de la Propiedad Intelectual (INDECOPI), Lima, Peru

PCH-2–87 (1988) State Building Committee of the Republic of Kyrgyzstan/Gosstroi of Kyrgyzstan: Возведение малоэтжных зданий и сооружений из грунтоцементобетона РСН-2-87

(Building of low storied houses with stabilized rammed earth), Republic Building Norms RBN-2-87, Frunse (Bischkek) Republic of Kyrgyzstan
Regione Piemonte L.R. 2/06 (2006) Norme per la valorizzazione delle costruzioni in terra cruda. B.U.R. Piemonte, p 3
SAZS 724 (2001) Standards Association of Zimbabwe: rammed earth structures. Zimbabwe Standard Code of Practice, Harare
SLS 1382 Part 2 (2009) Specification for compressed stabilized earth blocks—part 2: test methods. Sri Lanka Standards Institution, Colombo, Sri Lanka
SLS 1382-Part 3 (2009) Specification for compressed stabilized earth blocks, part 3: guidelines on production, design and construction. Sri Lanka Standards Institution, Colombo, Sri Lanka
TS 2514 (1997) Adobe blocks and production methods. Turkish Standard Institution, Ankara
TS 2515 (1985) Adobe buildings and construction methods. Turkish Standard Institution, Ankara
TS 537 (1985) Cement treated adobe bricks. Turkish Standard Institution, Ankara
UNE 41410 (2008) Bloques de tierra comprimida para muros y tabiques. Definiciones, especificaciones y métodos de ensayo (Compressed earth blocks for structural and partition walls. definitions, specifications and test procedures). AENOR, Madrid
US 849 (2011) Specification for stabilized soil blocks, 1st edn. Uganda Standard, Uganda

Chapter 2
Soils

2.1 Soil Formation

Soil is a mixture of solid and mineral particles of sizes ranging from less than 2 μ to 5 mm and sometimes containing other materials such as organic matter, salts and rarely some contaminants. The soil is formed due to geological cycle of events such as weathering, transportation and deposition continuously taking place on the earth surface. Basically, it is formed by the weathering of rocks.

The chemical properties of the soil are dictated by the minerals that constitute the soil particles and hence the parent rock. The physical properties of the soil are dictated by the size, shape and composition of the grains. Varieties of soils exist in nature, and their characteristics vary widely depending upon the nature of the parent rock, type of climatic influence and the age in geological terms. The rock masses are broken down due to the action of a variety of physical, chemical and biological processes. The physical processes reduce particle size, increase surface area and increase bulk volume. The chemical and the biological processes cause changes in both physical and chemical properties.

Mechanical weathering

The agents of mechanical weathering causing disintegration of rocks include: (a) glacial movements, (b) running water, (c) impact of wind-driven rain, (d) wind erosion, (e) expansion and contraction due to freeze–thaw effects, (f) biological actions, etc. The rocks are subjected to cyclic conditions of heating and cooling, which cause expansion and contraction resulting in the disintegration of the rocks into smaller pieces. The freezing of the water in the cracks and the crevices can also cause splitting and disintegration of the rocks. The movement of glacier ice, wind, running water, ocean waves, etc., can also disintegrate the rocks. In the mechanical weathering process, there is only size change without any change in the chemical composition. The particles due to the rock disintegration occur in wide range of sizes, from boulders to very fine powdery form.

B. V. V. Reddy, *Compressed Earth Block & Rammed Earth Structures*,
Springer Transactions in Civil and Environmental Engineering,
https://doi.org/10.1007/978-981-16-7877-6_2

Chemical weathering

The rocks are prone for chemical weathering. The original rock forming minerals get transformed into new minerals by chemical reactions. The water and the carbon dioxide from the atmosphere form carbonic acid which reacts with the existing rock minerals to form new minerals and soluble salts. The soluble salts present in the ground water and the organic acids formed from decayed organic matter can also cause chemical weathering. An example of chemical weathering of orthoclase to form clay minerals, silica and potassium is illustrated below.

$$H_2O + CO_2 \rightarrow \underset{\text{(carbonic acid)}}{H_2CO_3} \rightarrow H^+ + (HCO3)^-$$

$$\underset{\text{Orthoclase}}{2K(AlSi_3O_8)} + 2H^+ + H_2O \rightarrow 2K^+ + \underset{\text{Silica}}{4SiO_2} + \underset{\text{Kaolinite (clay mineral)}}{Al_2Si_2O_5(OH)_4}$$

Similarly, weathering of the basalt results into formation of the montmorillonite clay mineral. The weathering process changes solid rock masses into smaller fragments of various sizes ranging from the boulder to the clay particles. The uncemented aggregates of these small grains in various proportions form different types of soils. The clay minerals which are a product of chemical weathering of feldspars, ferromagnesian and mica give plastic properties to the soils. Sometimes the weathering products of rocks can get transported and deposited elsewhere. The transported soils are classified into several groups depending upon their mode of transportation. For example, alluvial soils are transported and deposited by the running water in the rivers and the streams. The soils formed at their places of origin by the weathered products are called the residual soils. Generally, the residual soils have fine grains in the top surface layers, and the grains become coarser with the depth.

2.2 Soil Particle Size

The soil consists of particles of different sizes, and these particles are named as gravel, sand, silt, and clay-based on their particle sizes. The soils are generally called gravely, sandy, silty, or clayey depending upon the predominant particle size in the soil. There are several soil classification systems. Table 2.1 gives the particles size range for various soil particles as per IS 1498 – 1970 and Unified Soil Classification (USC). The particles classified as clay based on the particle size may not necessarily contain clay minerals. The clays have been defined as those particles which develop plasticity when mixed with water. They are mostly in the colloidal size range.

Table 2.1 Soil particles sizes based on IS 1498–1970 and USC classification

Particle type	Particle shape	USC and IS-1498 classification Size range (mm)	Nature of particles
Gravel	Coarser pieces of rock	>4.75–76	Crystalline and inert
Sand	Fine pieces of rock	4.75–0.075	
Silt	Microscopic particles consisting of very fine quartz grains and some flakes of micaceous minerals	0.075–0.002	
Clay	Flake or needle shaped microscopic and sub-microscopic clay minerals	<0.002	Colloid and mineral

2.3 Clay Minerals

The clays are very tiny particles of one or more members of a small group of minerals. They are primarily hydrous aluminium silicates with the presence of other metals and alkalis. The clay minerals are generally derived from the chemical weathering of the silicate minerals and the rocks (Grim 1953). The clay particles are in the form of sheets or needle like tubular structure arranged in dispersed or aggregated or flocculated fashion. The clay minerals commonly found in the soils belong to the larger family of phyllosilicates. The structure of these silicates is in the form of thin structural sheets. Thus, the two basic structural units of the commonly occurring clays are the silica tetrahedral sheet and the alumina or magnesium octahedral sheet. The structure of the tetrahedron and the octahedron and their sheets are shown in Figs. 2.1 and 2.2, respectively.

The classification into different clay mineral groups and their characteristics are influenced by the stacking arrangements of the basic sheets of the structural units and the way the two successive sheet layers are held together. There are several

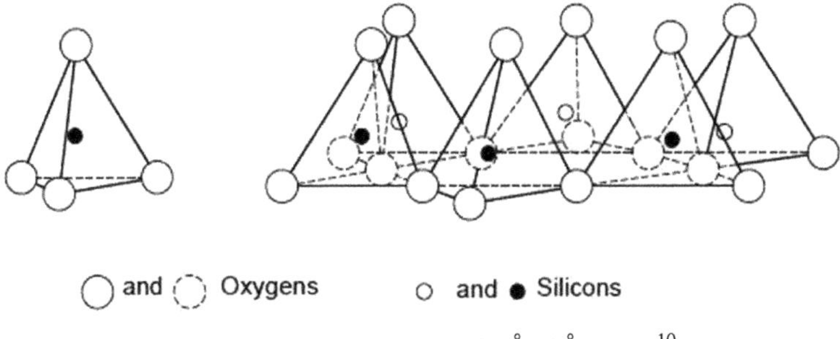

The thickness of the sheet = 4.63 Å (1Å = 10^{-10} m)

Fig. 2.1 Silica tetrahedron and silica tetrahedral sheet

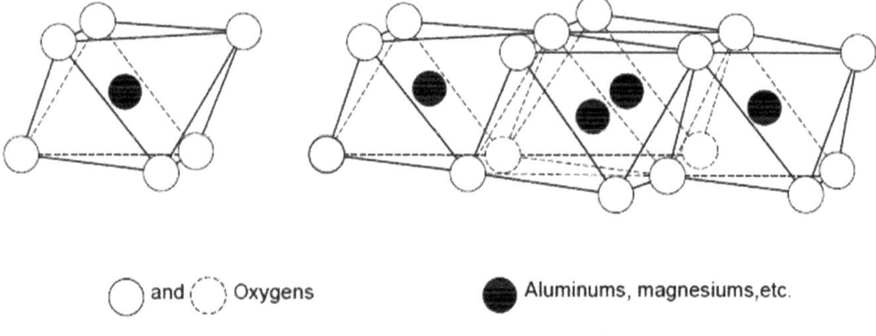

Fig. 2.2 Octahedral unit and octahedral sheet

families and groups of clay minerals. Some of the most commonly occurring soil clay minerals are kaolinite, illite and montmorillonite. There are a range of other clay minerals such as halloysite, beidellite, saponite, vermiculite, muscovite, biotite, attapulgite, chlorite, etc. Apart from the clay minerals, the soils can also contain the non-crystalline clay materials such as allophanes (have no definite composition), oxides and hydrous oxides of aluminium, silicon, and iron, etc. Some of the characteristics and the dimensions of the common clay minerals are highlighted in Table 2.2. The electron photomicrographs of kaolinite, illite and montmorillonite are shown in Figs. 2.3, 2.4 and 2.5, respectively.

Kaolinite

This clay mineral consists of alternating layers of silica and octahedral sheets. The bonding between the successive layers is both by the van der Waals forces and the hydrogen bonds having sufficient strength that there is no interlayer swelling.

Illite

The illite is like that of the montmorillonite except that some of the silicon is replaced by the aluminium, and the resultant charge deficiency is balanced by the potassium ions between the layers. Also, the illite structure is like that of the muscovite mica

Table 2.2 Characteristics and dimensions of clay minerals (Mitchell and Kenichi 2005)

Mineral type	Formula/unit cell	Crystal thickness	Specific gravity	Specific surface (m^2/gm)
Kaolinite	$(OH)_8Si_4Al_4O_{10}$	0.05 – 2 μm	2.60 – 2.68	10 – 20
Illite (hydrous mica)	$(K,H_2O)_2(Si)_8(Al,Mg,Fe)_{4,6}O_{20}(OH)_4$	0.003 – 10 μm	2.60 – 3.00	65 – 100
Montmorillonite	$(OH)_4Si_8Al_4O_{20}$ n H_2O	10 Å – 10 μm	2.35 – 2.70	700 – 840

Fig. 2.3 Electron photomicrograph of kaolinite clay mineral (Image reproduced from the 'Images of Clay Archive' of the Mineralogical Society of Great Britain & Ireland and The Clay Minerals Society, https://www.minersoc.org/images-of-clay.html)

Fig. 2.4 Electron photomicrograph of illite clay mineral (Image reproduced from the 'Images of Clay Archive' of the Mineralogical Society of Great Britain & Ireland and The Clay Minerals Society, https://www.minersoc.org/images-of-clay.html)

Fig. 2.5 Electron photomicrograph of montmorillonite clay mineral (Image reproduced from the 'Images of Clay Archive' of the Mineralogical Society of Great Britain & Ireland and The Clay Minerals Society, https://www.minersoc.org/images-of-clay.html)

and is sometimes referred to as the hydrous mica. The interlayer bonds are by cation bonds and are stronger than the bonds found in the montmorillonite.

Montmorillonite

In montmorillonite, the octahedral sheet is sandwiched between two silica sheets. The bonding between successive layers is by the van der Waals forces and by the cations may be present to balance charge deficiencies in the structure. No potassium ions. The large amount of water is attracted into the spaces between the layers. The interlayer bonds are weak and are easily separated by cleavage, and the water can enter between the layers leading to swelling.

The characteristics and behaviour of the clays greatly depend upon the quantum of moisture present. The clays tend to absorb significant amounts of water. The different groups of clay minerals exhibit a wide range of properties. The clay lumps are very hard in the perfect dry state and soften on absorption of the water. The type and quantity of the clay minerals present in the soils have controlling influence on the properties of the soils. The plasticity, cohesion, swelling, shrinkage, strength, permeability, compressibility, etc., of the soils are mainly attributed to the presence of the clay minerals. The characteristics of some of the commonly occurring soils are discussed in the following sections.

2.3.1 Identification of Clay Minerals

(a) **X-ray Diffraction analysis**

The clay minerals present in the soils can be identified through X-ray diffraction (XRD) analysis. The X-rays are one of the several types of waves in electromagnetic spectrum and have the wave lengths in the range of 0.01 –100 Å, (1 Å = 0.1 nm). In the X-ray tube, the incident X-rays on the clay powder get radiated. A relationship between the angle of incident rays and the diffracted rays (θ), the X-ray wavelength (λ), and the atomic planes spacing distance (d) is obtained, which is based on Bragg's law (Mitchell and Kenichi 2005). The X-ray diffraction data of the soil clay fraction is compared with the standard X-ray diffraction data of the specific clay minerals for identifying the clay minerals present in the soil. More information on the X-ray diffraction and the clay mineral identification can be found in Mitchell and Kenichi (2005).

(b) **Blue-methylene test**

This is called blue methylene "stain" test for the clay mineral characterisation, which yields a semi-quantitative evaluation of the activity of a soil based on the type and the quantity of clay minerals contained in it. The ASTM C837 and the BS EN 933–9 codes give procedures for methylene blue Test.

Table 2.3 Atterberg's limit values for the clay minerals (Mitchell and Kenichi 2005)

Clay mineral type	Liquid limit (%)	Plastic limit (%)	Shrinkage limit (%)
Montmorillonite	100–900	50–100	8.5–15
Illite	60–120	35–60	15–17
Kaolinite	30–110	25–40	25–29

(c) **Atterberg's limits**

The values of the Atterberg's limits (liquid limit, plasticity index and shrinkage limit) can be used to infer the broad group of the clay minerals present in the soil. The range of Atterberg's limit values for the three clay minerals is given in Table 2.3. The Atterberg's limits are greatly influenced by the associated cation in the clay mineral. This is the reason for wide range of values given in the Table.

2.4 Some Common Terms, Definitions and Soil Classification

The classification of the soils for the specific purposes requires understanding of certain engineering terms and definitions. The details of such terms and the definitions can be found in the books on soil mechanics and the literature on soils. A brief description of a few such terms and definitions are discussed below.

Atterberg's limits

The strength and the deformation characteristics of the soil depend upon the moisture content of the soil. The soil mixed with the water can result into a plastic paste which can be moulded into a requisite shape. As the moisture content is further increased, a state is reached where the soil deforms under its own weight and flows as a liquid. Thus, soil–water system passing from liquid state to solid state can be divided into four states (Fig. 2.6): (a) liquid state, (b) plastic state, (c) semi-solid state and (d)

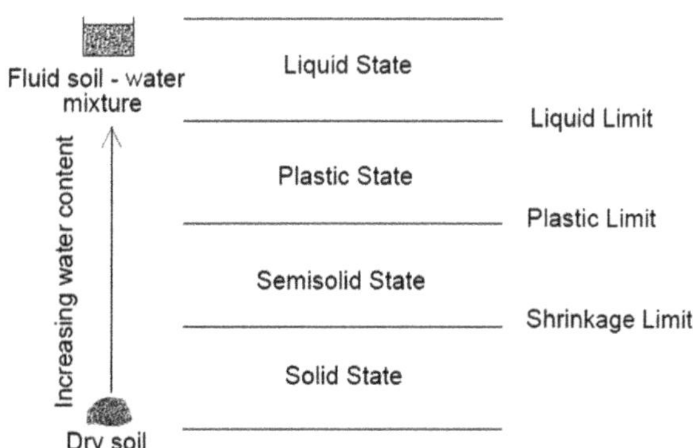

Fig. 2.6 Representation of Atterberg's limits and the related indices

solid state. The Atterberg, a Swedish agriculturist, set these arbitrary limits in terms of the water content of the soil–water system. These limits of water content are called Atterberg's limits. These limits are termed as liquid limit, plastic limit and shrinkage limit.

Liquid limit: This is the water content corresponding to an arbitrary limit between the liquid and the plastic state of soil. It is defined as the minimum water content at which the soil is still in the liquid state.

Plastic limit: This is the water content referring to an arbitrary limit between the plastic and the semi-solid state of the soil. It is the minimum water content at which the soil begins to crumble when rolled into a 3-mm diameter thread.

Shrinkage limit: The shrinkage limit is defined as the maximum water content at which a further reduction in water content will not lead to decrease in the volume of the soil mass. This represents the lowest water content at which the soil can be in completely saturated state.

Plasticity Index: This is the numerical difference between the liquid and the plastic limit of a soil. When the plastic limit cannot be determined for a soil, then it is reported as non-plastic.

The Atterberg's limits can be determined using the standard test procedures defined in various codes of practices such as IS 2720 (Part 5) (1985). The range of consistency within which the soil exhibits plastic properties is called plastic range and is indicated by the term plasticity index. The numerical values of the liquid limit and the plastic limit indicate the sensitivity of the soil to variations in the humidity and the moisture content. The Atterberg's limits and related indices are useful for the identification and the classification of soils.

Bentonite: It is a naturally occurring montmorillonite clay mineral. The bentonite has lot of affinity for water with very high swell–shrink properties.

Loam: It is a mixture of sand, silt or clay, or a combination of any of these. The term loam is originated and commonly used in the context of agricultural practices.

Murrum or moorum: It is a disintegrated rock or shale or indurated clay with or without boulder pieces. A mixture of the gravel and the red earth is also called murrum.

Indurate: Very strongly cemented and which does not soften under prolonged soaking in water.

Compressibility: The property of the soil pertaining to the volume reduction under compressive pressure is called compressibility. The quantum of reduction in the volume of soil under pressure depends upon the void ratio and the degree of saturation.

Compaction: It is defined as the process of packing soil particles closely together by mechanical means, thus increasing the dry density of the soil. The compaction leading to the density increase mainly happens due to reduction in the air voids under a loading of short duration.

2.5 Particle Size Distribution

The grain size distribution or the particle size distribution curve gives the details of the sizes and the percentages of various particles present in the soil. The particle size analysis of a soil is performed in two stages: (a) sieve analysis and (b) sedimentation analysis. The salient steps followed in the particle size analysis of a soil are as follows.

(a) A known weight of oven dried soil is washed through a 75-μ sieve. The material retained on the sieve is sand and gravel. Material passing through is silt and clay.
(b) The material retained is dried and sieved through a set of sieves having different aperture sizes. The percentage finer than each sieve size can be calculated.
(c) The natural soil particles passing through 75 μ sieve (washed) is dried at ambient conditions.
(d) A small representative portion of the dried silt and clay portion is subjected to sedimentation analysis to ascertain the percentages of various fine particle sizes.
(e) Percent finer than each particle size is plotted against particle size to obtain particle size distribution curve.

IS 2720 (part 4) (1985) and many such standard codes of practices give a detailed procedure to carry out the sieve analysis and the sedimentation analysis. Laser diffraction analysis is also widely used technique for particle size analysis.

The particle size distribution curves for kaolin clay, river sand, and three different types of soils are shown in Fig. 2.7. The kaolin clay was procured from a commercial supplier, which was probably ground to certain fineness, while processing the original solid lumps of kaolin. Hence, the curve for kaolin clay shows fractions more than 2 μm. The coarser fractions of the clay are also kaolin as indicated by the X-ray diffraction analysis. The curve for Soil-A is well graded. The soil is said to be well graded when it has good representation of particles of all sizes. The soils deficient in certain particle sizes or having excess of particles of certain sizes can be termed as poorly graded soils. The soils containing particles of same size are known as poorly graded or uniformly graded soils. The Soil-B is fine grained and uniformly graded. The Soil-C is uniformly graded and coarse grained. The poorly graded or the uniformly graded soils can pose problems in the compaction process and the green strength for the compressed earth blocks. The particle size distribution curve helps in determining the quantity of the sand to be added to a soil such that the resulting mixture could have optimum clay and sand percentages for the manufacture of the stabilised compressed earth blocks. The information from the particle size distribution curve along with the values of Atterberg's limits can be used for the classification of soils.

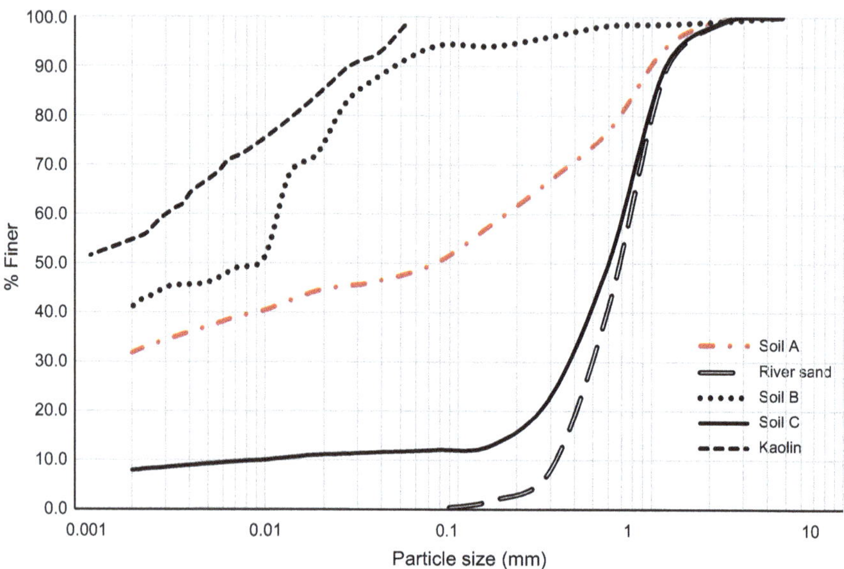

Fig. 2.7 Particle size distribution curves

2.6 Classification of Soils

The soils can be classified into some general groups according to their composition and other characteristics. Such classifications help in finding the suitability of a soil for the production of stabilised compressed earth blocks. There are several soil classification systems used for different engineering applications. Two types of soil classification systems are briefly discussed below, which could help in identifying a range of soils suitable for the stabilised compressed earth blocks.

(a) **Textural classification**

The soil consists of sand, silt and clay size particles in different percentages. The soil classification exclusively based on particle size distribution is known as textural classification. A triangular classification of United States Public Roads Administration based on the percentages of sand, silt and clay size fractions of a soil is shown in Fig. 2.8. To use this chart, lines are drawn parallel to the three sides of the equilateral triangle (as indicated by the arrows in the Figure), based on the percentages of sand, silt and clay size fractions of the soil. The lines so drawn will intersect at a point in the chart indicating the type of the soil. For example, 15% clay, 60% sand and 25% silt size fractions will lead to a region of sandy-loam soil. The textural classification is more useful for coarse-grained soils rather than clayey soils since the properties of the coarse-grained soils are less dependent on the particle size distribution.

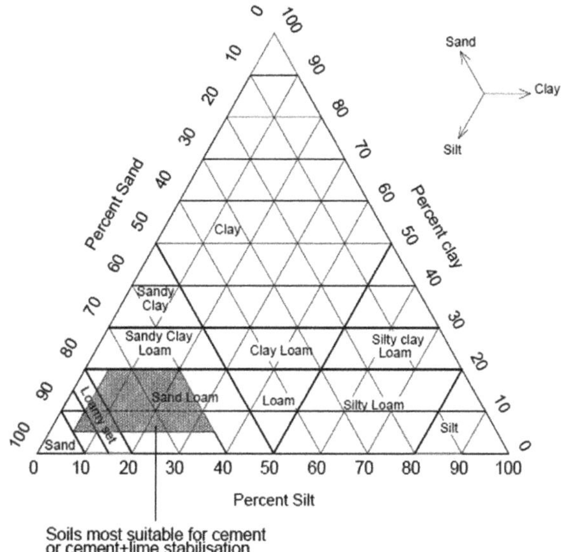

Fig. 2.8 Triangular textural classification chart

(b) **Unified soil classification system and the Indian Standard classification**

The Unified Soil Classification (USC) system was first developed by Casagrande. The USC system is based on both the particle size distribution and the plasticity properties (Atterberg's limits). A modified USC system as adopted by I.S. 1498–1970 code is given in Tables 2.4 and 2.5. The USC system is related to the physical properties inherent in the soils and is useful for a variety of engineering problems involved with soils.

Coarse-grained soils: Table 2.4 gives classification for the coarse-grained soils. The coarse-grained soils having broadly two symbols G and S are used in combination with other symbols (W, C, P or M) to designate the type of the coarse-grained soil. For example, GC means clayey gravel. The details of such combination symbols along with the description of the soil type are given in Table 2.5.

Fine-grained soils: Table 2.4 gives details of classification for the fine-grained soils. Various symbols (M, C, O, L, I and H) are used to describe these soils. The combinations of these symbols indicate the type of the fine-grained soil. For example, the symbol ML indicates inorganic silt with medium compressibility. Table 2.5 gives the descriptions of such possible symbols. A plasticity chart shown in Fig. 2.9 could be used to classify the fine-grained soils. The figure shows a line designated as A-line which divides the inorganic clays from the silt and the organic soil. Equation of the A-line is as follows.

$$\mathrm{PI} = 0.73(\mathrm{LL} - 20) \tag{2.1}$$

where PI = Plasticity Index (%) and LL = Liquid limit (%).

Table 2.4 Classification of soils (modified USC system as per IS 1498–1970)

Soils					
Coarse-grained soils		Fine-grained soils			Highly organic soils and other miscellaneous materials
Material retained on 75-μ sieve is > 50%		Material passing through 75-μ sieve is > 50%			
Gravels (**G**)	Sands (**S**)	Inorganic silts and very fine sands (**M**)	Organic clays (**C**)	Organic silts and clays, & organic matter (**O**)	
More than 50% of coarse fraction (retained on 75 μm sieve) is retained on 4.75 mm sieve	More than 50% of coarse fraction (retained on 75 μm sieve) is finer than 4.75 mm sieve	**L**: Silts and clays of low compressibility, having LL < 35% **I**: Silts and clays of medium compressibility, having LL > 35% & < 50% **H**: Silts and clays of high compressibility, having LL > 50%			Contain large amount of fibrous organic matter such as peat, decomposed vegetation. In addition, soils containing shells and other non-soil materials in sufficient quantities
W: Well graded **C**: Well graded with excellent clay binder **P**: Poorly graded **M**: Contain fine materials not covered in other groups					

The modified USC soil classification system as described in the above sections appears little cumbersome, especially for non-civil engineers. The textural classification and USC classification systems reveal the characteristics and the behaviour of soils as a first attempt to accept or reject soils suitability for the stabilised compressed earth blocks and the rammed earth. The range of soils suitable for the stabilised compressed earth blocks and the rammed earth especially using cement or a combination of cement and lime as stabiliser is marked in Figs. 2.8 and 2.9. More discussion is provided in the subsequent sections for making use of these classification systems for selecting ideal soils or modifications to the soils to make them suitable for the stabilised compressed earth blocks and the rammed earth.

2.7 Commonly Occurring Soils

(a) **Red loamy soil or red earth**

These are reddish to dark brown coloured soils. They generally contain predominantly less expansive clay minerals such as kaolinite and the clay size fraction (< 2 μm) ranges between 5 and 40%. The swelling and the shrinkage associated with such soils is very small when compared to the swelling and shrinkage of expansive soils such as black cotton soil.

Table 2.5 Group symbols and their details (IS 1498–1970)

Major division	Group symbol	Typical names and details
Gravels	GM	Silty gravels, poorly graded gravel-sand-silt mixtures
	GC	Clayey gravels, poorly graded gravel-sand-silt mixtures
Sands	SW	Well graded sands or gravely sands, little or no fines
	SP	Poorly graded sands or gravely sands, little or no fines
	SM	Silty sands, poorly graded sand-silt mixtures
	SC	Clayey sands, poorly graded sand-clay mixtures
Silts and clays	ML	Inorganic silts and very fine sands, silty or clayey fine sand or clayey silts with none to low plasticity
	CL	Inorganic clays, gravely clays, sandy clays, lean clays of low plasticity
	OL	Organic silts of low plasticity
	MI	Inorganic silts, silty or clayey fine sands or clayey silts of medium plasticity
	CI	Inorganic clays, gravely clays, sandy clays, silty clays, lean clays of medium plasticity
	OI	Organic silts and organic silty clays of medium plasticity
	MH	Inorganic silts of high compressibility, micaceous or diatomaceous fine sandy or silty soils, elastic silts
	CH	Inorganic clays of high plasticity, fat clays
	OH	Organic clays of medium to high plasticity

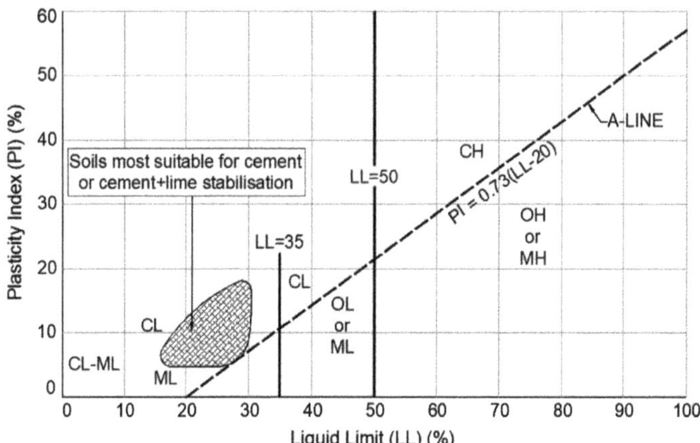

Fig. 2.9 Plasticity chart (IS 1498–1970)

(b) **Black cotton soils**

These soils contain considerable amount of expansive clay mineral such as montmorillonite. Such soils are characterised by the presence of very hard clay lumps in dry condition. The soils shrink badly (volumetric shrinkage 200–300%) on drying, leading to wide cracks in the ground. These soils possess high liquid limit, in excess of 50% depending upon the quantity of clay fraction and type of exchangeable ion associated with the clay mineral. It is difficult to stabilise the black cotton soils using Portland cement alone.

(c) **Ash-Modified Soils**

It is a special kind of soil often found in black cotton soil (BCS) zones. This soil is *called haalu mannu* (in Kannada), or as *ghad ki mitti* (in Hindi). Kannada and Hindi are local dialects in India. Such soil deposits are mined, and the soil is used for the construction of buildings, for walls and roofs. These soils contain 2 – 3% of organic matter in addition to 5 – 6% dissolved salts. They are known to be artificially made soils, by mixing organic matter and wood ash and allowing it for very long durations of natural cyclic wetting and drying cycles. These soils exhibit low swelling and shrinkage characteristics and are silty in nature. The haalu mannu is well suited for stabilised compressed earth block production using lime and cement. The properties of stabilised compressed earth blocks using ash-modified soils were investigated in greater detail by Reddy et al. (2003). The ash-modified soils (AMS) show excellent reactivity with stabilisers such as lime and cement when compared to the natural soils such as black cotton soil and red loamy soil (RS). The experiments have clearly shown that for any given density and stabiliser content, compressed earth blocks using AMS possess considerably much higher compressive strength than the ones using BCS and RS soils. The long-term behaviour of stabilised compressed earth blocks using AMS soils greatly depends upon the stabiliser to clay ratio of soil-sand mixture used for the compressed earth block production. The stabilised compressed earth blocks using AMS soil show stable long-term strength when the stabiliser-to-clay ratio $\geq$ 1.0 (Reddy et al. 2003).

(d) **Laterite soils**

The term laterite was first introduced by Buchanan (1807). The laterite was described as ferruginous, vesicular, unstratified and porous material with yellow ochres. The laterite can be cut into blocks and used for masonry. The Latin word "later" means brick. The laterite soils have low silica, high alumina in hydroxide form comparable to the form of bauxite (Gidigasu 1976). The laterite occurs up to several meters depth. Fermor (1911) defined various forms of laterites based on the relative contents of laterite constituents (Fe, Al, Mn and Ti) in relation to silica. Martin and Doyne (1927, 1930) defined three grades of laterite soils based on the silica–alumina ratio (SiO_2/Al_2O_3). The higher proportion of sesquioxides of iron (Fe_2O_3) and aluminium (Al_2O_3) relative to other chemical components is a feature of characteristic of laterite soils.

The laterite deposits are highly porous and pervious, and the laterite is a highly weathered material. The laterite soils are generally acidic in nature and require special attention while using them for stabilised compressed earth blocks. The pH of lateritic soils is generally in the range 4.8–6.8 (Evans 1958 and Dumbleton and Newill 1962). Some of the clay minerals occurring in lateritic soils include kaolinite and halloysite (Dumbleton and Newill 1962). In general, 2–3% hydrated lime may be used in addition to the Portland cement for stabilising the laterite soil. The lime is required to neutralise the soil's acidity.

The laterite hardens on dehydration and on exposure to air (Buchanan 1807; Alexander and Cady 1962). The studies on some Indian lateritic deposits by Nanda and Krishnamachari (1958) showed that the hardening of laterite may be due to (a) dehydration of hydroxides of iron and aluminium present in the soil, (b) further oxidation of the high iron content of the soil and (c) precipitation of these oxides as cementing material.

References

ASTM C837 – 09 (2019) Standard test method for methylene blue index of clay. American Society for Testing and Materials, West Conshohocken, PA, USA (Reapproved)

Alexander LT, Cady JG (1962) Genesis and hardening of laterite in soils. U.S. Dept Agric Tech Bull 1282

BS EN 933-9:2009+A1:2013, Tests for geometrical properties of aggregates Part 9: assessment of fines—ETHYLENE blue test. British Standards Institution, London, UK

Buchanan F (1807) A journey from Madras through the countries of Mysore, Canara and Malabar, 2. East Indian Company, London, pp 436–560

Dumbleton MJ, Newill DD (1962) A study of the properties of 19 tropical clay soils and the relation of these properties with the mineralogical constitution of the soils. British Research Lab Note LN/44

Evans EA (1958) A laboratory investigation of six lateritic soils from Uganda. British Road Research Lab Note 3241

Fermor LL (1911) What is laterite? Geol Mag 5(8), pp 453–566

Gidigasu MD (1976) Laterite soil engineering. Elsevier Publishing Company, Amsterdam, The Netherlands

Grim RE (1953) Clay minerology. McGraw-Hill, New York

https://www.minersoc.org/images-of-clay.html

IS 1498 – 1970 (2007) Classification and identification of soils for general engineering purposes, Bureau of Indian Standards. New Delhi, India (reaffirmed)

IS 2720 (Part 4) 1985 (2007) Methods of tests for soils—Part 4 Grain size analysis. Bureau of Indian Standards, New Delhi, India (reaffirmed)

IS 2720 (Part 5) 1985 (2006) Method of test for soils determination of liquid and plastic limit. Bureau of Indian Standards, New Delhi, India (reaffirmed)

Mitchell JK, Kenichi S (2005) Fundamentals of soil behaviour. Wiley, New Jersey, USA

Martin and Doyne (1927) Laterite and lateritic soils in Sierra Leone—1. J Agric Sci 17, pp 530–546

Martin and Doyne (1930) Laterite and lateritic soils in Sierra Leone—2. J Agric Sci 20, pp 135–143

Nanda RL, Krishnamachari R (1958) Study of soft aggregates from different parts of India with a view to their use in road construction. II. Laterites. Central Road Research Institute paper 15

Reddy BVV, Rao SM, Arun Kumar MK (2003) Characteristics of stabilised mud blocks using ash-modified soils. Indian Concr J February, pp 903–911

USC - ASTM D2487-17e1, Standard practice for classification of soils for engineering purposes (Unified soil classification system). ASTM International, West Conshohocken, PA, 2017, USA

Chapter 3
Soil Stabilisation

3.1 Introduction

The soils are used for the construction of roads, pavements, retaining structures such as bunds, tanks and dams and for the buildings. The building products especially used for superstructure are expected to possess certain desirable engineering properties (strength, low permeability, less erosion, low swelling and shrinkage characteristics, moisture buffering, etc.). The deficiencies in the expected engineering properties and the engineering performance of the soils can be improved by the process of stabilisation. In the broadest sense, the soil stabilisation is the alteration of any property of a soil to improve its engineering performance. The stabilisation techniques are employed in a variety of engineering applications. Some of them include construction of sub-base and base courses for roads and pavements, canal lining, ground improvement for foundations, reducing the permeability and the compressibility of the hydraulic works, retaining structures, production of the building materials (bricks, blocks, rammed earth, etc.), renderings to masonry walls and many other applications. Thus, the scope for the use of stabilisation techniques is very wide. The soil stabilisation techniques can be classified into two broad categories: (1) In-situ ground stabilisation (generally referred to as ground improvement) and (2) Stabilisation of excavated soils used for any specific purpose. The second type of stabilising technique is discussed here, especially for the production of the stabilised compressed earth blocks and the stabilised rammed earth for the construction of the superstructure of the buildings and other structural applications. The stabilisation techniques of excavated soils can be classified into three broad categories as shown in Fig. 3.1.

B. V. V. Reddy, *Compressed Earth Block & Rammed Earth Structures*,
Springer Transactions in Civil and Environmental Engineering,
https://doi.org/10.1007/978-981-16-7877-6_3

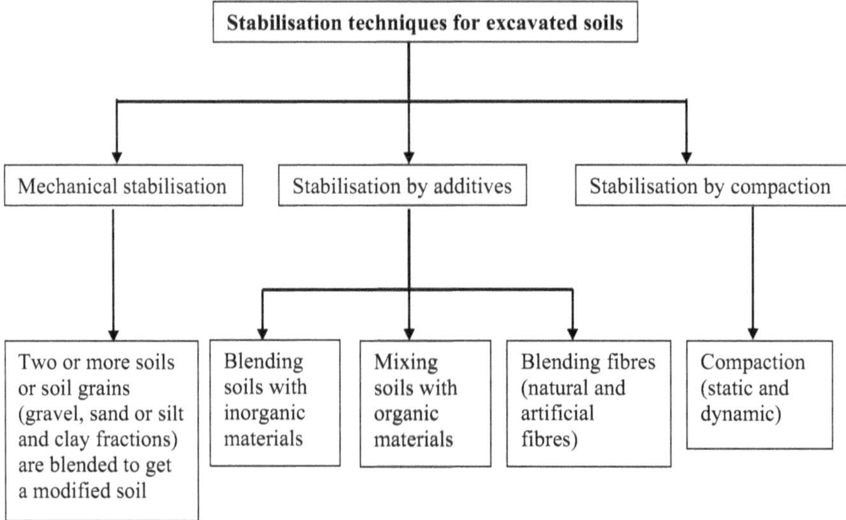

Fig. 3.1 Stabilisation techniques for excavated soils

3.2 Need for the Stabilised Earth Products in Superstructure

The superstructure of the load bearing buildings consists of the walls and the floors/roof. It is common to find use of the soil/earth in the construction of the traditional mud roofs and the vaults and domes. The cob, adobe, rammed earth and the compressed earth blocks represent the major types of earthen wall construction techniques, where bulk of the material used is soil or earth. The important requirements for an earthen wall are as follows:

1. Adequate strength to resist loads
2. Resistance to moisture ingress
3. Volume stability or dimensional stability, i.e. minimum swelling and shrinkage properties
4. Durability, i.e. resistance to rain erosion and mechanical damage.

Very rarely, the natural soil in its native state is suitable for the earth wall construction satisfying all the above-mentioned requirements. Hence, the properties of the natural soil must be tampered with or altered, i.e. stabilised to make the soil suitable for the earth wall construction and other superstructure applications.

3.3 Soil Stabilisation Techniques

There are many soil stabilisation techniques as applied to several engineering applications. The stabilisation techniques of the excavated soils can be classified into three broad categories (as shown in Fig. 3.1) apart from thermal stabilisation. Thermal stabilisation basically involves heating/firing the soil mass or the soil product. For example, heating at about 800 °C can convert the clay product into a burnt clay brick. This aspect of stabilisation is not the subject matter of this book. More information on thermal stabilisation can be found in the literature on ceramic science. There is vast amount of published material on the soil stabilisation and the soil stabilisation techniques. Some of these publications are listed under the section on references and bibliogaphy on soil stabilisation at the end of the chapter.

In majority of the soil stabilisation processes, more than one of the methods mentioned in Fig. 3.1 are adopted. For example, for stabilised rammed earth construction or the stabilised compressed earth block production, all the three methods mentioned in Fig. 3.1 are used.

3.3.1 Mechanical Stabilisation

This is the most common and inexpensive method of soil stabilisation. In simple terms, it is nothing but mixing of two or more soils or mixing sand and gravel to a soil. Generally, soils have either excess fines (clay/silt) or deficient in sand fraction. To make them suitable for the earth/mud wall construction, it is essential to either dilute the soil by the addition of sand/gravel or mix with another soil of a different composition. Thus, the original natural soil properties are altered to suit a specific earth wall construction technique. This type of soil stabilisation technique can be seen in a large number of applications as mentioned below.

(a) Blending of clayey soils with sandy soils for reducing the drainage through earthen bunds
(b) Mixing of sand or sandy soils with clayey soils to control shrinkage cracks in adobe block production and the earth-based plasters
(c) Production of compressed earth blocks using highly silty soils. Here, coarse sand or clayey soil is added to the highly silty soil in order to facilitate compaction and impart green strength for the blocks.
(d) Gravel and gravel rich soils are blended with normal soils to prepare base courses of water-bound Macadam roads.
(e) Controlling the clay content of the soils by blending with sand for making the mix suitable for the cement stabilised compressed earth blocks.

Many more such applications can be found where the blended soils are used for different engineering applications. For example, natural soil characteristics are altered by diluting the soil with the addition of sand as illustrated in Table 3.1. Here,

Table 3.1 Altering soil characteristics by changing composition

Type of soil	Composition (%, by mass)			Atterberg limits (%)		Compaction characteristics	
	Sand	Silt	Clay	LL	PI	OMC (%)	MDD (g/cc)
Natural soil	50.3	18.1	31.6	40	21	15.6	1.83
Reconstituted soil	72.6	11.6	15.8	27	17.5	9.4	2.00

the original sand, silt and clay fractions of the soil were changed by diluting the soil with equal amount of sand fraction. The index properties and compaction characteristics of the natural soil are drastically modified by diluting with the sand. The standard Proctor maximum dry density (MDD) of the reconstituted soil increases while reducing the liquid limit (LL), plasticity index (PI) and standard Proctor optimum moisture content (OMC) values, when compared with the properties of the natural soil.

3.3.2 *Stabilisation by Compaction*

Compaction is the well-known soil stabilisation technique to improve the strength and to reduce porosity of the compacted products. Here, the soil in a loose state is brought to a dense state by supplying compaction energy. Two types of generally employed compaction techniques are: static compaction and dynamic compaction. Improvement in strength, porosity and other properties depends upon the soil composition and the compaction energy supplied. Different types of compaction methods or techniques are employed in the compaction of soils for various engineering applications. Construction of the earthen bunds and dams, embankments, retaining walls, base course for roads, etc., employs vibratory rollers or other conventional rollers for the compaction of different soil layers to a specified density. The preparation of bases for the flooring in buildings, the foundations and the construction of rammed earth walls, the compaction is carried out by rammers dropping from a height or using vibratory rammers. The compaction processes such as pugging or kneading are employed in the stiff mud process of clay brick moulding. Partially saturated soil can be densified by employing a confined compaction process in a mould using a piston. This kind of compaction can be termed as static compaction. Manufacture of the high-density ceramic products, the brick manufacture using powdered partially saturated or dry soils and the stabilised compressed earth block (CEB) production employ the static compaction process. Details of the compaction techniques employed for the CEB production are discussed in the subsequent sections.

3.3.3 Stabilisation by Additives

In this type of stabilisation, both organic and inorganic additives either in solid or liquid form are added to alter the properties of the soil. Additives such as cement, lime, bitumen, polymers, certain types of salts, organic binders and organic and inorganic fibres are used. The selection of the type of additive depends upon the kind of alteration expected in the natural soil property and the properties of the ultimate product/application. For example, if it is required to control shrinkage cracks on drying, use of fibres is recommended or if there is a need to control the swelling of a soil, lime is added or improvement in erosion characteristics of adobe blocks can be achieved by adding bitumen.

There are many studies on the effect of chemical additives on the characteristics of the cement/lime stabilised mixtures (Mehra et al. 1955; Lambe et al. 1960; Davidson et al. 1960; Mateos and Davidson 1961; Moh 1962; Dan Marks and Allan Haliburton 1972; Ingles and Metcalf 1972). The chemical additives such as sodium-based salts and alkali metal compounds in small dosages have been explored. Sodium carbonate, sodium silicate, sodium metasilicate, sodium hydroxide, sodium sulphate, gypsum and magnesium sulphate represent some of these chemicals. The studies show small dosages of some of these chemicals, improve the strength of the cement/lime stabilised soil mixtures. Determining the optimum dosages of chemicals is difficult and tricky. Addition of dosages more than optimum values result in adverse effects on the stabilised soil or earth product.

3.4 Cement Stabilisation

The cement stabilisation is the most commonly employed and widely used technique. The cement stabilisation is ideal for stabilising coarse grained and gravely soils. Soils with high clay content are difficult to mix, and high cement content is required for an appreciable change in the properties of the stabilised earth product. The cement stabilisation is used for improving the strength (especially in saturated condition), the resistance against rain erosion and the mechanical damage of the earth-based products. The principle involved is illustrated below.

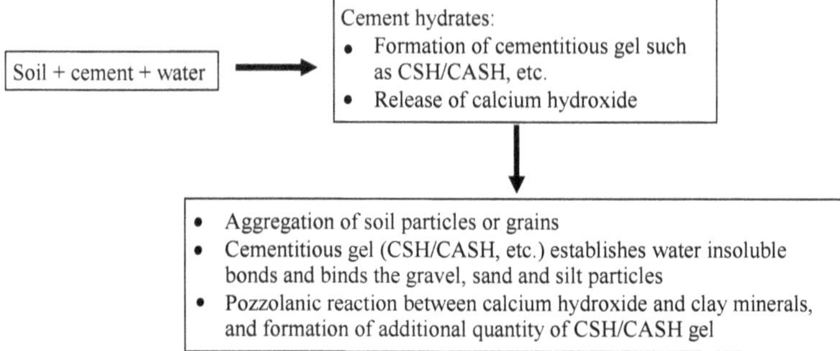

Herzog and Mitchell's (1963) investigations on the reactions accompanying the stabilisation of the pure clay minerals (kaolinite and montmorillonite) with cement proved that there is a pozzolanic reaction between the calcium hydroxide released during the hydrolysis and the hydration of the Portland cement with the clay minerals. This type of reaction is called the secondary reaction contributing additional strength to the cement–clay matrix. The quantity of additional cementitious material formed due to the secondary reactions is limited by the amount of calcium hydroxide released in the cement hydration process and may not be sufficient to react with all the clay particles present in the earth mix.

The cement stabilisation techniques are employed in the CEB production and the rammed earth wall construction. Major factors affecting the characteristics of the cement stabilised soils are as follows:

(a) Soil composition
(b) Percentage of cement added
(c) Degree of compaction achieved for the product
(d) Curing period and curing temperature.

The strength of the cement stabilised earth product increases with the increase in the cement content, density, curing period and the curing temperature. The characteristics of the cement stabilised CEB and the rammed earth are discussed in the subsequent chapters.

3.5 Lime Stabilisation

Lime: Naturally available lime stone (to a limited extent sea shells) is the source for lime. The lime stone is essentially the calcium carbonate. The following chemical reactions take place while processing the lime stone to obtain lime, i.e. calcium hydroxide.

3.5.1 Burning of Limestone

$$CaCO_3 \xrightarrow[\sim 1000\,^\circ C]{Heat} CaO + CO_2 \quad (3.1)$$

The raw limestone is burnt in kilns at about 1000 °C. In this process, the carbon dioxide is released leaving the calcium oxide. The calcium oxide (CaO) is called the quick lime and has lot of affinity for the water.

3.5.2 Slacking of Lime

The slacking is the process of adding water to quick lime to get hydrated lime, i.e. $Ca(OH)_2$.

$$CaO + H_2O \rightarrow Ca(OH)_2 \quad (3.2)$$

The hydrated lime must be stored in airtight bags to postpone the re-carbonation of the unstable product. Generally, the hydrated lime is used in the lime stabilisation process. The lime stabilisation techniques are generally adopted for heavy clay soils, to control the swelling and the shrinkage. The principle of lime stabilisation is as follows.

3.5.3 Lime-Clay Reactions

Mixing of the soil and the calcium hydroxide results in four basic reactions: cation exchange, flocculation, agglomeration and pozzolanic reaction. The first three reactions are responsible for the alterations in plasticity, shrinkage and the workability characteristics. The carbonation effect slightly contributes to the strength gain in soil–lime mixtures with time. The pozzolanic reaction is the main contributor to the strength development in the lime-soil mixtures. The earlier studies on the lime stabilised soils (Herrin and Mitchell 1961; Davidson and Handy 1960; Eades and Grim 1960; Glenn and Handy 1963; Mateos 1964; Thompson 1966) substantiate the above-mentioned summary.

The clay minerals are basically the complex aluminous-silicates. The lime (calcium hydroxide) reacts with the clay minerals. The lime-clay reactions (pozzolanic reactions) lead to the formation of the water insoluble gel of the silicate and the silicate-aluminates, and this gel finally crystallises into CSH/CASH gel with time. The cementing agent is like that found in the hydration of the ordinary Portland cement. The cementitious gel formed coats the soil particles and establishes bonds. The pace of lime-clay reactions and the formation of CSH/CASH products is slow. These reactions will be continuously taking place (in the presence of moisture) and can extend over to several years. The lime-clay reactions can be accelerated by raising the curing temperature. The lime stabilisation is a must for soils containing expansive clay minerals (such as montmorillonite), to control the swelling and the shrinkage problems associated with such soils. Also, it is common to employ the lime stabilisation techniques for clayey soils. The characteristics of the lime stabilised earth blocks are discussed in the subsequent chapters.

3.6 Soil Compaction

3.6.1 Soil as a Three-Phase System

A heap of processed moist soil consists of solids, water and air. The soil can be represented as a three-phase material as illustrated in Fig. 3.2. Some useful relationships among the unit weight (or density), the void ratio and the water content are as follows.

$$S\,e = w\,G_s \tag{3.3}$$

$$\gamma_d = (W_s \div V) = (G_s\gamma_w) \div (1 + e) = (G_s\gamma_w)\,(1 - n) \tag{3.4}$$

where, S: Degree of saturation; W_s: Weight of soil solids; e: Void ratio; w: water content; G_s: Specific gravity of solids; γ_d: Dry density; γ_w: Unit weight of water; V: Volume; n: Porosity.

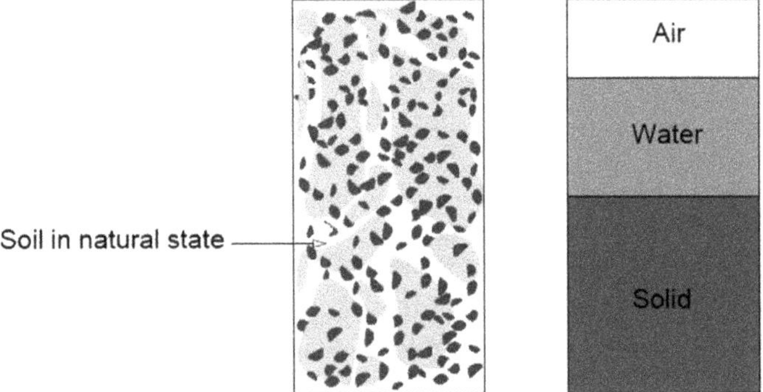

Fig. 3.2 Different phases of soil mixture

3.6.2 Soil Compaction Process

The densification is the most commonly employed method to improve the strength of the soil. The soil density can be increased through the process of compaction. The compaction is the process of reducing the air voids volume in the soil through the application of a short duration loading. For a given soil, the density achieved through compaction process is a function of the compaction energy supplied and the soil moisture content. The compaction energy can be supplied through the application of either the static or the dynamic force. During the compaction process, air present in the voids of the soil is expelled resulting in volume reduction and hence the increased density. The soil compaction requires the mechanical energy. On addition of the water to the soil, the clay particles present in the soil absorb water and develop plasticity and facilitate compaction process. During the compaction, the clay particles with a water layer around them facilitate slipping over of soil particles leading to a densely packed condition.

3.6.3 Dynamic Compaction

The moisture content of the soil during compaction and the density achievable are closely related. The moisture content–density relationships are dependent on the quantum of energy supplied and the method of compaction (static or dynamic compaction). The most commonly employed method to understand the compaction process in majority of the engineering applications is the standard Proctor test. In this type of test, the soil is compacted in a cylindrical mould in three layers with constant energy input (a hammer with fixed mass is dropped over a fixed height with 25 blows per layer). The moisture content and the dry density of the compacted soil mass are estimated. A relationship of the type shown in Fig. 3.3 is obtained for a normal soil. The curve shows a distinct optimum value of the maximum density achieved for specific moisture content. This dry density is referred to as the maximum dry density (MDD) and the corresponding moisture content is designated as the optimum moisture content (OMC). If the compaction energy is increased, the MDD increases, and the corresponding OMC reduces. A compaction curve for the modified Proctor test is shown in Fig. 3.3 along with the standard Proctor curve. The energy supplied in the standard Proctor test is 0.60 MJ/m^3, whereas in the case of modified Proctor test, the compaction energy supplied is 2.70 MJ/m^3. Both these tests are dynamic compaction tests.

Zero air voids line: Theoretical maximum dry density is obtained when there is no air in the voids, i.e. the degree of saturation is 100%. The line representing maximum dry density at various moisture contents gives zero air voids line.

$$\text{Zero air voids dry density} = \gamma_d = (G_s\gamma_w) \div (1 + wG_s) \quad (3.5)$$

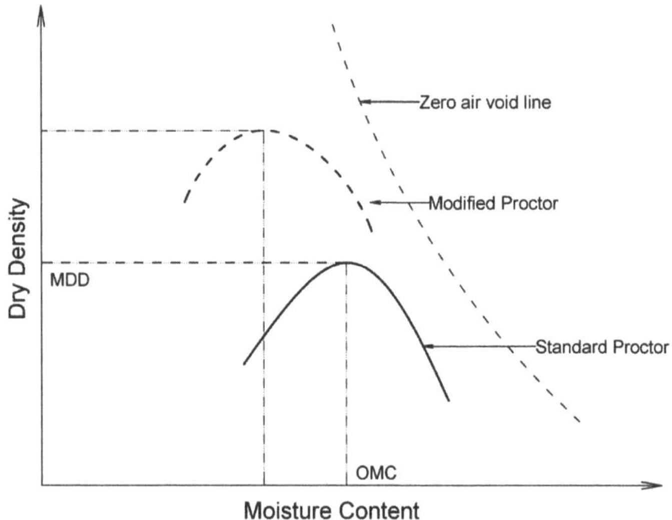

Fig. 3.3 Typical moisture density relationships

Figure 3.3 shows the zero air voids line. Compaction curves under no circumstances lie to the right side of the zero air voids line. Thus, the zero air voids line represents the limiting value of the dry density achievable for a given moisture content.

3.6.4 Factors Affecting Soil Compaction

The moisture content, the soil composition and the compaction energy are the major factors affecting the soil compaction. The moisture content of the soil mixture has significant influence on the soil compaction characteristics, as discussed in the previous sections. Increase in the compaction energy input leads to higher density and lower void ratio. The soil composition and the associated clay mineral affect the standard Proctor OMC and the MDD. Table 3.2 gives some typical results of the compaction tests on soils. The soil composition, the clay mineral type, the standard Proctor MDD and OMC values are detailed in the table. The soils with expansive clay mineral (montmorillonite) give higher OMC and lower MDD when compared with the values for soils having less expansive clay minerals such as kaolinite. Also, within the kaolinite clay group of soils, the soil with higher sand content shows lower OMC and higher MDD values, when compared to the values of the soils with more silt/clay content.

Table 3.2 Typical compaction characteristics of soils

Sl. No.	Soil composition (% by mass)			Type of clay	Standard proctor	
	Sand	silt	clay		OMC (%)	MDD (kg/m^3)
1	50.3	18.1	31.6	Kaolinite	15.6	1825
2	87.3	4.8	7.9	Kaolinite	8.5	1937
3	28.8	53.2	18.0	Kaolinite	21.4	2049
4	74.6	9.2	16.2	Kaolinite	12.3	2059
5	43.8	13.2	43.0	Kaolinite	16.5	1815
6	28.0	32.0	40.0	Montmorillonite	26.0	1501

3.6.5 Static Compaction of Soils

The static compaction techniques are employed in the production of compressed earth blocks. In the static compaction method, the soil is compacted through the gradual application of the static force. In the production of the stabilised compressed earth block using manually operated or automatic machines, the processed loose soil is confined in a metal mould, and the compaction is achieved by the gradual movement of a piston. The piston stroke length will depend upon the thickness of the CEB to be produced. The static compaction process as applied to the compressed earth block production can be classified into two categories.

(a) **Constant peak stress-variable stroke compaction**

In this method of compaction, the external stress is gradually varied at a definite rate (or a regime of different rates) till a specific peak stress is reached. The thickness of the compacted specimen will vary depending upon the soil moisture content and the soil composition. The static compaction tests of this type have been carried out by Turnbull (1950), Fitzmaurice (1958) and Olivier and Mesbah (1987). The compaction curves somewhat similar to the standard Proctor compaction test curves (Fig. 3.4) were generated by this type of static compaction tests. Increasing the peak stress during compaction, in general, leads to lower optimum moisture content. This is analogous to the dynamic compaction tests wherein an increase in compaction energy per unit volume leads to the lower optimum moisture content. These types of static compaction tests do not attempt to evaluate the compaction energy per unit volume.

(b) **Variable peak stress-constant stroke compaction**

In this type of compaction, the static force is gradually applied to the processed soil mass until a specific final thickness (volume) is achieved. The static force applied at the end of the compaction stroke length, can vary, depending upon the moisture content, the soil composition and the mass of the soil. This kind of soil compaction operation is very much like the one employed in the production of the compressed earth blocks using either manual or mechanised machine. The machines employed

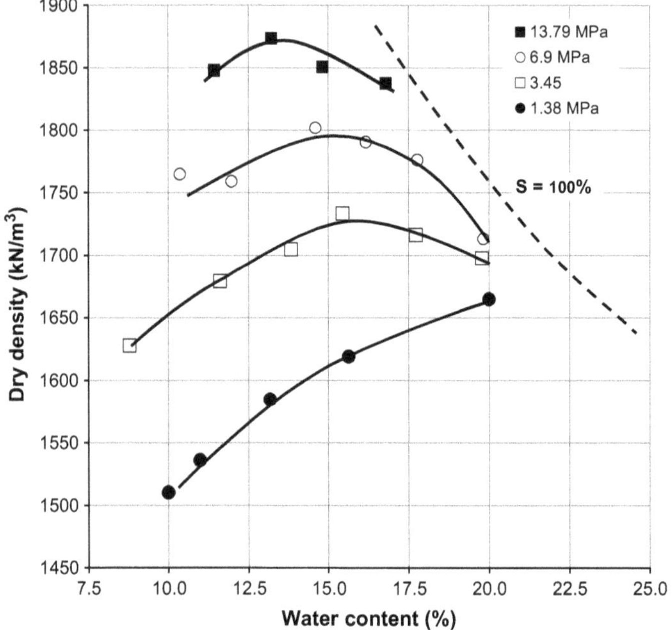

Fig. 3.4 Typical compaction curves for constant peak stress-variable stroke length (after Turnbull 1950)

for compressed earth block production are limited by the available compaction force and the piston displacement.

3.6.6 *Compaction of Soils to Produce Compressed Earth Block (CEB)*

The static compaction process is employed in the production of compressed earth blocks using either manual or automated machines as illustrated in Fig. 3.5. It is clear from this figure that the soil gets compacted in two stages: initially by the closing of the lid from top (Fig. 3.5b) and then by the piston movement from the bottom (Fig. 3.5c). The magnitude of compaction due to the lid closure and the piston stroke length will depend upon the final thickness and the density of the block produced. The compression ratio is an important parameter in controlling the block density depending upon the block thickness and can be defined as: ***Compression ratio = (Compaction stroke length + Final block thickness) ÷ (Final block thickness)***

The investigations of Reddy (1983, 2015) showed that the compression ratio should be at least 1.60 for a block thickness of 90–100 mm having a dry density of about 1800 kg/m^3. Understanding the magnitude of the static compaction force required

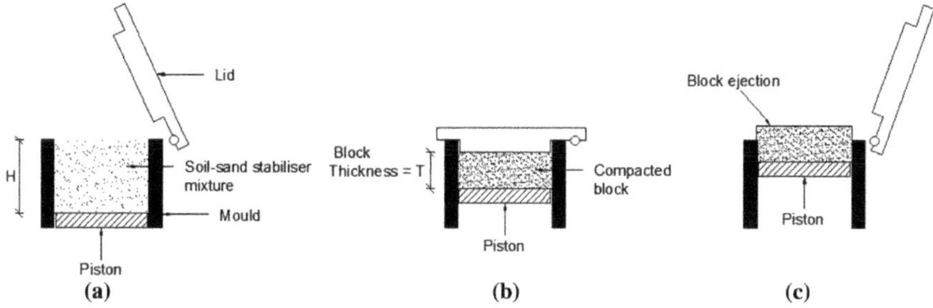

Fig. 3.5 Compaction process for compressed earth block production

to compact a CEB and its variation as the block compaction proceeds due to the piston movement becomes important in assessing a machine capacity to produce the compacted CEB's. The force–piston stroke length relationships for a sandy soil and clayey soil are shown in Fig. 3.6 (Reddy 1983). The clayey soil had a clay (kaolinite) fraction of 26%, and the sandy soil (with a clay fraction of 8%) was a reconstituted one obtained by diluting the clayey soil with the sand. The standard Proctor OMC values were 14.3% and 12.9% for the clayey soil and the sandy soil, respectively. The blocks (size: 305 × 144 × 100 mm) were compacted at the standard Proctor OMC to a dry density of 1850 and 1900 kg/m^3 for the clayey soil and the sandy soil, respectively. The force–stroke relationships reveal several features of the static soil compaction. The compaction forces increase slowly in the beginning but attain very large values at the completion of the stroke. The maximum force required to compact the soil blocks using the clayey soil (dry density: 1850 kg/m^3) and the sandy soil (dry density: 1900 kg/m^3) was 3.9 tonnes and 18.5 tonnes, respectively. The sandy

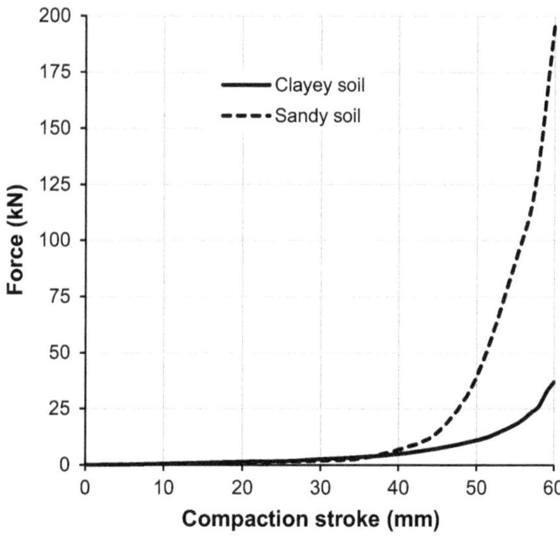

Fig. 3.6 Force–compaction stroke relationships for compressed earth block

soil needs nearly four times the maximum force needed for clayey soil. The sandy soils demand larger compaction forces than the clayey soils.

3.6.7 The Static Compaction Test

The standard Proctor test gives a relationship between the density and the moisture content for the constant energy input and reveals OMC and MDD values which can be used in the field for the soil compaction. The energy input is a variable quantity in the static compaction operations especially in the production of the compressed earth blocks using a machine. Therefore, it is desirable to devise a compaction test, wherein the quantum of compaction energy input and static nature of loading are simulated in the laboratory. The relationship between the compaction energy, the dry density and the moisture content can be obtained by the compaction of a soil into a small cube at different moisture contents while monitoring the energy input to the soil cube.

(a) **Test procedure**

This type of test can be classified under variable peak stress-constant stroke compaction category. Figure 3.7 shows the experimental set-up for the static compaction test. The strain or the displacement controlled machine with facilities to acquire the piston displacement and the corresponding force is essential for the test. The mould size is such that finally a compacted soil cube of size 76 mm is obtained. Brief test procedure is as follows.

(i) Dried soil sample is powdered and then mixed with a specific quantity of water (well below the saturated moisture content) and then stored in an airtight bag.
(ii) Different quantities of moist soil (say 850, 875, 900, 925, 950 gm, etc.) can be compacted in the mould.
(iii) The moist soil of known quantity is charged into the mould and is subjected to static compaction until the requisite cube thickness (76 mm) is achieved.

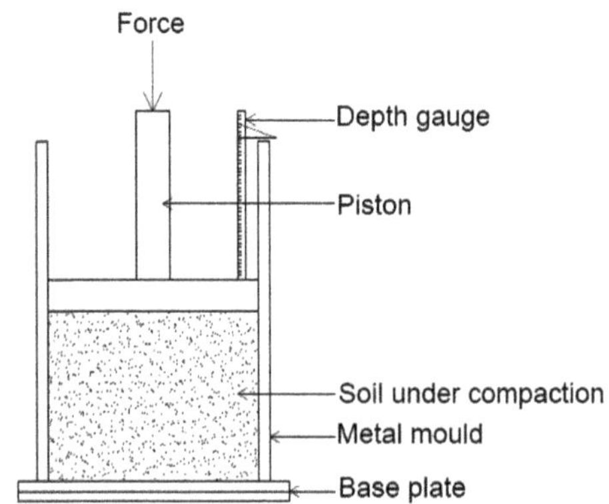

Fig. 3.7 Experimental set-up for the static compaction test

(iv) The compaction force and the corresponding displacement (stroke length) are recorded. The force–compaction stroke relationship is obtained.
(v) Each test is repeated to ascertain the mean force–stroke relationship.
(vi) The test can be repeated for other bulk weights of the moist soil and for other moulding moisture contents so that a family of force–stroke curves are generated.

Figures 3.8 and 3.9 show force–stroke relationships for the bulk densities of 1940 and 2050 kg/m^3 generated by Reddy (1991). Such relationships can be obtained for different moulding moisture contents and densities. The characteristics of the soil used in these static compaction studies are given in Table 3.3. Area under the force–stroke curve gives the total energy supplied to the soil mass during the compaction process. The compaction energy per unit volume can be estimated by dividing the total compaction energy by the final compacted volume of the soil or the earth block. The moulding moisture has a dominant influence on the force–stroke relationships. Increase in the moisture content results in a decrease in the magnitude of the compaction force as well as the compaction energy per unit volume.

(b) **Static compaction curves**

A family of force–stroke curves can be generated for a specific soil considering wide range of values for the moulding moisture and the density. The compaction energy

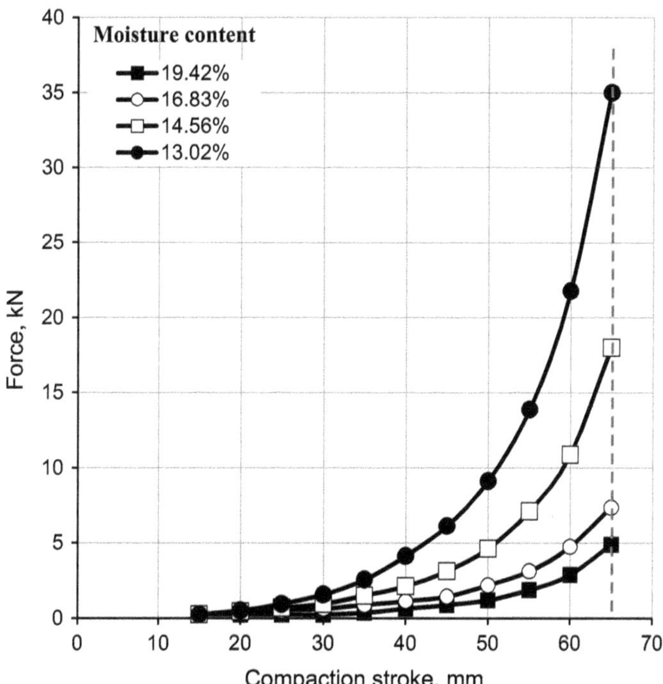

Fig. 3.8 Force–compaction stroke relationships for a bulk density of 1940 kg/m^3

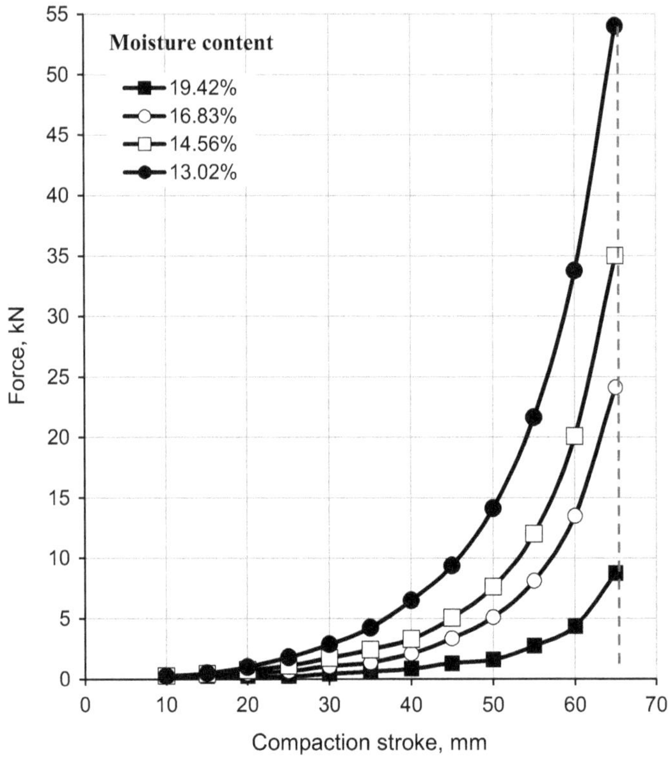

Fig. 3.9 Force–compaction stroke relationships for a bulk density of 2050 kg/m^3

Table 3.3 Characteristics of the soil

Soil property	Values
Sand (4.76–0.074 mm) (%)	48.8
Silt (0.074–0.002 mm) (%)	22.4
Clay (<0.002) (%)	28.8
Liquid limit (%)	42.0
Plastic limit (%)	19.7
Plasticity index (%)	22.3
Predominant clay minerals	Kaolinite and traces of montmorillonite
Specific gravity	2.69
pH	7.71
Organic matter (%)	0.72

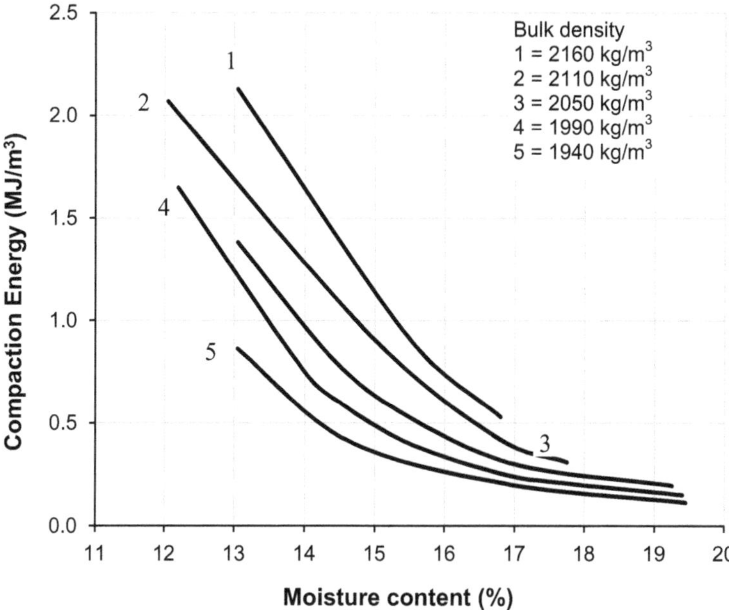

Fig. 3.10 Effect of moisture and density on compaction energy (Reddy 1991)

per unit volume can be estimated from these curves for each bulk density and the moulding moisture content. The resulting information, relating moisture content and density to compaction energy is shown in Fig. 3.10 (Reddy 1991). The figure shows five different curves representing five different bulk densities for the earth block. These curves indicate that compaction energy per unit volume of the soil or the earth block decreases as the moisture content increases irrespective of the block density. The lower ends of these curves (except the one corresponding to 1940 kg/m^3, which probably can be extended slightly further until 21% moisture content) represent limiting moisture contents, beyond which further compaction will lead to squeezing of the water from the soil mass. Any further attempt to carry out compaction at this stage would utilise the input energy to squeeze out the water (consolidation) rather than causing compaction. Using the energy and density information from the force–stroke curves, it is useful to construct the compaction energy and moisture content curves at constant dry density values as displayed in Fig. 3.11. These curves can be termed as static compaction curves. These curves provide a relationship between compaction energy per unit volume and the moisture content for a specified density. For a specific energy input, the appropriate moisture content (OMC-static) can be obtained by drawing a horizontal line at the given energy level to read the OMC-static value on the abscissa. These curves clearly show that for any specified dry density, an increase in the compaction energy leads to a reduction in the OMC-static value. Each of the curves shown in Fig. 3.11 terminate at a specific moisture content value indicating that the given density cannot be achieved at moisture content greater than

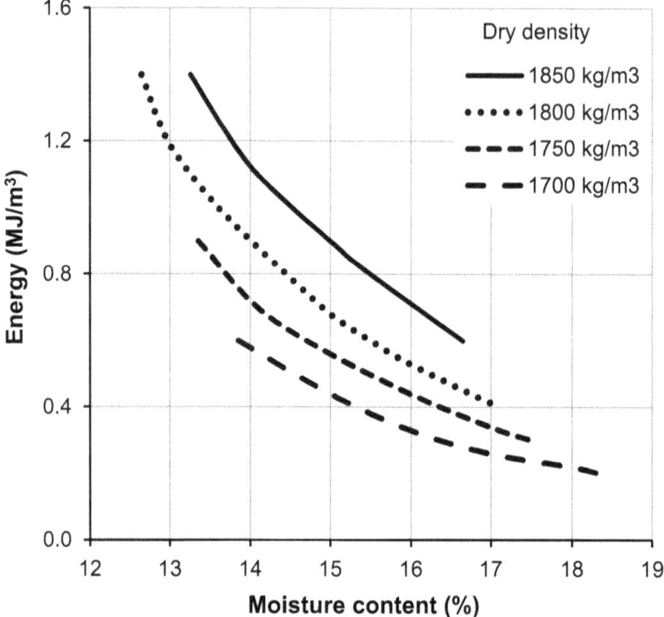

Fig. 3.11 Compaction energy-moisture relationships (Reddy 1991)

this value. These curves give a complete picture of the static compaction for a specific soil for various dry density values.

A single static compaction curve can be generated by joining the lower end points, (corresponding to the limiting moisture contents shown in Fig. 3.11) as shown in Fig. 3.12. The bold curve indicates the relationship between the energy and the OMC-static to achieve a dry density of 1850 kg/m^3. This curve shows that the limiting energy to achieve this dry density is 0.6 MJ/m^3. If the energy supplied is less, the corresponding OMC-static can be read off the dashed line curve. The dry density attainable at any energy input less than 0.6 MJ/m^3 reduces progressively as the energy is reduced. This curve may now be referred to as the static compaction curve for the given soil for a maximum dry density of 1850 kg/m^3.

3.6.8 Comparison Between Static Compaction Test and Standard Proctor Test

It is useful to compare the results of the static compaction curves with those of the standard Proctor compaction test curves. The results of the static compaction tests discussed in the previous sections have been reorganised to yield curves relating dry density and moisture content for constant energy input values as displayed in

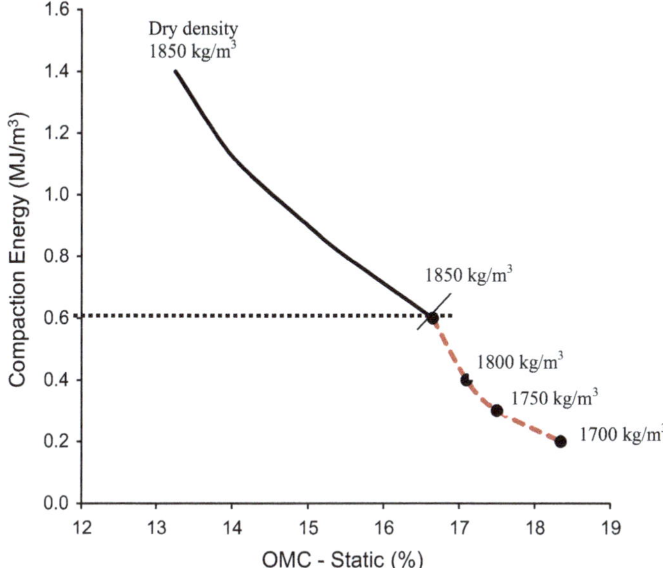

Fig. 3.12 Static compaction curve (Reddy 1991)

Fig. 3.13. The standard Proctor curve for the soil has also been superposed on static compaction curves. The static compaction curves show only the "rising" portion of the compaction curve. The drooping portion of the compaction curve noticed in standard Proctor compaction curve is absent in the static compaction curves. For any given static compaction energy, the dry density increases monotonically as the moisture content is increased till a limiting point is reached. The static compaction curves end up near zero air voids line without showing the downward slope as seen in the standard Proctor curve. For each energy input, one can define the maximum dry density possible and the corresponding moisture content.

The behaviour of the moist soil during compaction (static and dynamic) especially when moisture content is beyond OMC needs careful examination. In the case of the standard Proctor compaction test, the energy of the dropping weight is not used effectively for the compaction, as the moist soil bulges around it. This is attributed to the fact that the diameter of the falling weight (50 mm) is much smaller than the diameter of the soil specimen (100 mm). In the case of static compaction test, the entire soil mass is subjected to positive displacement. As the compaction proceeds, the water phase becomes continuous readily and leads to the beginning of consolidation, and compaction is no longer possible as a quick or instantaneous response (the beginning of consolidation phase is independent of any specific air void ratio). The point at which this situation arises for a given energy input gives the OMC-static and the possible dry density. The compaction process in the standard Proctor test avoids this stage, because the dynamic nature of the test precludes the development of a

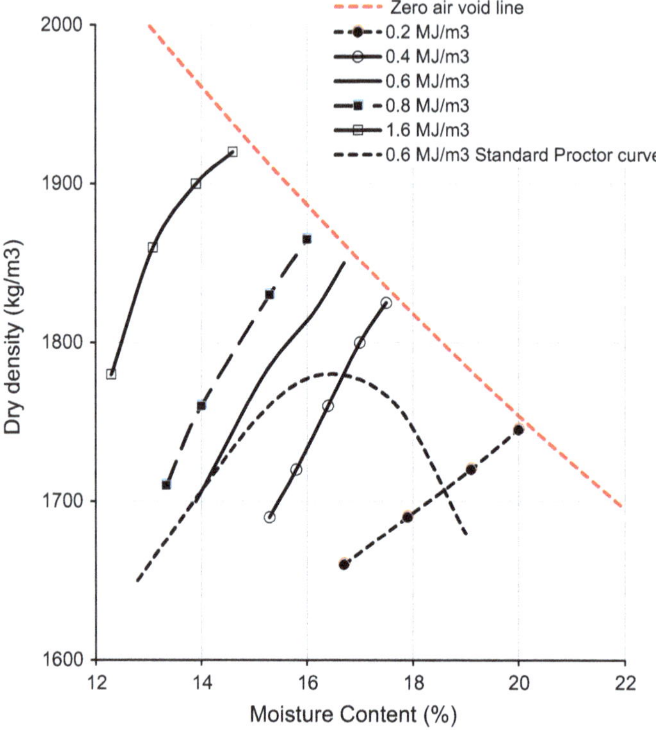

Fig. 3.13 Density versus moisture curves for static compaction and standard Proctor tests (Reddy 1991)

consolidation phase. Thus, the static compaction curves generated differ from those of the standard Proctor compaction curves especially at higher moisture contents.

A comparison between the static and the dynamic compaction curves clearly shows that for the same energy input and OMC value, the static compaction produces much higher dry density. The compaction energy supplied in the standard Proctor test is 0.60 MJ/m^3, which achieves a dry density of 1750 kg/m^3 at an OMC of about 16.5%, whereas in the case of the static compaction with similar energy input and moisture content, a dry density of 1860 kg/m^3 can be achieved. These results seem to indicate that the static compaction process is more energy efficient than the standard Proctor method. This can be attributed to higher energy loss during the impact of falling weight in the standard Proctor test.

Investigations of Reddy (1991) and Reddy and Jagadish (1993) give more details on the static compaction of soils and the static compaction test. These investigations reveal the influence of the specimen size and the rate of compaction on the static compaction characteristics of the soil. The static compaction test discussed here appears too cumbersome.

References

Davidson DT, Mateos M, Barnes HF (1960) Improvement of lime stabilisation of montmorillonitic clay soils with chemical additives. Bulletin No 262, Highway Research Board, National Research Council, Washington, D. C.

Davidson DT, Handy RL (1960) Lime and lime-pozzolana stabilization. Highway Engineering Handbook, McGraw-Hill Book Co., New York, N. Y., pp 21–98

Dan Marks B, Allan Haliburton T (1972) Acceleration of lime-clay reactions with salt. J Soil Mech Found Div Proc ASCE 98(SM4), pp 327–339

Eades JL, Grim RE (1960) Reaction of hydrated lime with pure clay minerals in soil stabilization. Bulletin 262, Highway research board, Washington, DC

Fitzmaurice RF (1958) Manual on stabilised soil construction for housing. U. N. Technical Assistance Programme, New York

Glenn GR, Handy RL (1963) Lime-clay mineral reaction products. Highway Research Record No 29, Highway Research Board, National Research Council, Washington, D. C.

Herzog A, Mitchell KK (1963) Reactions accompanying the stabilization of clay with cement, Highway Research Record No. 36, Highway Research Board, National Research Council, Washington, D. C.

Herrin M, Mitchell H (1961) Lime soil mixtures. Bulletin No 304, Highway Research Board, National Research Council, Washington, D. C.

Ingles OG, Metcalf JB (1972) Soil stabilisation—principles and practice. Butterworth's publisher, Australia

Lambe TW, Michaels AS, Moh ZC (1960) Improvement of soil-cement alkali metal compounds. Bulletin No 241, Highway Research Board, National Research Council, Washington, D. C.

Mateos M (1964) Soil-lime research at Iowa State University. J Soil Mech Found Div Proc ASCE 90(SM2):127–153

Mateos M, Davidson DT (1961) Further evaluation of promising chemical additives for accelerating hardening of soil-lime-fly ash mixtures. Highway Research Record No 304, Highway Research Board, National Research Council, Washington D.C.

Mehra SR, Chadda LR, Kapur RN (1955) Role of detrimental salts in soil stabilization with and without cement: the effect of sodium sulphate. Indian Concr J 29(10), pp 336–337

Moh ZC (1962) Soil stabilization with cement and sodium additives. J Soil Mech Found Div Proc ASCE 88(SM6), pp 81–105

Olivier M, Ali M (1987) Influence of different parameters on the resistance of earth, used as a building material. In: Proceedings of international conference on mud architecture, Trivandrum, India

Reddy BVV (1983) On the technology of pressed soil blocks for wall construction. MSc(Engg) thesis, Department of Civil Engineering, Indian Institute of Science Bangalore, India

Reddy BVV (2015) Design of a manual press for the production of compacted stabilised soil blocks. Curr Sci 109(9), pp 1651–1659

Reddy BVV (1991) Studies on static compaction of soils and compacted soil-cement blocks for walls. PhD thesis, Department of Civil Engineering, Indian Institute of Science Bangalore, India

Reddy BVV, Jagadish KS (1993) The static compaction of soils. Geotechnique 43(2), pp 337–341

Thompson MR (1966) Lime reactivity of Illinois soils. J Soil Mech Found Div Proc ASCE 92(SM5), pp 67–92

Turnbull WJ (1950) Compaction and strength tests on soil. Presented at annual meeting ASCE (January)

Bibliography on Soil Stabilisation

Bell FG (1996) Lime stabilization of clay minerals and soils. Eng Geol 42:223–237
Bofinger HE (1964) The structure of soil-cement. Australian Road Res J 2(1):46–19
Bryan AJ (1988) Criteria for the suitability of soil for cement stabilization. Build Environ 23(4):309–319
Chaston FM (1967) A practical approach to soil stabilization. J Inst Eng Australia 29(10–11):241–248
Consoli NC, Dalla RA, Corte MB, Lopes LS (2011) Porosity-cement ratio controlling strength of artificially cemented clays. J Mater Civ Eng 23(8):1249–1254
Consoli NC, Foppa D, Festugato L, Heineck KS (2007) Key parameters for strength control of artificially cemented soils. J Geotech Geoenviron Eng 133(2):197–205
Clare KE, Cruchley AE (1957) Laboratory experiments in the stabilization of clays with hydrated lime. Geotechnique 7:97–111
Clare KE, Pollard AE (1954) The effect of curing temperature on the compressive strength of soil-cement mixtures. Geotechnique 4:97–107
Clare KE, Sherwood PT (1956) Further studies on the effect of organic matter on the setting of soil-cement mixtures. J Appl Chem 6:317–324
Clare KE, Sherwood PT (1954) The effect of organic matter on setting of soil-cement mixtures. J Appl Chem 4(2):625–630
Croft JB (1968) The problem in predicting the suitability of soils for cementitious stabilization. Eng Geol 2(6):397–424
Croft JB (1967a) The structures of soils stabilized with cementitious agents. Eng Geol 2:63–80
Croft JB (1967b) The influence of soil mineralogical composition on cement stabilization. Geotechnique 17:119–135
De Sousa Pinto C, Davidson DT, Laguros JG (1962) Effect of lime on cement stabilization of montmorillonitic soils. Bulletin 353, Highway Research Board, National Research Council, Washington, D. C., pp 64–83
Diamond S, Kinter EB (1965) Mechanisms of soil-lime stabilization: an interpretive review. Highway Research Record No 92, Highway Research Board, National Research Council, Washington, D. C., pp 83–102
Felt EJ, Abrams MS (1957) Strength and elastic properties of compacted soil-cement mixtures. ASTM Special Publication 206
Felt EJ (1955) Factors influencing physical properties of soil-cement mixtures. Bulletin No 108, Highway Research Board, National Research Council, Washington, D. C., pp 138–162
Hilt GH, Davidson DT (1961) Isolation and investigation of lime-montmorillonite crystalline reaction product. Highway Research Record No 304, Highway Research Board, National Research Council, Washington, D. C.
Ingles OG (1968) Advances in soil stabilization 1961–1967. Rev Pure Appl Chem 18(291)
Ladd CO, Moh ZC,Lambe TW (1960) Recent soil-lime research at MIT. Bulletin No 262, Highway Research Board, National Research Council, Washington, D.C.
Lambe TW (1962) Soil stabilization, Chap 4. Foundation Engineering, Edited by Leonard's Mc. Graw Hill, pp 351–347
Remus MD, Davidson DT (1961) Relation of strength to composition and density of lime-treated clayey soils. Bulletin No 304, Highway Research Council, Washington, D. C., pp 65–75
Robertson JA, Blight GE (1978) Stabilized earth fill dams. In: Proceedings of symposium on soil reinforcing and stabilizing techniques in engineering practice University of New South Wales, Sydney, Australia
Reddy BVV, Latha MS (2014) Retrieving clay minerals from stabilised soil compacts. Appl Clay Sci 101:362–368
Winterkorn HF (1955) The science of soil stabilization. Bulletin No 108, Highway Research Board, National Research Council, Washington, D. C., pp 1–24

Part II
Stabilised Compressed Earth Blocks and Masonry

Chapter 4
Stabilised Compressed Earth Block Production

4.1 Introduction

Stabilised compressed earth blocks (CEBs) can be manufactured using a mixture of soil-sand and the stabiliser. A uniform mixture of soil-sand-stabiliser is compressed into a high-density CEB at an optimum moisture content using a machine. The production of the stabilised CEB involves the following critical steps.

(a) Selection of a suitable soil, excavation and transportation
(b) Processing the stabilised soil mixture
(c) Compaction, ejection and stacking the CEB
(d) Curing

The stabilised CEBs can be produced at a construction site, where the building or the structure is being built or in a factory. The CEBs can be produced by employing totally manual process or semi-mechanised or completely mechanised process. The activities involved and the machinery employed will depend upon the type of the production process employed. The basic raw materials required to produce the stabilised CEB include soil, sand or inert industrial by-product such as crushed rock dust, stabilisers (lime, cement, etc.) and potable water. The crushed rock dust, an inert material, can be used as a substitute for sand in diluting the natural soil for adjusting the grain size distribution and controlling the clay fraction in the mix.

4.2 Soil Selection, Excavation and Transportation

4.2.1 Soil Selection

A soil deposit or quarry site must be identified prior to the commencement of excavation of the soil. Identification of a soil deposit or the quarry site requires some

B. V. V. Reddy, *Compressed Earth Block & Rammed Earth Structures*,
Springer Transactions in Civil and Environmental Engineering,
https://doi.org/10.1007/978-981-16-7877-6_4

experience and testing to assess the suitability of the soil for the soil-based constructions. There are varieties of tests which can be carried out on the soil. These tests can be broadly divided into two types: (a) field tests and (b) laboratory tests. The field tests provide a preliminary assessment of the suitability of the soil, and it should be followed by the laboratory tests for further confirmation on the suitability of the soils to produce stabilised CEB.

4.2.1.1 Field Tests

Examining texture, colour, dry strength of the soil lump, soil deposit's surface crack pattern and crack width, touch sensation test and wash test represent some of the field tests for the preliminary assessment of the soils suitability. These tests give definite clues to the suitability of the soil for the CEB or rammed earth construction. The details of these field tests are outlined below.

(i) **Soil texture and colour**

Soil texture in dry state is a good indicator of the type of soil and its suitability for the CEB or rammed earth. If the soil in dry condition is powdery and fine grained or coarse grained, it is generally suitable for stabilised earth products. Too much of fine's fraction (silty) can pose problems in compaction and the green strength for the CEB. However, such soils can be reconstituted with coarse-grained materials (sand and gravel) to make them suitable for the stabilised compressed earth products.

Natural soils are available in a very wide range of colours. For example, soils can have colours like dark red, brown, yellow, white, grey, dark brown, black, etc. The colour of the soil is indicative of the nature or type of the clay mineral present or its constituents. For example, dark red or brown coloured soils contain non-expansive kaolinite-type minerals. The black or greyish coloured soil can be suspected to contain expansive clay minerals such as montmorillonite, and this can be substantiated by the presence of very hard dry soil lumps.

(ii) **Surface crack pattern and crack widths**

Examine the exposed surface of the soil deposit for surface cracks and their pattern. The deposits with heavy clay and fine-grained soils show irregular cracking pattern. The dry cracked surface gives a fine soapy feeling on touching the surface. The best examples for this type of deposits are the tank beds where the fine-grained heavy clay soils are deposited over a period of time. Figure 4.1 shows a typical tank bed deposit with surface cracks. Generally, heavy clay soils are difficult to handle for the manufacture of stabilised earth products such as CEB and rammed earth. Such soils possess very hard lumps in dry condition posing problems for crushing/powdering, uniform mixing of the stabiliser, needing large dosage of stabiliser, etc.

The black cotton soil (rich in montmorillonite mineral) deposits are easy to identify. Their colour is generally dark grey, and the deposit surface will have very wide and deep cracks in dry condition (Fig. 4.2). The presence of large sized soil lumps (in dry condition) upon excavation and the lumps are very hard to crush. Wetting

Fig. 4.1 Surface cracks in a tank bed soil deposit

Fig. 4.2 Typical surface cracks in a black cotton soil ground

such lumps makes them extremely sticky. These soils are called expansive soils with high swell–shrink potential. Such soils are unsuitable for cement stabilisation, and lime is essential to control the swelling and shrinkage.

(iii) **Strength of the dry soil lumps**

The soil lumps of different sizes can be found in the excavated soils as shown in Fig. 4.3. Hold the dry lump between the fingers (Fig. 4.4) and attempt to crush it. If it can be crushed easily, then it can be inferred that the soil contains lower percentage of clay or it is a sandy soil. Also, a dry soil lump of fist size dropped from shoulder height on to a hard surface can either break easily into small pieces or may not break at all. If the lump breaks into several small pieces, it indicates that the soil is a sandy soil. Such soils are suitable for stabilisation especially using additives such as cement. If the lump is very hard and does not break, the soil is expected to contain

Fig. 4.3 Soil lumps in the excavated soil

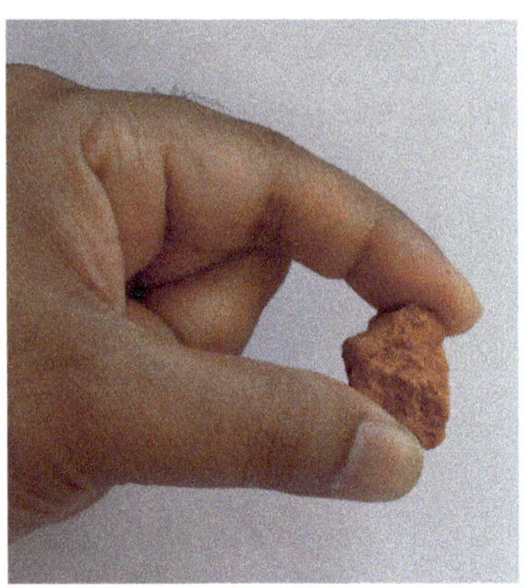

Fig. 4.4 Crushing test on a soil lump

either higher clay content or the soil contains expansive (such as montmorillonite) type of clay mineral.

(iv) **Touch sensation test**

Remove the larger/coarser grains such as gravel from the soil and place a small quantity of the sample in the middle of the palm. Add water and saturate it. Rub the saturated soil sample on the palm using bear fingers (Fig. 4.5). The sandy soils give a rough sensation with little plastic cohesion. The clayey soil gives smooth and sticky feeling.

(v) **Wash test**

Take a small quantity of soil and place it on the palm. Saturate the soil and rub it using the fingers. Try to wash the palm by keeping the palm under a running tap. Sandy and silty soils can be easily washed off or rinsed clean without much difficulty, whereas clayey soil needs some effort to wash the palm free of sticking soil, i.e. mainly clay/silt particles.

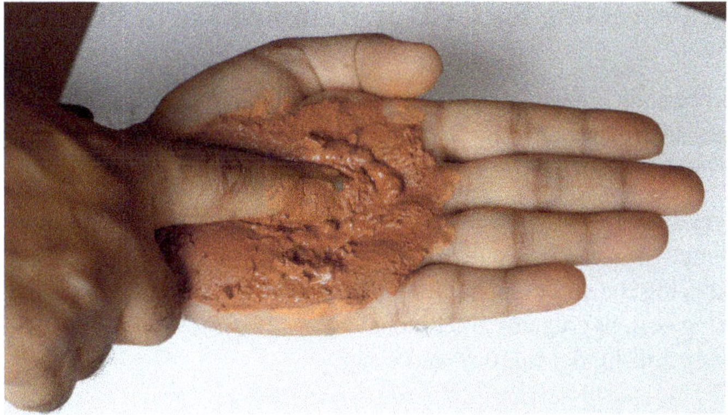

Fig. 4.5 Touch sensation test on the soil

4.2.1.2 Laboratory Tests

The soil samples can be collected from the soil deposits selected in the preliminary assessment survey for further detailed laboratory testing before the soil is excavated and transported. Avoiding the laboratory tests could result in problems related to the strength and the durability of the stabilised CEB masonry and the rammed earth walls. Important laboratory tests to assess the soils suitability include: (a) Grain size distribution, (b) X-ray diffraction to identify the clay minerals, and (b) Atterberg's limits. The X-ray diffraction can be dispensed with if some indirect methods are used for assessing the swell–shrink potential of the soil. Detailed soil testing procedures were discussed in Chap. 2.

4.2.2 Excavation and Transportation

The soil can be excavated either by using mechanical equipment/machinery or manually using hand tools. The quantum of output required and the availability of labour including the cost of labour dictates the choice of choosing either mechanical equipment/machinery or manual process. If the quantity of the excavated output required is large, and for quicker operations, use of mechanical equipment/machinery becomes essential. If the manual labour is available at an affordable cost, then excavating smaller quantities of soil can be carried out employing manual process. The mode of transportation employed depends upon the distances involved in carting the excavated soil. For larger hauling distances, use of trucks will be convenient and economical. For shorter distances, small capacity vehicles such as tractors or mini trucks may be ideally suited. Transported soil dumped at the CEB manufacturing site needs further processing involving crushing and sieving.

4.3 Soil Processing for CEB or Rammed Earth Production

The soil processing is an important step in the production of the stabilised compressed earth blocks and the rammed earth. The soil processing and the preparation involves the following activities:

(a) Pulverisation
(b) Sieving
(c) Modifying the soil composition
(d) Mixing soil, aggregates and binders
(e) Mixing soil-binder mixture and water.

4.3.1 Pulverising the Soil

The excavated soils contain lumps, and the size of the dry lumps mainly depends upon the type and quantity of the clay minerals present in the soil. The size of the soil lumps can be sometimes as big as 300 mm or more. These lumps must be pulverised for utilising the soil either for the stabilised CEB or the rammed earth construction. The soil pulverisation (i.e. breaking the lumps) becomes essential to blend the natural soil with the fine aggregates and the stabilisers (such as cement or lime). Powdery or pulverised soil is effective in arriving at uniform mixture and better distribution of the fine aggregates and the stabilisers.

Pulverising of soil lumps is a difficult task. Breaking the soil lumps and the pulverisation can be effective when the soil is in dry state. The soil lumps can be pulverised either by using rollers or a mechanical mill or through manual process of beating. Simple ways to crush the dried soil lumps are by spreading onto a hard platform or surface in thin layers of about 100 mm or less and then allow a heavy roller to pass over the soil lumps, number of times. Instead of heavy roller, a heavy automobile wheel such as a tractor wheels can be utilised. Special purpose soil pulverises such as hammer mill can be used to pulverise the dry soil. If the use of manual labour is feasible, the soil lumps can also be broken into smaller pieces by beating the lumps with long wooden sticks or rammers manually.

4.3.2 Sieving

Sieving the pulverised soil is necessary to remove bigger size particles and unwanted organic matter in the form of roots, leaves, etc. Generally, 4–5 mm mesh is employed for the soil sieving. There are varieties of sieving techniques. Rectangular fixed sieve, rotary sieve and vibrating screen represent some of the simple gadgets employed for the soil sieving.

Figure 4.6 shows a fixed screen kept at an inclination and the soil being thrown on to the sieve. The soil grains smaller than the mesh size get heaped up, on the bottom side of the inclined screen. The bigger soil lumps retained on the sieve can be

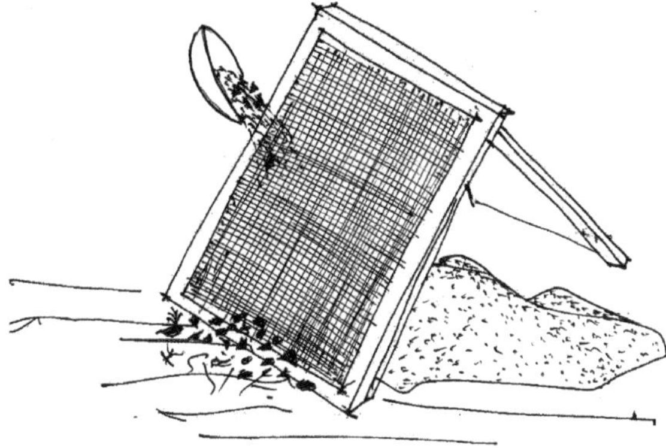

Fig. 4.6 Soil sieving using inclined fixed sieve

Fig. 4.7 Manually operated rotary sieve

further pulverised and passed through the screen again. This is the simplest manual process for sieving the soil. The quantity of soil sieved by a couple of persons may not exceed 2.0 m^3 in a day.

Rotary sieve is another simple device for sieving the soil. Figure 4.7 displays a manually operated rotary sieve. The cylindrical mesh of any required opening size can be chosen. Also, rotary sieves can be powered using a machine for increasing the output. A stack of mechanically operated vibrating sieves can also be adopted to sieve the soil.

4.3.3 Modifying the Soil Composition

Rarely natural soils in their native state possess optimum composition found suitable for the stabilised compressed earth block or rammed earth. Hence, in majority of

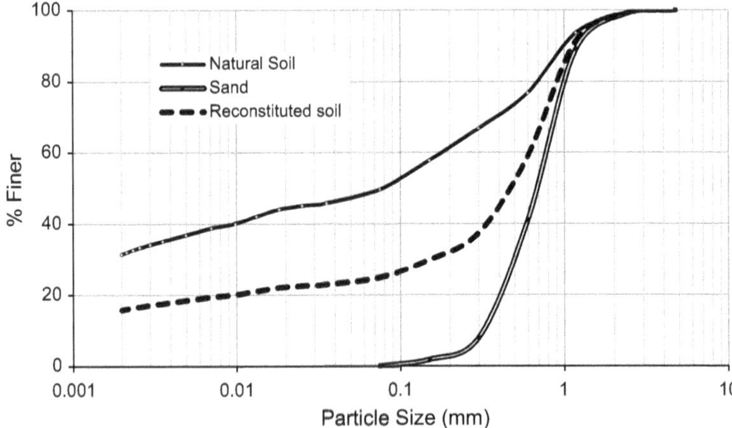

Fig. 4.8 Grain size distribution curves of natural soil, sand and the reconstituted soil

the cases, the natural soils are blended with fine aggregates, or sometimes soils of two different compositions are blended such that the resulting mixture is suitable to produce stabilised CEB or rammed earth. Fine aggregate is basically sand, and it is used to modify the natural soil composition. For example, a natural soil containing 31.6% clay fraction can be reconstituted by mixing with the sand in the ratio of 1:1 (soil: sand, by mass) such that the resulting mixture has optimum composition (especially optimum clay content), which is the essential requirement for the production of the stabilised CEB. Figure 4.8 shows the grain distribution curves for a natural soil, sand and the reconstituted soil. The natural soil has 49.7% sand, 18.7% silt and 31.6% clay size fractions. When this soil was blended with sand in the ratio of 1: 1 (soil: sand, by mass), the resulting composition of the reconstituted soil will be 74.8% sand, 9.4% silt and 15.8% clay. Thus, soils can be reconstituted to obtain optimum grain size distribution curves for the manufacture of the stabilised CEB or rammed earth.

4.3.4 Mixing Soils, Aggregates and Binders

Uniform mixing is essential for effectively exploiting the stabilisation potential in earth-based construction products. Soils, aggregates (such as sand) and stabilisers can be effectively mixed in dry state rather than in partially saturated state. Avoid mixing the soil, sand and stabilisers in wet or saturated condition. The simplest and the easiest way of mixing all the ingredients is by manual operation using a spade/shovel. First, the soil and aggregates are mixed and then mixed with the stabilisers such as cement/lime. The manual mixing process involves the following steps.

Fig. 4.9 Rotating drum-type mixer

(i) Spread the soil into a thin layer and then spread the aggregates (sand) on top of this layer. Using a spade/shovel, the two layers are mixed till a uniform mixture is obtained.
(ii) Spread the cement/lime on the uniformly mixed soil-sand layer, and carryout the mixing using a spade/shovel to attain uniform distribution of the stabiliser in the mix.

This type of dry mixing can also be carried out using a motorised rotating drum-type concrete mixer (Fig. 4.9) or a paddle mixer or a pan mixer (Fig. 4.10). Many such mechanical mixers can be explored for mixing of the various ingredients in dry condition.

4.3.5 Mixing Soil-Binder Mixture and Water

The stabilised soil mix should be compacted into a high-density CEB while producing the stabilised compressed earth blocks. Compacting into a high-density CEB is possible only when the stabilised soil mix is partially saturated. The water and the clay fraction present in the mix play a crucial role in reducing the compaction force as well as attaining enough green strength for handling the fresh CEBs. Adding the right amount of moisture and mixing it uniformly with the soil-binder mixture is a very important operation in the CEB production process.

Fig. 4.10 Pan mixer

The soil-sand-stabiliser mixture contains clay minerals, and hence, the mix becomes sticky forming lumps (Fig. 4.11) after adding moisture to the dry mix. These lumps must be broken for uniform moisture distribution. Uniform distribution of moisture in the entire mix is essential for maintaining the stabilised CEB quality. Mixing water with the dry stabiliser-soil mix can be carried out either manually or using a mixer.

(a) **Manual mixing**

Figures 4.12, 4.13, 4.14 and 4.15 illustrate various steps involved in the manual mixing process in obtaining a uniform distribution of the moisture in the stabilised soil mix. The dry mixture of soil, sand and the stabiliser is spread into a thin layer (100 mm thickness), and then, water is sprayed using shower rose-can such that the entire layer is wetted. Using a spade, the soil is chopped and churned. The soil lumps formed during the spade work should be crushed such that a uniform mix is obtained.

The manual mixing process is generally carried out in small batches. The wetted mix containing stabilisers such as cement should be compacted into a CEB in about an hour. The delayed compaction affects the strength development. The percentage of water to be added depends on the soil composition and the stabiliser percentage. The role of moulding water content on the characteristics of stabilised CEB will be discussed in the subsequent chapters.

Fig. 4.11 Lumps formed during mixing soil and water

Fig. 4.12 Thin layer of dry soil-stabiliser mix being sprayed with water using a shower rose-can

(b) Mixing using mechanical mixers

The mechanical mixers can be employed for effective mixing as well as for achieving higher productivity. The rotating drum-type mixers such as concrete mixer can be adapted to carry out the mixing of partially saturated soil-stabiliser mix. The simple adaptation of the concrete mixer is by loading certain amount of charge (coarse

Fig. 4.13 Chopping and mixing the thin layer of the mix with water using spade

Fig. 4.14 Breaking lumps of wet soil-stabiliser mix

Fig. 4.15 Partially saturated soil-stabiliser mix ready for compaction into a CEB

aggregates of size 40–50 mm) along with the soil-stabiliser mix. Generally, the aggregate charge is about 10–15% (by mass) of the mix to be handled in the mixer. The requisite quantity of water should be sprayed into the rotating drum at certain intervals. The wetted mix is discharged onto a mesh (size: 25 mm), where the mesh retains the aggregate charge which can be reloaded into the mixer. Figures 4.16 and 4.17 show the aggregate charge and the wetted mix discharged onto the mesh.

The pan mixer (Fig. 4.10) or a paddle mixer can also be used for mixing water with the dry soil-stabiliser mix. The mechanical mixers are more effective in ascertaining uniform distribution of moisture in the mix apart from higher productivity. A helical screw mixer (Fig. 4.18) is also quite effective in obtaining a uniform mixture of soil-sand-stabiliser and water.

Fig. 4.16 Aggregate charge and the dry materials in the rotating drum-type mixer

Fig. 4.17 Uniformly mixed partially wet mix discharged onto a mesh

Fig. 4.18 Helical screw mixer and the mix discharged

4.4 Stabilised Compressed Earth Block Production

4.4.1 CEB Production Process

The principle of CEB compaction using static compaction process was discussed in Sect. 3.6.6 as illustrated in Fig. 3.5. The production of the stabilised CEB from the processed stabilised soil mixture using a manual or semi-automatic machine involves the following steps.

(a) Weighing the mix
(b) Feeding the machine mould
(c) Block compaction
(d) Ejecting the CEB, and
(e) Stacking.

The production of the stabilised CEB illustrating the above-mentioned steps using a manually operated machine is illustrated in the Fig. 4.19a–f. The details of these steps are as follows.

(a) Processed soil is taken in a scoop and weighed using a pan balance (Fig. 4.19a). Weigh batching is essential to control the CEB density. Since the density is a significant parameter in controlling the CEB characteristics, and therefore, weigh batching is the fool proof technique to control CEB density.
(b) The weighed mix is fed into the mould followed by closing the lid and positioning the lever.
(c) Then, the lever is pulled down till it touches a knob, and further pulling of the lever is not possible. Thus, the piston stroke length is controlled. To achieve a proper density, effort from two persons is essential for the CEB compaction to complete.
(d) Bring the lever to rest position and then open the lid. The CEB is ejected out of the mould by pulling the lever down. The ejected CEB is carried along with a thin base plate and stacked for curing.
(e) The CEBs are stacked one above the other up to a height of 1.5 m as illustrated in Fig. 4.20. The stacking should be such that there are no gaps in between the CEBs (to minimise the water loss due to evaporation from the exposed CEB surfaces), and the CEBs should be staggered to achieve a good stability for the stack.

Manufacturing of the CEB using semi-mechanised and fully mechanised machines will be slightly different from that of the manually operated machines. For example, the fully mechanised machine will have a conveyor belt attached to a mixer at one end. Definite amount of mix falls into the moulds, then the processes of lid closing, compaction, ejection and conveying the CEB happen automatically without much human intervention. Figure 4.21 shows a fully mechanised system of CEB production. Here, the operations of feeding the mix, lid closing, CEB compaction,

Fig. 4.19 Different steps involved in the CEB production using a manually operated machine. [**a** Weighing the mix; **b** Machine mould; **c** Feeding the mix; **d** Pulling lever—compaction; **e** CEB ejection; **f** Removing the CEB

Fig. 4.20 Stacking of stabilised CEBs

Fig. 4.21 CEB making process in a mechanised machine (Courtesy: David Easton, Watershed Materials LLC, USA)

CEB ejection and pushing the CEB on to the conveyer are carried out by a hydraulically operated system. A typical flow diagram of the fully automated CEB production process is illustrated in Fig. 4.22.

4.4.2 Analysis of CEB Production Process

A time and motion study with clearly defined activities will help in assessing the optimum output in any CEB production system. In a manually operated CEB production system, the production is a function of the number of persons deployed in various activities. The detailed study of the manual CEB production system was carried out by Reddy (1983). Table 4.1 gives the details of the time and motion study conducted in a CEB production system using a manually operated CEB making machine (Reddy 1983). In the manual CEB production system, soil preparation, i.e. basically mixing soil, sand, stabiliser and water, is carried out in small batches. Generally, the batch size is restricted to a mix sufficient for about 25–30 CEBs (CEB size is 2.5–3 times the volume of normal burnt clay brick), i.e. about 250 kg of processed mix. This kind of batch size is chosen to complete the CEB pressing within the initial setting time of

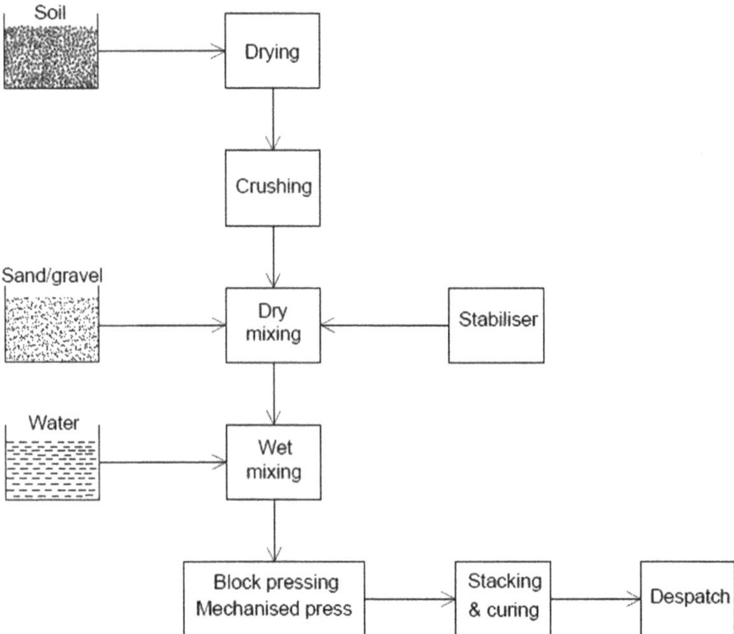

Fig. 4.22 Fully automated production process for stabilised compressed earth blocks

the cement stabiliser. Also, it is difficult to achieve uniform mix with large quantities of soil in a batch using manual mixing operation. Table 4.1 lists various activities along with the average time taken for each activity and the standard deviation values. The activity of soil mixing/preparation can be a parallel activity along with the other activities of the CEB production. Figure 4.23 shows a view of typical stabilised earth block production system at a construction site where all the operations involved are carried out manually.

A batch operation involves two parallel activities: (a) preparing the soil for a set of CEBs and (b) making the CEBs in the machine. The CEB production per day or per shift depends upon the number of persons working at the machine and the number of persons involved in preparing the soil. Figures 4.24, 4.25 and 4.26 show the activity networks and labour utilisation diagrams for various combinations of persons working with the machine. Figures 4.24 and 4.25 show the activity network and labour utilisation diagrams for making one CEB with two and three persons, respectively. When only two persons are producing CEBs, there are no parallel activities, and the critical path will be along a straight line. In this case, total time needed for making a CEB is more (51.7 s). For the case, where three persons were involved in CEB making, there are some parallel activities and the time required for producing a CEB reduces to 30.1 s. Figure 4.26 shows the activity network and labour utilisation diagrams when four persons are working at the machine in various CEB making operations. There are parallel activities in this case also, but the total time needed

Table 4.1 Duration of various activities in manual CEB making system

Sl. No.	Activity	Average time required	Standard deviation (n = 100)
1	Soil preparation		
	(a) one person	31 min	
	(b) two persons	26 min	
	(c) three persons	22 min	
2	Loading the scoop and weighing	7.7 s	3.34
3	Filling the mould	14.7 s	4.71
4	Lid closing and compaction	8.1 s	3.41
5	Lid opening and CEB ejection	7.3 s	3.57
6	Stacking the stabilised CEB	14.0 s	3.33

Fig. 4.23 Typical view of activities in a manual CEB production system

for making a CEB is 30.1 s. The labour utilisation with four persons at the machine is not as efficient as with three persons.

Activity networks for making a batch of 25 CEBs (size: 230 × 190 × 100 mm) when two and three persons working at the machine and preparing the soil are displayed in Figs. 4.27 and 4.28, respectively. Networks like that of Fig. 4.28 can

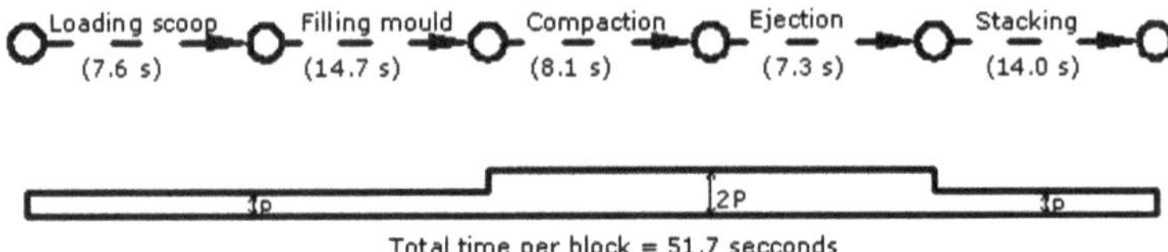

Fig. 4.24 Activity network and labour utilisation diagrams for making one CEB with two persons (P: person) (Reddy 2015)

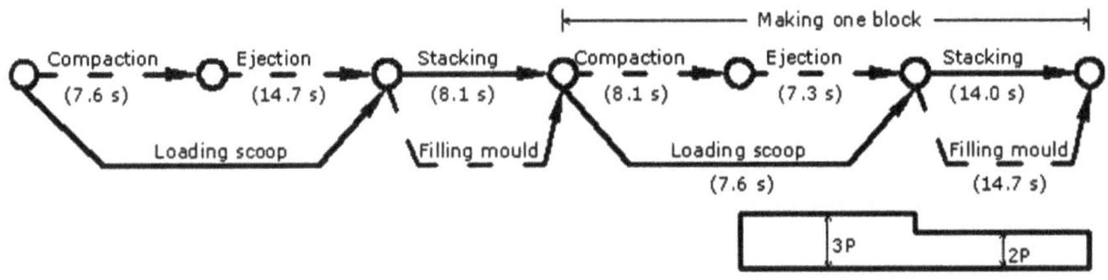

Fig. 4.25 Activity network and labour utilisation diagrams for making one CEB with three persons (P: person) (Reddy 2015)

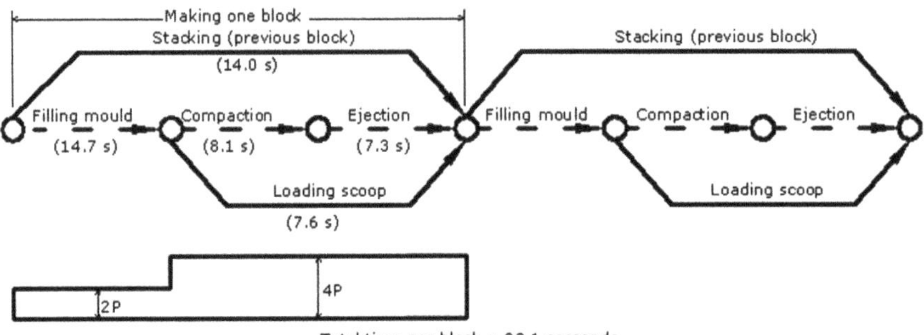

Fig. 4.26 Activity network and labour utilisation diagrams for making one CEB with four persons (P: person) (Reddy 2015)

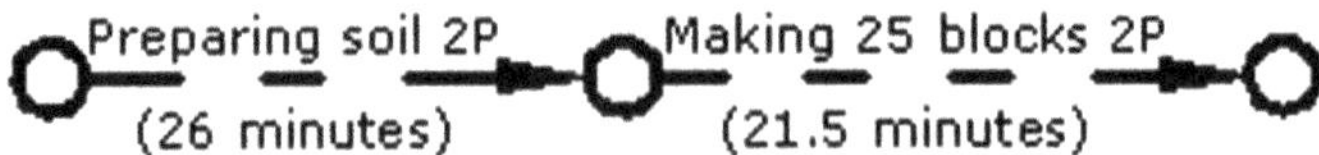

Fig. 4.27 CEB making activity network with two persons (P) (Reddy 2015)

be drawn when more than three persons are working in the CEB making operations. Table 4.2 summarises analysis of the CEB production when the number of persons working in various activities is in the range of 2–6. Analyses of these networks and data given in Table 4.2 show that the time needed for making a batch of 25 CEBs is

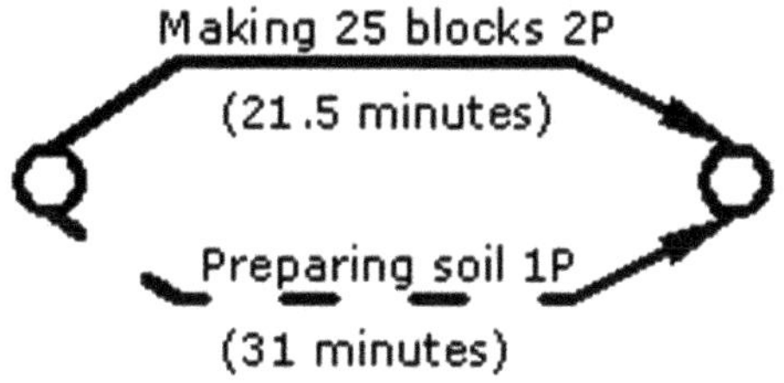

Fig. 4.28 CEB making activity network with three persons (P) (Reddy 2015)

Table 4.2 Analysis of manual CEB production

Total number of persons	Number of persons for		Time for making a batch of 25 CEBs (minutes)
	Soil preparation	CEB making	
2	2	2	47.5
3	1	2	31.0
4	2	2	26.0
5	3	2	22.0
6	3	3	22.0

less than the time required for preparing the soil (~225 kg) for that batch. The total time needed for completion of the batch operation is controlled by the soil preparation activity. Making use of these networks time taken for per batch (25 CEBs) production for an 8-h shift and productivity per person can be estimated. Figure 4.29 shows a plot of total CEB production per 8-h shift and production per person versus the number of persons employed in the CEB production activities. Some useful observations

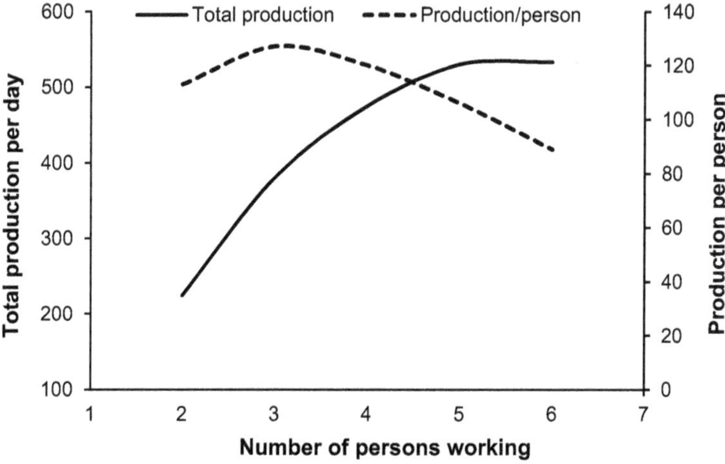

Fig. 4.29 Labour utilisation and CEB production (Reddy 2015)

Table 4.3 CEB production in different types of CEB making systems

Type of production system	CEB production in 8-h shift
1. Manual machines	400–700
2. Semi-automatic machines	1500–3000
3. Fully automatic machines	4000–8000

with reference to the production of stabilised CEBs using manual operations are as follows.

1. CEB production increases as the number of persons employed in various activities are increased.
2. The optimum time needed to produce a CEB is 30.1 s (Figs. 4.25 and 4.26), and therefore, theoretically a manually operated machine can produce about 950 CEBs in an eight-hour shift when 3–4 persons are working at the machine.
3. The major bottleneck hampering production in the manual production system is soil preparation activity. The productivity per shift can be increased by hastening the soil preparation activity. This can be achieved by either employing more persons in the soil preparation activity or by employing a motorised mixer machine.

It has been observed in several manual CEB manufacturing units in India that employing six persons, the daily production was about 600–700 CEBs. The CEB production per shift in employing semi-automatic and fully automatic production systems can be different depending upon the designed capacities of mixing devices and the CEB making machines. Table 4.3 gives an idea of daily CEB production using different types of CEB making systems.

4.5 Curing

The stabilised compressed earth blocks must be cured to facilitate formation of hydration products of cementitious binders (cement) and pozzolanic reactions between the lime and the clay minerals, responsible for the strength development. The stabilised CEBs can be cured by either moist curing or steam curing.

4.5.1 Curing by Keeping the Stack of CEBs in Moist Condition

The stabilised CEBs are stacked on a level ground one above the other without any gaps between the CEBs, and the curing is commenced a day after the casting. The stack is covered with either a layer of straw or gunny cloth, and then, water is sprinkled (avoid concentrated jetting of water) three to four times daily for four

Fig. 4.30 Stacking and curing

weeks (Fig. 4.30). There is no need for the stabilised CEBs to be soaked in water for the curing purposes. The curing duration is generally 28 days for CEBs with binders such as the Portland cement. For binders such as the lime and the lime-pozzolana cements, the duration can be longer than 28 days. It is preferable to cure the stabilised CEBs under shaded area, so that the moisture loss due to evaporation is minimised. After completion of curing, the stabilised CEBs are allowed to dry in the stack before being used for the construction.

4.5.2 Steam Curing at Atmospheric Pressure

To reduce the curing period (especially for lime stabilised CEBs) and to achieve higher strengths, steam curing at low temperature (80 °C) and at atmospheric pressure for 14–16 h can be explored. The lime-clay reactions are slow at ambient temperatures, and hence, there is a need for longer curing period (>28 days) to achieve meaningful strengths. Such lime stabilised CEBs can be cured using steam curing technique. More information on the steam curing can be found in the studies of Reddy and Hubli (2002). Figure 5.23 (Chap. 5) shows a steam curing plant, where a truck load of CEBs (10–12 tonnes) can be cured. This plant operates on biomass fuel, and there is no pressurised boiler in this system. The boiler is a simple tray holding water on top of which the trolley-containing lime stabilised CEBs is positioned.

4.6 Machines to Produce Stabilised Compressed Earth Blocks

The CEBs can be manufactured using varieties of machines commercially available in the global market. Mukerji (1986) compiled the details of more than 100 varieties of machines used for the CEB production. Earliest examples of the use of machines for the CEB production can be seen in the late forties (ILO/UNIDO 1987). Then onwards, many different types of machines were developed. Machines available for the CEB production can be grouped into two categories: (a) manual machines and (b) mechanised machines. The basic principle involved in the stabilised CEB production using a machine is subjecting the processed moist (partly saturated) mixture of soil and stabiliser to confined compaction in a mould to obtain a dense CEB having sufficient green strength for handling and stacking in layers. This means a loose mixture of stabiliser and soil (with a bulk density of about 1400 kg/m^3) is compressed into a dense CEB having a fresh bulk density of more than 2000 kg/m^3.

4.6.1 Manual Machines

Majority of the manually operated machines produce constant volume CEBs. This means that the plan size of the CEB as well as piston stroke length is fixed. The first manual press was developed by the Chilean Engineer Ramirez at the Inter-American Housing and Planning centre (CINVA) in Bogota, Columbia, during 1956. This machine has since been known as CINVA-Ram. This was a portable lightweight machine weighing about 60 kg capable of producing 2 MPa compaction pressure. The machine has a fixed mould producing one CEB at a time. Several variants of this machine were developed and used to produce stabilised compressed earth blocks. Brepak, Ceta-Ram, Tek-Block, Zora, GEO-50, Ceraman, CTA Block press, Mardini, Ellson block master, Astram, Itgevoth, Auram, Balram, etc., are some of the other manual machines. Figure 4.31 shows ASTRAM-10 manual machine. Some of the manual machines use fixed mould, i.e. CEB size cannot be changed, whereas some machines have the facility to interchange the mould, so that the CEBs of desired shape and size can be produced. Some of the machines have special attachments to produce specially shaped CEBs such as rounded corner, cornice and filler CEBs.

4.6.1.1 Compaction Mechanisms in Manual Machines

The compaction of soils to produce the compressed earth blocks was discussed in Sect. 3.6.6. The manually operated machines employ the static compaction process to produce compressed earth blocks. Figure 3.6 (in Chap. 3) illustrates the force–compaction stroke relationships to produce a CEB of size: 305 × 144 × 100 mm. The task of a soil compaction machine is to produce the desired force variation.

Fig. 4.31 ASTRAM-10 manual compressed earth block making machine

Thus, the force–deflection relationship of the soil mass is a prerequisite in designing the machine. The force–stroke relationships for the CEB (Fig. 3.6) show that large forces are required towards the end of the compaction stroke. This means that the mechanism for manually operated machine should be capable of providing gradually increasing force amplification as the compaction proceeds. The toggle mechanism is ideally suited for this purpose as it provides gradually varying mechanical advantage reaching a large value at the end (infinity) of the stroke. Toggle mechanisms are used extensively for manually operated machines and tools where a large force is required (Tuttle 1967).

Line diagrams of the toggle mechanisms are illustrated in Fig. 4.32. Figures 4.32a and 4.32b which show the toggle mechanism and the reverse toggle mechanism, respectively. The links AB and BC are the two main links of the mechanism, and AD is the lever through which the force is applied manually to generate large force at C. When the lever AD occupies the position AD^1, the point C moves to C^1, thereby AB and BC will be in one straight line. The point C is hinged to the bottom or top of the mould in the machine. As the lever, AD is pulled, the point C moves upwards, and the stroke length CC^1 is responsible for the compaction of the soil in the mould.

The mechanical advantage of the toggle mechanism is the ratio of the angular velocities of point C and point B.

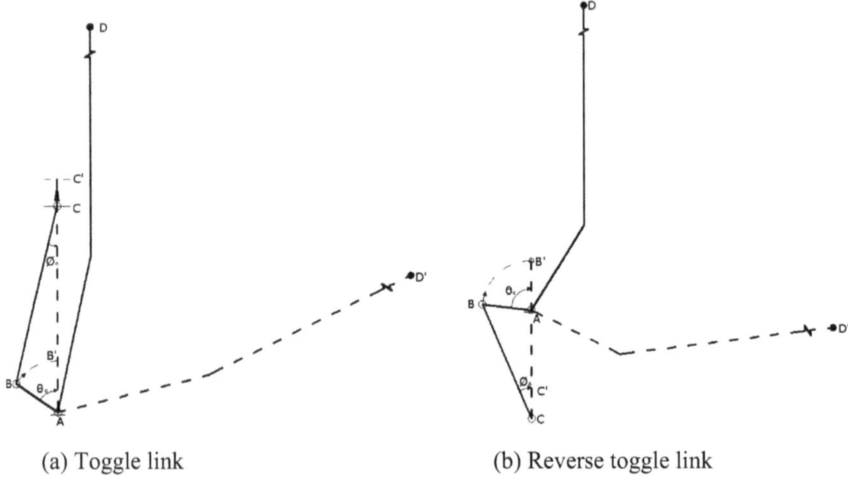

Fig. 4.32 Toggle mechanisms of CEB machines. (AB = r; BC = l; CC′ = X_m; AD = L)

(a) **Mechanical advantage of the Toggle mechanism**

Let Φ_o and θ_o be the initial angles (Fig. 4.32). The maximum value of the stroke length CC′ depends on the initial value of angle θ. As θ approaches zero, the point C moves to C′.

Let CC′ = X_m = maximum stroke length,

X = Any intermediate value of stroke corresponding to the value of θ.

The expression for mechanical advantage is

$$A_T = \left| \frac{L\dot{\theta}}{\dot{X}} \right| \tag{4.1}$$

Then,

$$A_T = \frac{L}{r\mathrm{Sin}\theta \left[1 + \frac{\left(\frac{r}{l}\right)\mathrm{Cos}\theta}{\left\{1-\left(\frac{r}{l}\right)^2\mathrm{Sin}^2\theta\right\}^{1/2}} \right]} \tag{4.2}$$

Asymptotic behaviour of A_T

As $\theta \rightarrow 0,\ \mathrm{Sin}\theta \rightarrow \theta,\ \mathrm{Cos}\theta \rightarrow 1.0$

$$\lim_{\theta \rightarrow 0} A_T = \frac{L}{r\theta\left(\frac{r}{l} + 1.0\right)} \tag{4.3}$$

Again as $\theta \rightarrow 90°,\ \mathrm{Sin}\theta \rightarrow 1.0,\ \mathrm{Cos}\theta \rightarrow 0.$

and

$$A_T \to \frac{L}{r} \tag{4.4}$$

The value of θ varies between the initial angle θ_o and zero. It is clear from Eq. (4.3) that, to achieve maximum amplification towards the end of the stroke (i.e. when $\theta \to 0$), the r/l ratio should be close to zero. Hence, to keep "A_T" large for a given value of θ, L/r should be large, and r/l should be close to zero. In practice, it is difficult to have exceedingly small r/l ratios.

(b) **Mechanical advantage of the Reverse Toggle mechanism**

The mechanical advantage of reverse toggle link is $A_{RT} = \left|\frac{L\dot{\theta}}{\dot{X}}\right|$

Then,

$$A_{RT} = \frac{L}{r\mathrm{Sin}\theta\left[\frac{\left(\frac{r}{l}\right)\mathrm{Cos}\theta}{\left\{1-\left(\frac{r}{l}\right)^2\mathrm{Sin}^2\theta\right\}^{1/2}} - 1.0\right]} \tag{4.5}$$

Asymptotic behaviour of A_{RT}

As $\theta \to 0,\ \mathrm{Sin}\theta \to \theta,\ \mathrm{Cos}\theta \to 1.0,\ \mathrm{Sin}^2\theta \to 0$

$$\lim_{\theta \to 0} A_{RT} = \frac{L}{r\theta\left(\frac{r}{l} - 1.0\right)} \tag{4.6}$$

Again as $\theta \to 90°,\ \mathrm{Sin}\theta \to 1.0,\ \mathrm{Cos}\theta \to 0.$

and

$$A_{RT} \to \frac{L}{r} \tag{4.7}$$

In this case, also the value of θ varies between the initial angle θ_o and zero. Equation (4.6) describes the behaviour of A_{RT} as θ tends to zero. For a specific value of θ, "A_{RT}" will be large if L/r is large and if r/l is close to 1.0.

Equations 4.3 and 4.6 clearly indicate that, for specific values of θ, L and r/l, the reversed toggle will always have a better amplification than toggle mechanism. As a typical illustration if "r" and "L" are the same for both the mechanisms and $l_{\text{reverse toggle}} = 0.5\ l_{\text{toggle}}$ and $r = 0.5\ l_{\text{reverse toggle}}$, then for small values of θ the amplification of the reverse toggle is 2.5 times that of the toggle mechanism.

The force amplification versus ram-stroke curves are different for different *r/l* ratios as illustrated in Figs. 4.33 and 4.34 for toggle and reverse toggle mechanisms, respectively. As the "*r/l*" ratio is reduced, the magnitude of the mechanical advantage

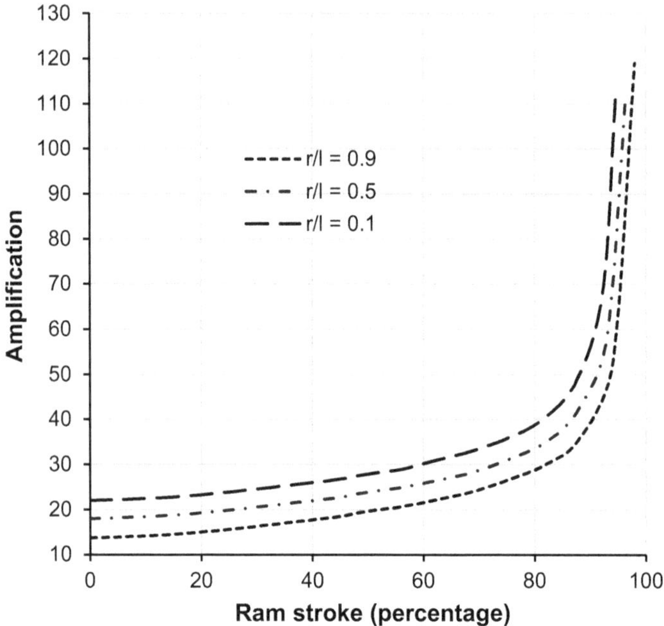

Fig. 4.33 Effect of "*r/l*" ratio on force amplification in toggle mechanisms (Reddy 1983)

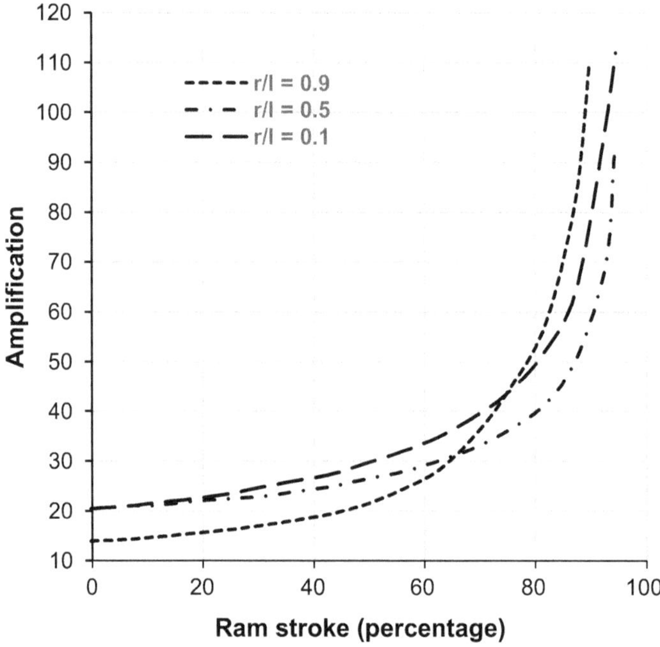

Fig. 4.34 Effect of "*r/l*" ratio on force amplification in reversed toggle mechanisms (Reddy 1983)

Table 4.5 Variation of θ_o with maximum ram-stroke (X_m)

Type of mechanism	r (mm)	r/l ratio	X_m (mm)					
			20	30	40	50	60	70
			θ_o (degrees)					
Toggle mechanism	90	0.1	37.25	46.20	54.00	61.10	67.85	74.40
	90	0.5	31.90	39.60	46.30	52.50	58.40	64.15
	90	0.7	29.85	36.95	43.10	48.80	54.15	56.75
	70	0.1	42.50	52.90	62.10	70.70	78.95	87.10
	70	0.5	36.40	45.35	53.40	60.90	68.15	75.50
	70	0.7	34.00	42.25	49.60	56.40	62.95	69.50
Reversed toggle mechanism	100	0.1	38.70	47.70	50.50	62.60	69.10	75.30
	100	0.5	49.50	59.70	68.20	75.50	82.10	88.10
	100	0.7	59.10	69.53	77.65	84.50	90.50	95.95
	70	0.1	49.60	57.60	67.30	76.20	84.65	92.90
	70	0.5	58.50	70.40	80.25	89.00	97.00	104.50
	70	0.7	68.20	79.70	88.25	96.65	103.80	110.50

(force amplification) increases for the toggle mechanism, whereas for the reversed toggle mechanism the mechanical advantage increases as the "*r/l*" ratio tends to unity. Figure 4.34 shows the variation in the force amplification with the ram-stroke for three "*r/l*" ratios of 0.1, 0.5 and 0.9 of reversed toggle mechanism. It is clear from this figure that for "*r/l*" ratio of 0.9, the force amplification is relatively less until 75% of the ram-stroke, and then, it increases rapidly for the remaining 25% of the stroke length compared to the curve for "*r/l*" = 0.5. These curves are useful in choosing the "*r/l*" ratio for a CEB compaction machine.

The initial angle θ_o, maximum ram-stroke length X_m and r/l ratio are all interrelated. Table 4.5 gives the variation of initial angle θ_o for different values of maximum ram-stroke (X_m) and "*r/l*" ratio. The crucial factors affecting the force amplification and the maximum ram-stroke of the mechanism are the initial angle θ_o and the "*r/l*" ratio. These factors must be considered in the design of a soil compaction ram for the compressed earth block production.

4.6.1.2 Human Effort Requirement in Manual Machines

The magnitude of human effort required to produce a compressed earth block of specific size and the density depends upon the type of the toggle link used in the machine. The force–stroke relationship for the manufacture of compressed earth block of size 305 × 144 × 100 mm (with a dry density 1900 kg/m^3) using a sandy soil at OMC is shown in Fig. 3.6 of Chap. 3. To produce compressed earth block

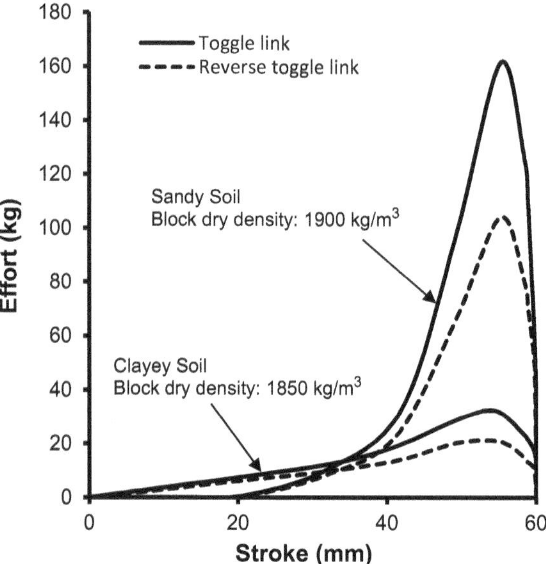

Fig. 4.35 Human effort variation during pressing of a CEB of size: 305 × 144 × 100 mm, (Reddy 2015)

of thickness 100 mm, with a dry density of 1800 kg/m^3 or more, using a manual machine, the compaction stroke length of about 60 mm is essential. The human effort required to produce this CEB can be calculated by dividing the force required by the force amplification provided by the toggle mechanism of the machine. Figure 4.35 shows the effort-ram-stroke relationships for the machines using the toggle link and the reverse toggle link. These relationships clearly indicate that the effort required gradually increases with stroke reaching a maximum when 90–95% of the stroke is completed. The effort diminishes rapidly during the last 5–10% of the stroke. The maximum effort needed for the CEB production varies with the type of the soil and the dry density needed. For example, with 1800 kg/m^3 dry density and using clayey soil, the maximum compaction effort needed is about 18 and 30 kg using a machine with the reverse toggle and the toggle link, respectively (Fig. 4.35). The effort needed to produce a CEB with 1900 kg/m^3 dry density and with a sandy soil is about 100 and 170 kg (Fig. 4.35) using a machine with the reverse toggle and the toggle links, respectively. The effort reduces by about 40% when the machine has a reverse toggle link.

The laboratory tests conducted (Reddy 1983) to elicit the effort produced by a person reveal that the Indian construction worker can generate an effort of 50 kg over short durations (of the order of few seconds). One can expect a lower value for the maximum effort when sustained compaction operations have to be carried out. Force requirement of the order of 10–12 tonnes cannot be produced by the effort of a single person in a manual machine. Also, it has been found (Reddy 1983 and 1991) that the soils with 15% clay fraction can be compacted to a CEB (size: 305 × 144 × 100 mm) with dry density of 1800 kg/m^3 with a maximum compaction force of 10 tonnes. Manual machines generate ultimate compaction pressures in the range of 2–3 MPa on the block/brick surface. These discussions suggest that two persons effort is required to achieve dry density of >1800 kg/m^3 (for CEB of size: 305 × 144 × 100 mm).

4.6.2 Mechanised and Semi-mechanised Machines

The major difficulty in the manual machines is the supply of human effort for the compaction which is a strenuous operation causing a lot of fatigue to the human beings involved in the compaction operations. The obvious choice is to supply energy through a motor. This leads to the concept of mechanising the CEB production system. Both the soil processing and CEB production can be mechanised. Various activities in the CEB production process (feeding, compaction, ejection and conveying) can be facilitated using mechanical driven systems or hydraulic systems. The Winget CEB making machine represents the first mechanised machine developed in 1948. This machine was tested in Tanzania to produce stabilised compressed earth blocks (ILO/UNIDO 1987). The mechanised machines can be grouped into two broad categories viz. (1) semi-automatic and (2) fully automatic. In the semi-automatic machines, many of these operations: mixing, feeding, compaction and ejection of the CEB, are mechanised. For example, the simplest form of semi-automatic machine could be mechanising the compaction and ejection process of the CEB. In fully mechanised systems, all the operations starting from soil sieving to conveying the CEB to the curing crate are mechanised. Figure 4.36 shows a semi-automatic machine, and Fig. 4.37 shows a fully automatic machine. In the mechanised systems, the ultimate compaction pressures exerted on the CEB can range between 5 and 20 MPa.

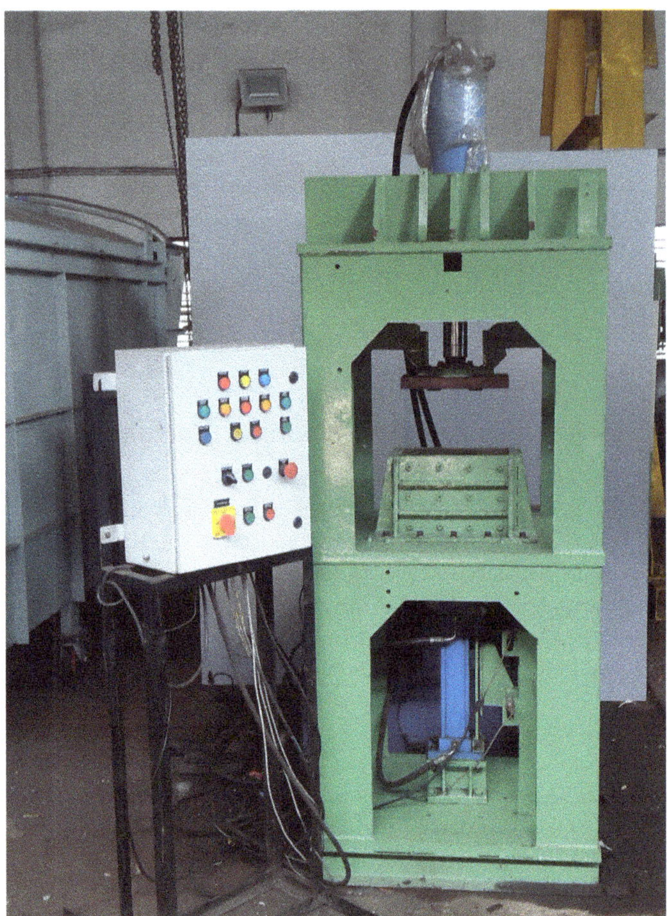

Fig. 4.36 ASTRAM-P20 semi-automatic machine

Fig. 4.37 A fully automated compressed earth block machine (Courtesy: David Easton, Watershed Materials LLC, USA)

References

ILO/UNIDO (1987) Technical memorandum No. 12—small scale manufacture of stabilised soil blocks. International Labour office, Geneva, pp 4–5

Mukerji K (1986) Soil block presses, Report on global survey. German Appropriate Technology Exchange, Dag-Hammerxjold-weg, 1, 6236, Eschborn, Germany

Reddy BVV, Hubli SR (2002) Properties of lime stabilised steam-cured blocks for masonry. Mater Struct 35, pp 293–300

Reddy BVV (1983) On the technology of pressed soil blocks for wall construction. MSc (Engg) thesis, Department of Civil Engineering, Indian Institute of Science Bangalore, India

Reddy BVV (1991) Studies on static compaction of soils and compacted soil-cement blocks for walls. PhD thesis, Department of Civil Engineering, Indian Institute of Science Bangalore, India
Reddy BVV (2015) Design of a manual press for the production of compacted stabilised soil blocks. Curr Sci 109(9), pp 1651–1659
Tuttle SB (1967) Mechanisms for machine design. Wiley

Chapter 5
Characteristics of Stabilised Compressed Earth Blocks

5.1 Introduction

The stabilised compressed earth block (CEB) is used for the structural masonry. Apart from the physical form (proper shape, size and edges), the specific characteristics of the stabilised CEBs to be addressed are listed as follows.

1. Strength
   a. Initial and long-term strength
   b. Strength in wet state
2. Absorption characteristics
   a. Water absorption
   b. Rate of water absorption
   c. Initial rate of absorption (IRA)
3. Development of bond with the mortar
4. Durability
   a. Dimensional stability
   b. Mass loss due to cyclic wetting and drying
   c. Erosion due to rain impact
5. Deformation characteristics and elastic properties
   a. Stress–strain relationships
   b. Modulus, ultimate strain, Poisson's ratio
6. Thermal characteristics
   a. Thermal conductivity and thermal performance

B. V. V. Reddy, *Compressed Earth Block & Rammed Earth Structures*,
Springer Transactions in Civil and Environmental Engineering,
https://doi.org/10.1007/978-981-16-7877-6_5

The structural requirements of a load bearing masonry building require information on the characteristics of the masonry unit such as CEB, mortars, type of bonding, strength and deformation properties of the masonry as well as the masonry materials. Understanding the absorption and the durability characteristics of CEBs becomes essential to throw light on the durability of the stabilised CEB buildings. Assessing the issues on thermal comfort in CEB buildings necessitates the understanding on the thermal behaviour and the thermal characteristics of the CEB and the CEB masonry.

5.2 Factors Affecting the Strength Characteristics of Stabilised Compressed Earth Block

Soil, sand, gravel, stabilisers and water, form the main ingredients used to manufacture the stabilised CEBs. The factors influencing the strength characteristics of the stabilised CEB are:

1. Soil composition
    a. Particle size distribution (gravel, sand, silt and clay fractions)
    b. Type and quantity of clay minerals
    c. Chemical properties of the soil-sand mix
    d. Organic matter
2. Density
    1. Porosity and pore structure
    2. Mass density
3. Moulding moisture content
4. Type and quantity of stabiliser
    a. Type of stabiliser (lime, cement, lime-pozzolana, etc.)
    b. Quantity of stabiliser.

The roles of these factors in controlling the characteristics of the stabilised CEB are discussed in detail in the subsequent sections.

5.3 Influence of Soil Composition on the Strength of Stabilised Compressed Earth Block

The soil composition includes the particle size distribution (fractions of clay, silt, sand and gravel), organic matter and the presence of salts and the pozzolanic materials. The particle size distribution of the soil and the clay minerals present in the soil-sand mix play significant role on the strength of the stabilised CEB. The type of stabiliser needed to produce CEB will depend upon the types of clay minerals present in the mix used for the CEB production. The type and the percentage of stabiliser needed will depend upon the clay mineral type and the desired strength for the brick or the block.

The compressive strength of the stabilised CEB is very sensitive to its degree of saturation. The stabilised CEB can be tested at the two extreme moisture conditions, i.e. either in the dry state or in the wet state. It is difficult to achieve 100% saturation for the stabilised CEB through soaking in cold water at ambient temperature conditions. Soaking for even 72 h in cold water results in about 75% saturation for the cement stabilised CEB (Reddy and Latha 2014b). The compressive strength measured at near saturated condition (48–72 h soaking in cold water at ambient temperature) can be termed as the wet compressive strength, and the strength measured in oven dry (~60 °C) condition can be designated as the dry compressive strength.

5.3.1 Soil Reconstitution

The soils used for the stabilised CEB production are sieved through either a mesh (5–6 mm) or ground to break the clay lumps. The ground or the crushed material is then passed through a mesh. Therefore, the mix used for the stabilised CEB production is generally free from large percentage of gravel-sized particles. Generally, the natural soils do not have optimum soil grading suitable to produce the stabilised compressed earth blocks or the stabilised rammed earth. They are either excess or deficient in sand, silt and clay size fractions meeting the requirements of an optimum grain/textural composition. The composition of such soils can be brought to an optimum level by reconstituting with sand or silt or by mixing two different soils. This kind of reconstitution is nothing but the mechanical stabilisation of soils. Let us consider a natural soil designated as NS. Figure 5.1 shows the grain size distribution curve for the soil. Table 5.1 gives the engineering characteristics of the natural soil.

The soils can be reconstituted by the addition of gravel, sand, silt or clay size fractions such that the final soil mix contains optimum clay, silt and sand size fractions. The grain size distribution curves for the natural soil and the two reconstituted soils are shown in Fig. 5.1. Table 5.1 gives the clay, silt and sand size fractions of the three different types of soils. The natural soil (NS) contains 50.3, 18.3 and 31.6% of sand, silt and clay size fractions, respectively, upon reconstitution with sand in the ratio of 1:1.25 (NS: sand, by mass) the reconstituted soil (RS1) contains 77.7, 8.3 and 14.0% sand, silt and clay size fractions, respectively. Similarly, when the natural soil (NS) upon diluting with silt in the ratio of 1: 1.25 (NS: silt, by mass), the reconstituted soil (RS2) contains 22.4, 63.6 and 14% sand, silt and clay size fractions, respectively. When the textural composition of the soil is altered, other engineering characteristics of the soil also change. For example, the liquid limit and the plasticity index of the NS soil change from 40% and 21% to 26.9% and 17.5%, respectively, when reconstituted with sand in the proportion of 1:1.25 (soil: sand, bay mass). In addition, the compaction characteristics get altered. Similarly, for reconstituted soil RS2 using silt fraction, the liquid limit and the plasticity index get changed to 28%

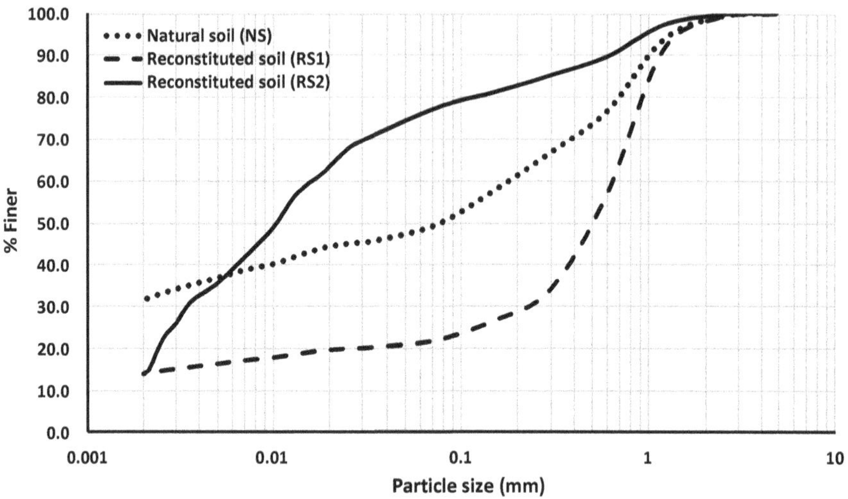

Fig. 5.1 Grain size distribution curves of natural and reconstituted soils

and 10.2%, respectively. Thus, the grain size fractions of clay, silt and sand in a soil can be easily altered by mixing with gravel, sand, silt or a different soil to arrive at an optimum soil grading. This technique of reconstituting the natural soils is commonly practised in the production processes of the stabilised CEBs.

5.3.2 Optimum Soil Grading for Stabilised Compressed Earth Bricks/blocks

The clay minerals have controlling influence on the behaviour of a soil. The clay minerals influence the strength, Atterberg's limits and the compaction characteristics of the soil. Among the different soil particles, the type and percentage of clay minerals present in the soil have controlling influence on the strength of stabilised CEB. It is desirable to define the optimum clay fraction in the mix, which yields maximum strength for the stabilised CEB. Obviously, it cannot be a unique value for the clay fraction; a narrow range can be defined. The optimum clay fraction is again dependent upon the type of the clay mineral and the type of cementitious stabiliser additive used. The Portland cement and the lime are the commonly used stabilisers in the stabilised CEB production.

5.3.2.1 Cement Stabilised Compressed Earth Specimens or Bricks/Blocks

The cement stabilisation is ideally suited for coarse-grained soils with non-expansive clay minerals (Mitra 1951; Fitzmaurice 1958; Ingles and Metcalf 1972; Houben and Guillaud 2003; Walker and Stace 1997; Reddy and Gupta 2005). However,

Table 5.1 Textural composition and properties of natural and reconstituted soils

Soil type	Reconstituted soil proportion (by mass)			Textural composition (%)			Atterberg's limit			Standard Proctor test	
	Soil	Sand	Silt	Sand	Silt	clay	Liquid limit (%)	Plastic limit (%)	Plasticity Index (%)	OMC (%)	MDD (kg/m^3)
Natural soil (NS1)	1	0	0	50.3	18.1	31.6	40.0	19.0	21.0	15.6	1830
Reconstituted soil (RS1)	1	1.25	0	77.7	8.3	14.0	26.9	9.4	17.5	9.15	1962
Reconstituted soil (RS2)	1	0	1.25	22.4	63.6	14.0	28.0	17.8	10.2	20.6	2053

both coarse and fine-grained soils are used to produce stabilised CEBs. Hence, it is necessary to understand the strength of the cement stabilised compressed earth block using both the coarse and the fine-grained soils containing the non-expansive clay minerals.

The influence of clay content on the strength and the durability characteristics of the stabilised CEBs has been explored by Mitra (1951), Fitzmaurice (1958), Bokhari (1976), Olivier and Mesbah (1987), Reddy and Jagadish (1995), Walker and Stace (1997), Reddy and Walker (2005), Reddy et al. (2007), and Reddy and Latha (2014b). Investigations by Mitra (1951) revealed that soils containing high silt and clay fractions are not suitable for the cement stabilised CEB production and emphasised the use of sandy soils. Fitzmaurice (1958) recommended the use of soils with low clay fraction and high sand/gravel fraction for the cement stabilised CEB.

Bokhari's (1976) studies showed pronounced strength increase for cement stabilised CEBs, when the sand was added to the soils, especially for the cement contents more than 3%. Olivier and Mesbah's (1987) investigations were restricted to the artificially reconstituted soils using gravel, sand, silt and kaolinite clay. The compressive strength of the cement (4%) stabilised compressed earth cylindrical specimens was examined considering wide range of sand (30–75%) and clay (13–60%) fractions. The results showed that the strength increased with increase in the sand and the gravel fraction, and the best results were for the soils with 70% sand and 20% clay fraction.

Considering the soil grain size and the clay mineral as variables, Reddy and Jagadish (1995) investigated the wet compressive strength and the durability of the cement stabilised compressed earth cubes and blocks. The study revealed that the wet compressive strength of the compressed soil–cement specimen was sensitive to the amount of clay fraction in the mix. The wet strength increased with the decrease in the clay content of the soil–cement mix. For cement stabilised CEBs, the study recommended using soils with predominantly non-expansive clay minerals containing 70 ± 5% sand and less than 15% clay. Walker and Stace (1997) observed drastic reduction in the strength of compressed soil–cement with the increase in the clay content of the soil mix. These studies emphasise the importance of the clay minerals in controlling the strength of the cement stabilised compressed earth specimen. Reddy et al. (2007) made a detailed study on arriving at optimum soil grading limits for the manufacture of the cement stabilised CEB. The studies were conducted using a coarse-grained soil and its reconstituted variants. Based on both the strength and the durability characteristics, the study concluded that the optimum clay fraction in the mix is about 16%. Though the earlier studies specify optimum clay content in the soil yielding maximum strength, the logical reasoning for this behaviour is not stated.

Reddy and Latha (2014b) carried out a comprehensive study considering a range of soils and cement contents. The study revealed the optimum clay content in the soil mixes to produce the cement stabilised CEB. Controlled experiments considering both coarse and fine grain soils (containing kaolinite clay mineral) and a range of cement contents (4–10%) were performed. The dry density of the compacted specimens was controlled at 1800 kg/m^3 at the time of casting. The moulding moisture

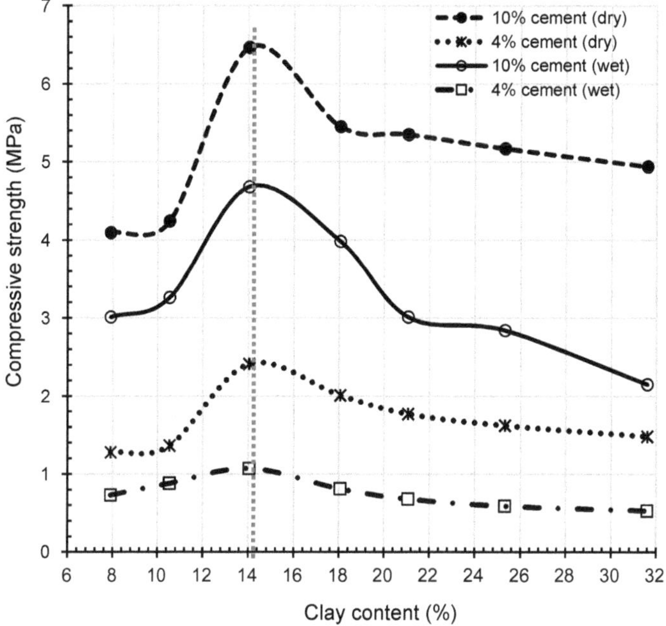

Fig. 5.2 Variation in the compressive strength with the clay content of the mix for compressed earth cylindrical specimen using coarse-grained soil

content was 12.5% and 18% for the specimens using coarse and fine-grained grained soils, respectively.

Figures 5.2 and 5.3 display relationships between the strength and the clay fraction of the mix used in the cement stabilised compressed earth specimen for the dry and the wet cases, respectively (Latha 2015). The figures clearly show that the strength of the cement stabilised compressed earth specimen increases with the increase in the clay content, reaches a peak value and then drops as the clay content increases further. These results are from controlled experiments on the compressed earth cylindrical specimens (38 mm diameter × 76 mm height). The dry density and the moulding moisture content of the specimens have been controlled to prevent their interference in the measurement of the strength of the cement stabilised compressed earth specimen. The clay content of the soil used in these studies was in the range of 5–32%. There was considerable difference in the strength from the peak to the extreme values. The optimum clay content yielding maximum strength was in the range of 10–15% considering both the fine and the coarse-grained soils. The optimum clay content was about 10% and 14% for the specimens using the fine and the coarse-grained soils, respectively. Figure 5.4 shows the soil grading curves yielding maximum strength for the coarse and the fine-grained soils.

Void ratio–strength relationships: The investigations of Latha (2015), and Reddy and Latha (2014b) explored further in giving logical explanations for the optimum clay content yielding maximum strength and better durability characteristics. Even

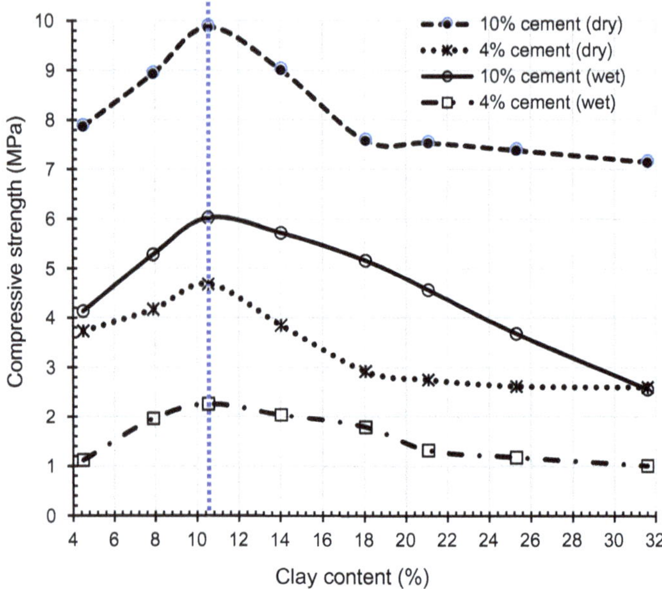

Fig. 5.3 Variation in the compressive strength with the clay content of the mix for compressed earth cylindrical specimen using fine-grained soil

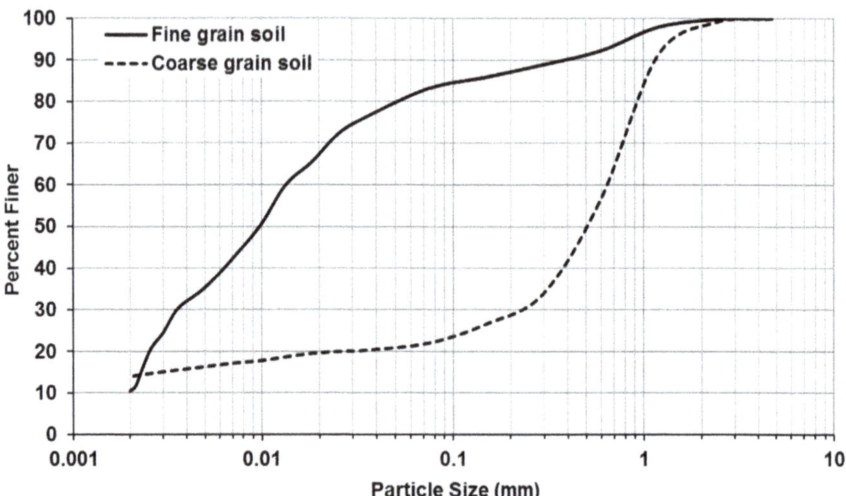

Fig. 5.4 Optimum soil grading curves for the coarse- and the fine-grained soils

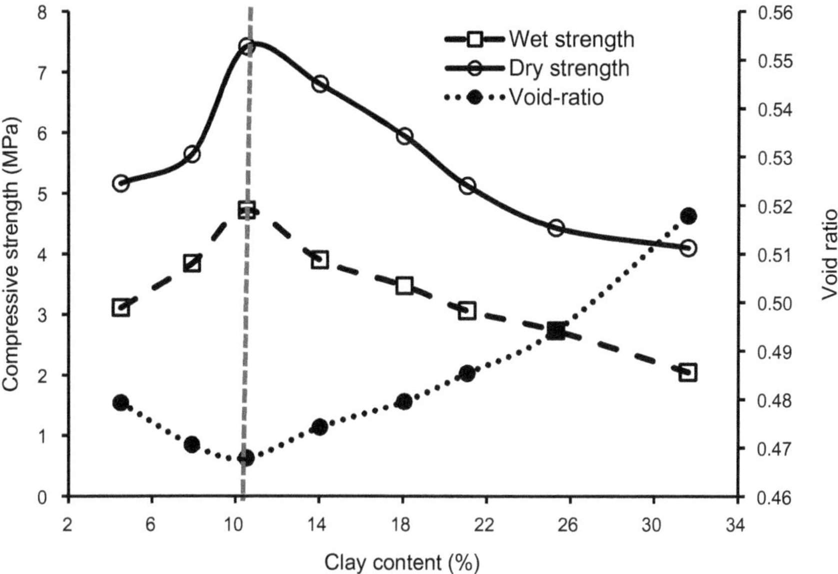

Fig. 5.5 Strength–void ratio–clay content relationships (7% cement) (Reddy and Latha 2014b)

though the initial dry density and the moulding moisture content of the specimens was controlled, the dry density at the time of testing marginally varies across the clay content range. This was mainly attributed to the variation in the pore structure and porosity across the range of clay contents used. This has led to variation in the void ratio of the specimen as the clay content varies. Figure 5.5 displays a typical variation in the strength and the void ratio with the clay content of the soil mix. At the optimum clay content, the void ratio is minimum. Figure 5.6 illustrates wet compressive strength and void ratio relationships measured on compressed earth cylindrical specimens (38 mm diameter × 76 mm height). As the void ratio increases, the strength decreases and the relationship is linear.

Surface porosity-strength-clay content: Fig. 5.7 shows the scanning electron microscope (SEM) images of the broken surfaces of the cement stabilised compressed earth specimens for few typical cases. The SEM images show the surface pore structure. The particle size and the surface pore size appear larger for the specimens using the coarse-grained (CG) soils when compared to the ones with the fine-grained (FG) soils. Figure 5.8 shows the plot of surface porosity–strength–clay content relationships for the cement stabilised compressed earth specimens. Table 5.2 gives the surface porosity of different samples assessed through SEM image analysis. In addition, the table gives wet compressive strength values for the 7% cement stabilised compressed earth cylindrical specimen.

SEM images show distinctly different pore structure for the CG and the FG soil compressed earth specimens. The pores and the grains are larger in size in the CG soil specimen when compared to those in the FG soil specimen. The surface porosity

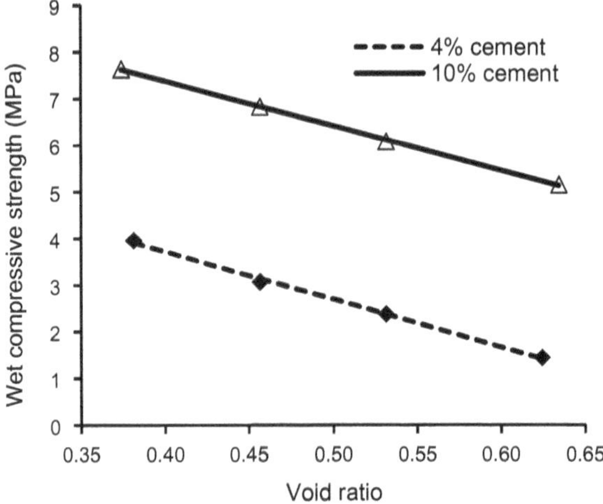

Fig. 5.6 Strength versus void ratio (Latha 2015)

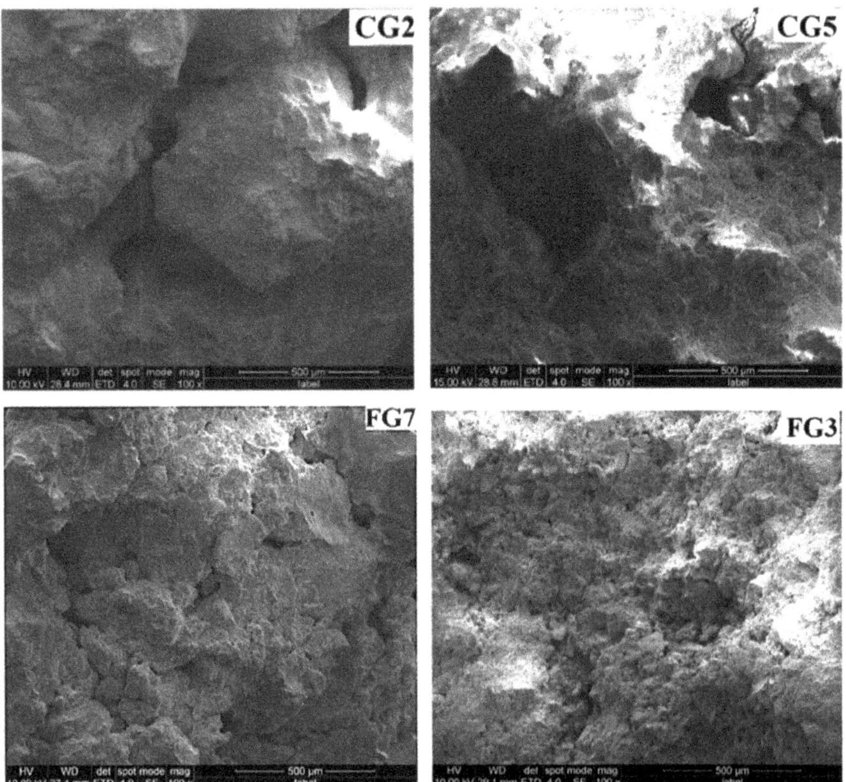

Fig. 5.7 Typical SEM images of the inside surfaces for cement stabilised compressed earth specimens (CG2 & CG5—Coarse-grained soil; FG7 & FG3—Fine-grained soil) (Latha 2015)

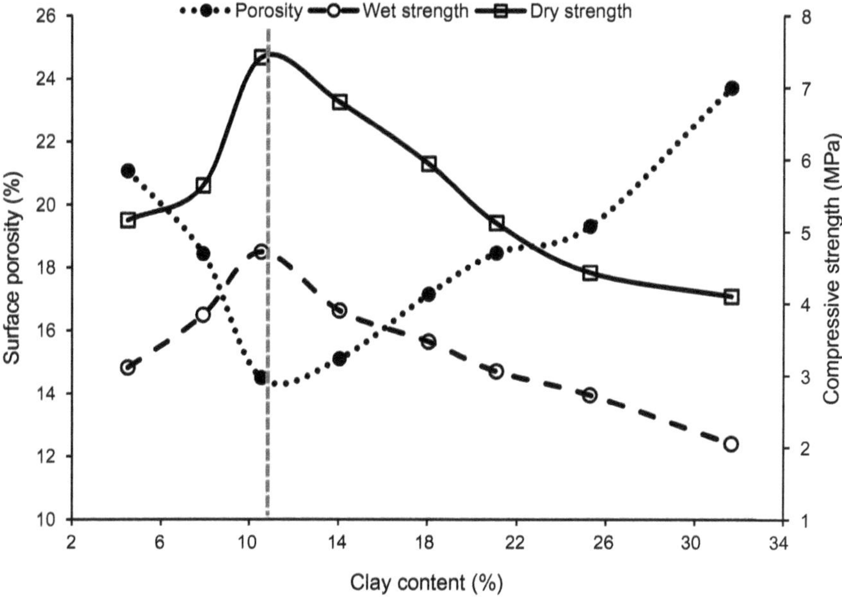

Fig. 5.8 Surface porosity–strength–clay content relationships (Reddy and Latha 2014b)

is minimum at the optimum clay content yielding maximum compressive strength for both the soil types.

For any given clay content, the FG soil specimen possesses smaller surface porosity when compared with the porosity of the CG soil specimen. For example, at 10.5% clay content the porosity of FG soil specimen is 14.5% while that of CG soil one is 20.5%. For 7% cement and 1800 kg/m^3 initial dry density, FG soil compressed earth specimens possess higher compressive strength when compared to the strength of CG soil ones. This can be attributed to lower porosity in the FG soil compressed earth specimen when compared to the porosity of the CG soil compressed earth specimen.

Summary: Considering soils containing predominantly non-expansive clay minerals as best suited for cement stabilisation; soil grading, especially the clay content of the soil, has controlling influence on the strength of the cement stabilised compressed earth bricks/blocks. The optimum clay content of the soil resulting in maximum compressive strength is in the range $10 \pm 2\%$ and $14 \pm 2\%$ (by mass) for the fine grain and the coarse grain soils, respectively. At the optimum clay content, the compressed earth specimens have least void ratio and least surface porosity. For a given density and cement content the fine-grained soils, show higher strength for the cement stabilised CEB when compared to the strength of the ones with the coarse-grained soils. Relationships between optimum clay content and the durability characteristics of cement stabilised CEB are discussed in the subsequent sections.

Table 5.2 Surface porosity and strength for cement stabilised compressed earth specimen (Cylindrical specimen: 38 mm diameter × 76 mm height; 7% cement; 1800 kg/m^3 initial dry density)

Soil				Surface porosity (%)	Wet compressive strength (MPa)
Designation	Composition (% by mass)				
	Clay	Silt	Sand		
Coarse-grained					
CG1	7.9	4.8	87.3	22.7	1.34
CG2	10.5	6.3	83.2	20.5	1.50
CG3	14.0	8.3	77.7	19.6	2.51
CG4	18.0	10.6	71.4	23.7	1.65
CG5	21.0	12.3	66.7	25.3	1.52
CG6	25.3	14.5	60.2	26.2	1.36
CG7	31.6	18.1	50.3	27.0	1.12
Fine-grained					
FG1	4.5	88.3	7.2	21.1	3.11
FG2	7.9	79.6	12.6	18.4	3.84
FG3	10.5	72.7	16.8	14.5	4.72
FG4	14.0	63.6	22.4	15.1	3.90
FG5	18.0	53.2	28.8	17.2	3.47
FG6	21.0	45.4	33.6	18.5	3.06
FG7	25.3	34.5	40.2	19.3	2.73

5.3.2.2 Lime Stabilised Compressed Earth Bricks/Blocks

The mechanism of strength development in the lime stabilised compressed earth was discussed in Sect. 3.5 (Chap. 3). The strength developed in the lime stabilised compressed earth is mainly attributed to the formation of the cementitious products due to the lime–clay reactions. The purity of lime, the type of lime and the type and the quantity of the clay minerals present in the soil play crucial role in the strength development. Generally, stabilisation of soils using lime is mainly aimed at controlling the swell–shrink characteristics of the stabilised soil. The lime stabilised soils have been extensively used in the construction of roads, pavements, retaining walls, slope protection, canal lining, drains, pile foundations, etc. (Dogra et al. 1960; Lambe 1962; Herrin and Mitchell 1961; Diamond and Kinter 1965; Ingles and Metcalf 1972; Mitchell 1981; Anon 1985, 1990; Sherwood 1993). There are a number of investigations on the lime stabilised compressed earth bricks/blocks for the masonry walls (Reddy and Jagadish 1984; Reddy and Lokras 1998; Reddy and Hubli 2002; Houben and Guillaud 2003; Oti et al. 2008a, b, 2009a, b, Oti et al. 2010; Oti and Kinuthia 2012; Inzemmouren et al. 2015).

Lime-clay and strength relationships

The lime-clay reactions are the main contributors for the strength attained in the lime stabilised compressed earth specimens. The calcium hydroxide (lime) reacts with the alumina–silica present in the clay minerals resulting in the formation of the cementitious products such as calcium silicate hydrate (CSH), calcium aluminate silicate hydrates (CASH), etc. (Clare and Cruchley 1957; Eades and Grim 1960; Herrin and Mitchell 1961; Croft 1964; Thompson 1966; Bell 1996; Oti et al. 2009a; Dash and Hussain 2012; Reddy and Latha 2014a). There is a limiting capacity for a given clay mineral to entice into lime-clay reactions. For a given clay content, there is an optimum lime content giving maximum strength. Figure 5.9 illustrates the relationships between the strength and the lime content for different types of clay minerals present in the soil. Table 5.3 gives the details of the optimum lime content, the clay mineral type and the percentage of clay mineral. These results indicate that the strength of the lime stabilised compressed earth increases with the increase in the lime content, reaches a peak value and then drops. There is a considerable difference in the strength between the optimum lime content (OLC) and the extreme values of lime contents. These relationships indicate that the lime beyond OLC remains as a weak filler in the stabilised compressed earth leading to fall in strength. The investigations of Bell (1996) attribute that the lime does not have appreciable friction nor cohesion, and hence, an excess amount of lime acts as lubricant between the soil particles, thereby decreasing the strength. The OLC corresponding to lime stabilised compressed earth for a given soil mix and depends upon the type and the quantity of

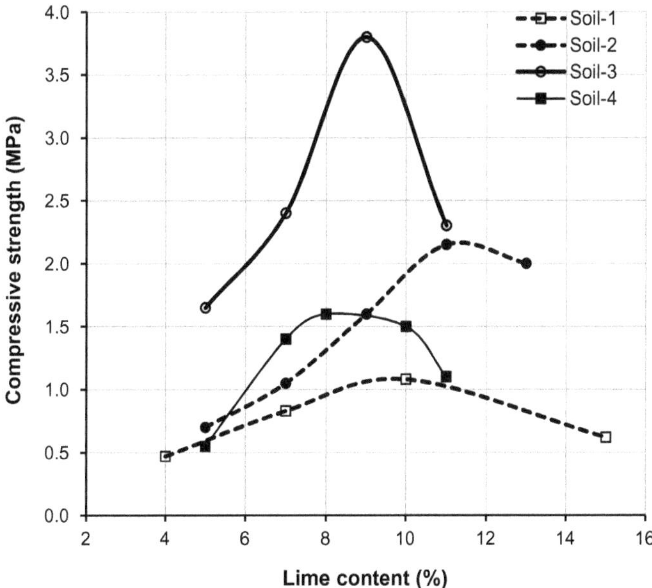

Fig. 5.9 Strength and lime content relationships for lime stabilised compressed earth specimen

Table 5.3 Details of optimum lime content and type of clay mineral in the soil

Soil designation	Type of clay	Clay content (%)	Optimum lime content (%)	References
Soil-1	Kaolinite	15.8	12.5	Latha (2015)
Soil-2	Montmorillonite	12.0	11.5	Reddy and Jagadish (1984)
Soil-3	Kaolinite	10.5	9.0	Reddy and Jagadish (1984)
Soil-4	Kaolinite and montmorillonite	6.0	8.0	Reddy and Jagadish (1984)

the clay mineral in the mix. As illustrated in Table 5.3, the OLC is about 0.8 times the clay content for the mixes with kaolinite clay mineral, whereas for the soils with montmorillonite clay, the OLC is in the range of 1.0–1.3 times the clay content. The expansive clay minerals such as the montmorillonite can absorb more lime than the non-expansive clay minerals such as kaolinite.

Unlike in the cement stabilised compressed earth specimens, there is no optimum clay content in the lime stabilised compressed earth specimen. Instead, the lime stabilised compressed earth specimens will have an OLC which is dependent upon the type and quantity of the clay mineral in the soil mix. The higher strength values can be achieved by handling soils with more clay content provided there is sufficient lime supplied for the lime-clay reactions. Depending upon the clay mineral type and OLC, the lime-clay ratio for the lime stabilised compressed earth will be in the range 0.75–1.25. Adding higher quantities of lime results in higher costs for the lime stabilised CEBs. In many situations, reconstituting the soil through addition of inert materials such as sand and gravel can bring down the clay content such that the lime content corresponding to OLC can be in reasonable limits. For example, if a soil contains 24% clay, it can be reconstituted with sand/gravel in the ratio of 1:1.5 (soil: sand/gravel, by mass). The resulting soil-sand mix will have 9.6% clay. This mix can be stabilised with 7–10% lime, which will be at its OLC.

5.3.3 *Chemical Properties of Soils and the Strength Development in Stabilised CEB*

The chemically contaminated soils should be avoided to produce stabilised earth products. It is difficult to stabilise the soils containing chemicals such as high concentration of salts (sodium, potassium, carbonates, etc.), phosphates, chlorites and sulphates. Soils with low pH (<6.5) are difficult to stabilise with the Portland cement. The low pH interferes with cement hydration and setting of the Portland cement. This is particularly true with the lateritic soils. In such situations, addition of small quantities of lime (~2% by mass) with Portland cement will facilitate in the strength development.

The presence of sulphates generally in the form of sodium sulphate and magnesium sulphate will have detrimental effects on the stabilised compressed earth products. The sulphates tend to undergo volumetric changes in the varying atmospheric conditions. The presence of these salts in the soil can lead to the formation of decahydrates by combining with the water resulting in volume increase. The effect of sodium and magnesium sulphates on the characteristics of the soil–cement blocks was examined by Chadda and Raj (1955), and Uppal and Kapur (1957). These studies showed that the use of soils containing sodium sulphate and magnesium sulphate for the cement stabilised CEBs resulted in strength loss and disintegration with age. The extent of damage to CEB varies with the concentration of these salts in the soil.

5.3.4 Organic Matter and Soil Stabilisation

Some of the soils contain organic matter. The organic matter in soils is derived from plants and animals/organisms. The organic matter can be in the form of fibres or decomposed form. The decomposed form of organic matter will be in a finely divided state (~1 μm). Generally, the amount of organic matter in the topsoils will be in smaller quantities (<2%), except in the highly vegetated/forest regions. The soils under the heavy vegetation such as forests may contain high percentage of decomposed organic matter. Such soils, which are very rich in organic matter, are termed as peat. The organic matter influences the plasticity, density, strength and the compressibility characteristics of the soil (Odell et al. 1960; Coutinho and Lacerda 1987).

The soil organic matter comprises of lignin, cellulose, glucose, alginic acid, pectin, casein, nucleic acid and many such elements (Clare and Sherwood 1954, 1956). The active organic compounds present in the soils affect the hardening of the cement in the soil–cement mixtures and hence leading to lower strengths for the cement stabilised CEB. The retardation of cement setting in the soils with the active organic compounds is attributed to the adsorption of calcium ions (liberated during cement hydration) by the active organic matter. The presence of active organic matter in the soils is characterised by the lower pH values. The use of small dosage of calcium chloride (~2% by weight of soil) or rapid hardening cement can overcome the retarding effect of the active organic matter in the cement stabilised compressed earth specimens (Clare and Sherwood 1954, 1956). Caution should be exercised while stabilising the soils using inorganic chemicals. The percentage of chemicals added beyond a certain limit can have adverse effects on the properties of the stabilised compressed earth product.

5.4 Density, Strength and Moulding Moisture Content

5.4.1 Density and Strength Relationships

The density is a sensitive parameter in controlling the strength of the stabilised CEB. Two types of densities can be recognised: (a) bulk density (γ_b) and (b) dry density (γ_d).

$$\text{Bulk density} = \gamma_b = (\text{m}) \div (\text{v}) \tag{5.1}$$

where m = mass, v = volume

$$\gamma_d = (\gamma_b) \div (1 + \text{w}) \tag{5.2}$$

where w = water content.

The density is an indirect measure of porosity of the densified soil specimen. The porosity and the dry density relationship of the compressed or compacted soil/earth specimen is as follows:

$$\text{Dry density} = \gamma_d = (\text{G}\gamma_w) \div (1 + \text{e}) \tag{5.3}$$

where γ_w is density of the water and "**e**" is the void ratio

$$\text{Void ratio} = \text{e} = (\text{V}_v) \div (\text{V}_s) \tag{5.4}$$

where V_v = Volume of voids and V_s = Volume of solids.

The density of the stabilised CEB has significant influence on its compressive strength. Typical relationships between the compressive strength and the density for the cement stabilised CEBs (230 × 108 × 75 mm) are shown in Fig. 5.10. Figure 5.11 shows the density and the strength relationships for the lime stabilised compressed earth specimen (36 mm diameter × 76 mm height cylindrical specimen) for the two types of soils. RS soil contains kaolinite clay mineral (clay: 15.8%, silt: 11.6%, sand: 72.6%, liquid limit: 27%, plasticity index: 17.5%), while BC soil contains montmorillonite clay mineral (15.8% clay, liquid limit: 57.2%, plasticity index 24.8%). Each one of the relationships shown in Figs. 5.10 and 5.11 corresponds to the results where the density was varied by keeping the moulding moisture content constant. The wet compressive strengths shown in Fig. 5.10 are for the CEBs produced in a manual press using static compaction process at the moulding moisture content of 12%. This was a fixed stroke length machine producing constant thickness bricks. The compaction pressure in such machines can vary if the moulding moisture content (MMC) is varied. Also, the results shown in Fig. 5.11 were generated by employing

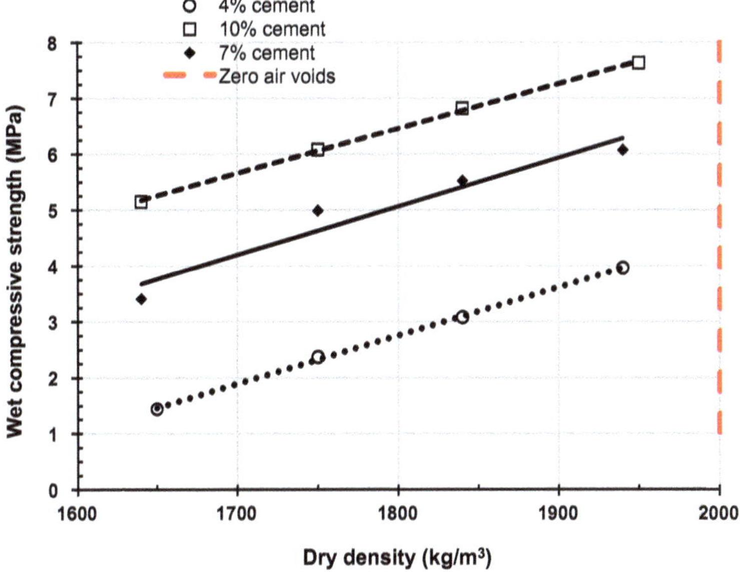

Fig. 5.10 Density–strength–moulding moisture relationships for cement stabilised compressed earth bricks

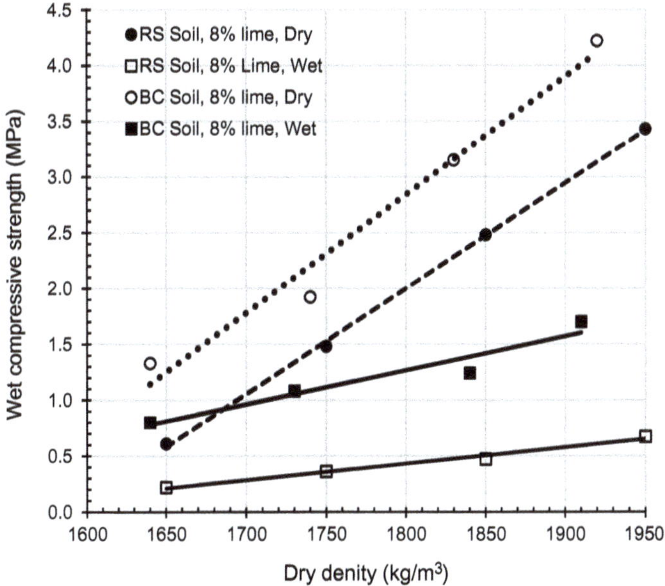

Fig. 5.11 Density–strength relationships for lime stabilised compressed earth specimen (38 mm diameter × 76 mm height)

static compaction process and maintaining fixed stroke length. The MMC for the specimens with BC soil mixes was 10%.

The compressive strength of the stabilised CEBs increases with the increase in the density irrespective of the stabiliser type and the stabiliser content. The compressive strength is very sensitive to the variations in the dry density. A marginal increase in the dry density results in large increase in the compressive strength. The strength increase caused by the increase in the dry density can be attributed to the reduction in the void ratio of the compacted specimen, resulting in better contact among the particles facilitating better bonding due to cementitious products. The maximum possible dry density was 2000 kg/m^3 at a moulding moisture content of 12% for the CEBs using the RS soil. This moisture represents the zero-air voids; any attempt to achieve dry density beyond this limit will result in squeezing the water from the compressed earth specimen. The void ratio and strength relationships for the cement stabilised compressed earth specimens was discussed in Sect. 5.3.2.1 The density and the porosity relationships for the cement stabilised CEB have been discussed in Chap. 7. Linear or nearly linear relationship between the compressive strength and the dry density for the cement stabilised compressed earth products has been reported in the investigations by Fitzmaurice (1958), Ingles and Metcalf (1972), Spence and Cook (1983), Reddy and Kumar (2011).

5.4.2 Density, Moulding Moisture and Strength Relationships

The clear distinction between static and dynamic compaction of soils was explained in Chap. 3. The machines used to produce the stabilised compressed earth blocks employ static compaction process. Two types of static soil compaction processes can be envisaged with reference to the machines used for CEB production:

(a) Constant compaction pressure–variable piston stroke length (CP-VS)
(b) Constant piston stroke length–variable compaction pressure (CS-VP).

CP-VS system: In this case, constant pressure is applied to a definite bulk mass of the processed partially saturated soil in the metal mould. Keeping the bulk mass constant and by varying the moisture content, compressed earth specimens of different thicknesses can be obtained (i.e. volume of the compressed earth specimen varies). In this system, pressure is constant, but the dry density varies as the moisture content is varied.

CS-VP system: In this case, the stroke length of the energy-supplying device is constant. For a given mould size, the dry density as well as the moulding moisture content can be varied, while producing constant volume compressed earth specimen. The compressed earth specimen volume remains constant.

Figure 5.12 gives the generic relationships for the strength and the moulding moisture content (MMC) for the cases of CP-VS and CS-VP type of compaction systems. For the CP-VS system, there is an optimum moisture content (OMC) yielding maximum strength, but the specimen volume changes at each moisture

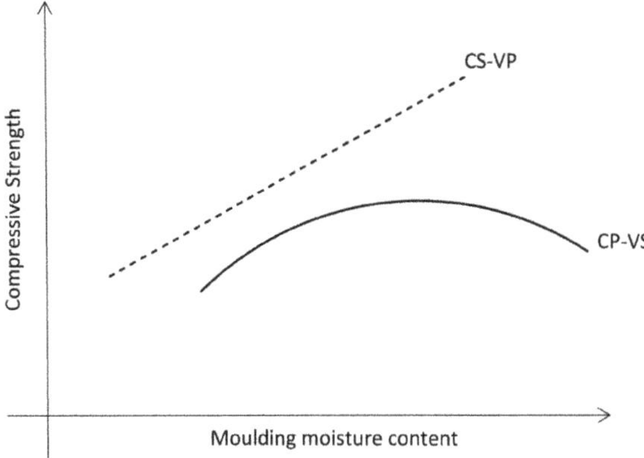

Fig. 5.12 Typical strength–moulding moisture relationships for cement stabilised compressed earth specimen

content and hence the dry density. In the case of CS-VP system, the volume of the compressed earth specimen remains constant, but the dry density varies because MMC varies. The compaction energy per unit volume varies with the variation in the MMC in both the types of compaction systems.

For CS-VP type of compaction system, the experimental results of Kumar (2009) on the strength and the moulding moisture content relationships for three different dry densities are shown in Fig. 5.13. In this case, the compressed earth specimens (cylinders) were made using 8% Portland cement, and the moulding moisture contents chosen were above, close to and below the standard Proctor OMC value. The standard Proctor OMC was 11%. The compressive strength increased with the increase in MMC irrespective of the dry density. The compressive strength increase is in the range of 20–40% as the MMC was varied between 8.5 and 14.5%. Both, the cement and the clay particles in the dry soil–cement mixture, have affinity for water. When the dry soil–cement mixture encounters moisture, the moisture is shared by both the cement and the clay particles. At lower MMC (say 8.0%), there could be insufficient supply of water for proper hydration of the cement. As the moulding water content increases (say 14.5%), more water is available for the cement hydration. This could explain for increased strength when higher percentage of MMC was used.

The results generated by Fitzmaurice (1958) for the two compaction pressures of 4.0 and 7.0 MPa are shown in Fig. 5.14. These curves represent the case of CP-VS compaction system. The dry density along the curves varies as the MMC varies. The density variation occurring here can be attributed to both volume change in the specimen as well as MMC.

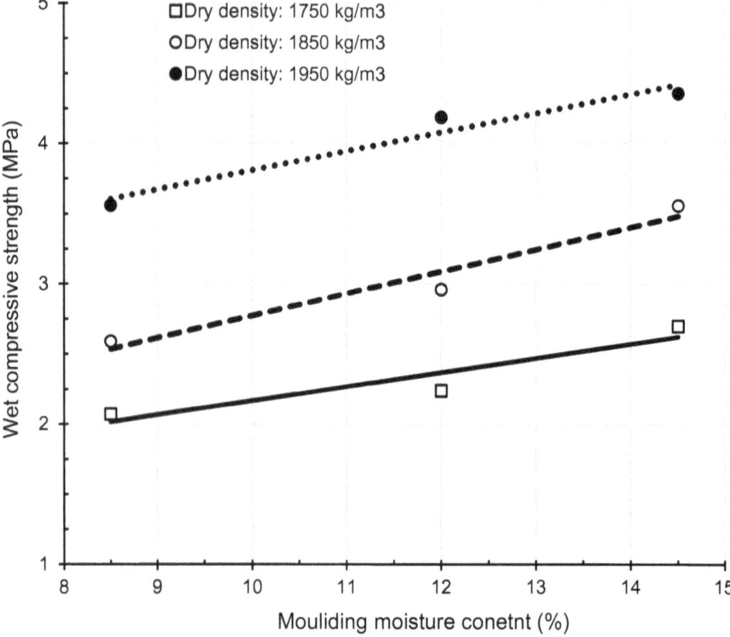

Fig. 5.13 Strength and moulding moisture relationships for cement stabilised compressed earth specimens

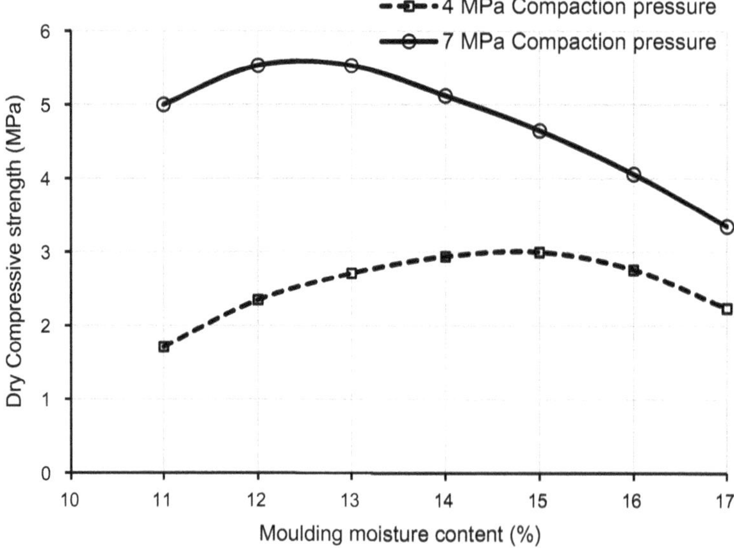

Fig. 5.14 Strength and moulding moisture relationships for CP-VS type of compaction (Fitzmaurice 1958)

5.5 Strength and Stabiliser Content

Varieties of stabiliser additives are used in the production of the stabilised compressed earth bricks/blocks as discussed in the earlier chapters. The lime and the Portland cement are the commonly used stabiliser additives for the production of stabilised compressed earth blocks. The selection of the stabiliser type (lime or cement) to produce CEB depends upon the types of clay minerals present in the soil as well as the strength and the durability requirements of such blocks. Ordinary Portland cement (OPC) is an ideal stabiliser for stabilising the soils with non-expansive clay minerals such as kaolinite and illite, whereas the expansive soils with clay mineral such as montmorillonite need lime as a stabilising additive to control the swell–shrink phenomenon of such a clay mineral.

Figure 5.15 illustrates the generic relationships between the compressive strength and the stabiliser (cement and lime) content for the stabilised CEBs. The compressive strength increases with the increase in the cement content. In the case of lime stabilised CEBs, the strength reduces with the increase in lime content beyond an optimum lime content.

5.5.1 Cement Stabilised Compressed Earth Products

For the cement stabilised CEB, Fig. 5.16 shows the relationships between the compressive strength and the cement content. The relationships shown in Fig. 5.16 represent the information culled from the experimental studies of Walker (1999), Kenai et al. (2006), Consoli et al. (2007), Kumar (2009) and Latha (2015). The dry density of the compressed earth specimen has been maintained constant along each of these lines representing the relationship between the strength and the cement

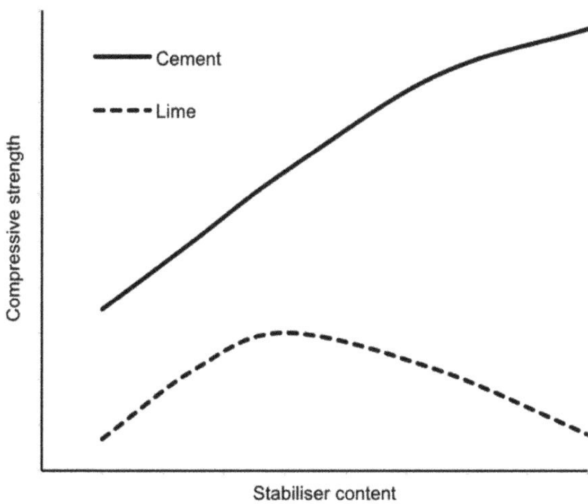

Fig. 5.15 Generic relationships for compressive strength and stabiliser content of stabilised compressed earth specimen

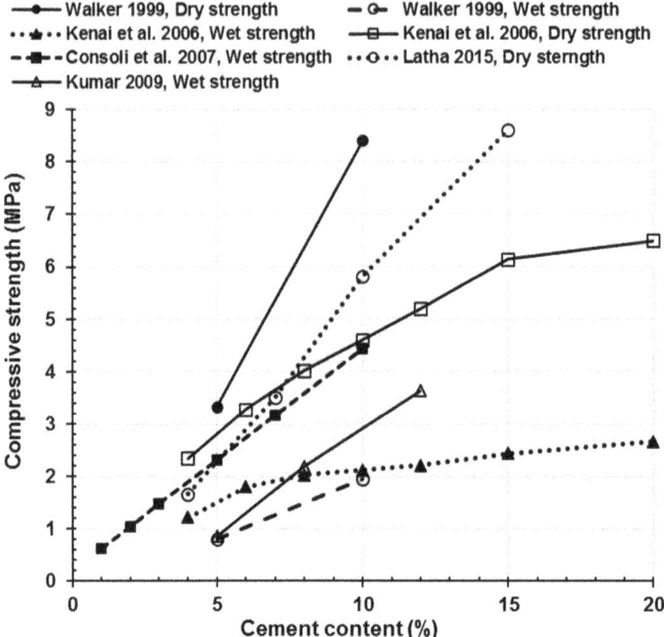

Fig. 5.16 Compressive strength and cement content relationships

content. The relationships are for different types of soils and of different densities. Table 5.4 gives the details of the cement content, density, clay content of the soil and the strength for the results shown in Fig. 5.16. Figure 5.17 shows the relationships

Table 5.4 Details of cement content, density, clay content and strength for the results shown in Fig. 5.16, (*Clay + silt)

Details of the references	Specimen size (mm)	Clay content (%)	Dry density (kg/m^3)
Walker (1999), dry strength	295 × 140 × 120	11	1750
Walker (1999), wet strength	295 × 140 × 120	28	1600
Kenai et al. (2006), wet strength	100 dia. × 100 height	62*	1760
Kenai et al. (2006), dry strength	100 dia. × 100 height	62*	1760
Consoli et al. (2007), wet strength	50 dia. × 100 height	5	2000
Latha (2015), dry strength	38 dia. × 76 height	15.8	1750
Kumar (2009), wet strength	38 dia. × 76 height	15.8	1700

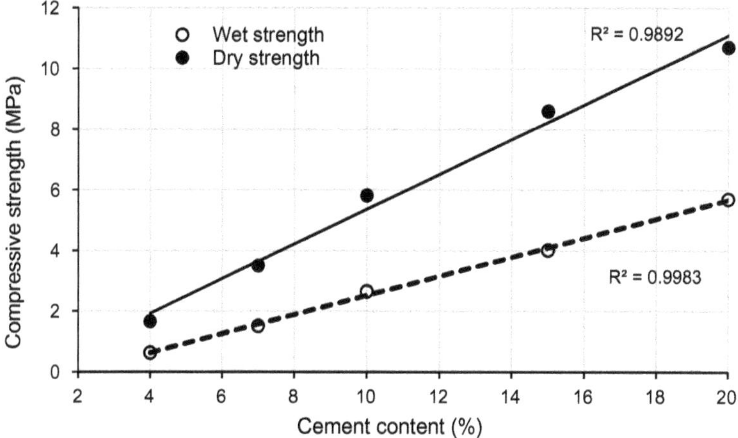

Fig. 5.17 Variation in compressive strength with cement content for cement stabilised compressed earth specimen (Specimen: 38 mm dia. and 76 mm height; dry density: 1750 kg/m^3)

between the cement content and the compressive strength obtained from the more controlled experiments using a soil having optimum clay (kaolinite) fraction (clay: 15.8%, silt: 11.6%, sand: 72.6%, liquid limit: 27%, plasticity index: 17.5%). The following interesting observations can be made from the results shown in Figs. 5.16 and 5.17 and the details in Table 5.4.

- The compressive strength of the cement stabilised CEB increases with the increase in the cement content irrespective of the soil type, clay content, specimen size and dry density of the specimen.
- The strength and the cement content relationships are linear or bilinear. For the wet and the dry cases, the slope of the linear relationships differ, though the dry densities are similar.
- The percentage increase in strength bears no relationship with various parameters controlling the strength (such as the cement content, clay content and density).
- For the specific results shown in Fig. 5.17, as the cement content increases from 4 to 10%, the compressive strength of the compressed earth specimen increases by 250% and 300% for the dry and wet cases, respectively. The compressive strength increases further as the cement content increases. The strength and the cement content are linearly related.

The selection of the percentage of cement required for the manufacture of the cement stabilised CEBs is dependent upon the desired compressive strength for the brick or the block. This in turn is dependent upon the wall strength required for a given situation in a building or the structure. The brick or the block strength required can be ascertained from the masonry design calculations for any specific case of a load bearing masonry structure.

5.5.2 Lime Stabilised Compressed Earth Blocks

The lime stabilisation becomes essential for handling the soils with expansive clays, mainly to control the high swell–shrink characteristics. Generally, lime in the form of calcium hydroxide is used for the production of the stabilised CEBs. The lime stabilisation technique is also used for the production of the stabilised CEBs using soils having non-expansive clays. Figure 5.9 shows the variation in compressive strength with lime content for the lime stabilised compressed earth specimens using different types of soils. Table 5.3 gives details of the optimum lime content and the corresponding clay content of the soil. Influence of the lime content on the strength of the lime stabilised compressed earth products was discussed in Sect. 5.3.2.2. There are many investigations on the strength and lime content relationships for the stabilised soils (Clare and Cruchley 1957; Mateos and Davidson 1961; Mateos 1964; Dan Marks and Allan 1972; Gidigasu 1976; Reddy and Jagadish 1984; Bell 1996, Oti et al. 2009a, b; Consoli et al. 2011; Dash and Hussain 2012; Armin and Behzad 2013; Latha 2015). One major issue with majority of these investigations is the dry density of the stabilised compressed earth product, and the moisture content during testing was not controlled. Therefore, strength and lime content relationships differ among the studies and may not show OLC precisely. Another major issue is achieving meaningful wet compressive strength for the lime stabilised CEBs. With 8–10% lime, the wet compressive strength of the lime stabilised CEB rarely crosses 2.0 MPa. Therefore, it becomes necessary to add other pozzolonaic additives such as fly ash, ground granulated blast furnace slag (GGBS), etc., and resort to increasing the curing temperature (50–80 °C).

5.6 Effect of Curing Conditions on the Strength Development in Stabilised CEB

Both the cement stabilised and the lime stabilised compressed earth products need curing for the strength development. The strength development greatly depends upon the curing type and the curing duration. The following types of curing techniques are possible.

(a) Moist curing at ambient temperature and pressure
(b) Steam curing at atmospheric pressure
(c) Autoclave curing at above atmospheric pressure.

5.6.1 Moist Curing

Moist curing is the most commonly employed curing technique for the stabilised compressed earth products. In this type of curing, moisture is supplied for the hydration of cementitious materials such as the Portland cement or the blended Portland cement. During the hydration process, the cementitious products such as silicate or aluminate hydrates are formed, which are responsible for bonding the inert materials such as silt and sand, present in the stabilised compressed earth product. During the Portland cement hydration process, calcium hydroxide (lime) is released, which can react with clay minerals in the compressed earth forming additional cementitious products. If the blended cements are used for stabilisation, the calcium hydroxide released during the hydration reacts with the pozzolanic materials in the blended cement forming additional cementitious products. Generally, the blends for Portland cement include pulverised fuel ash (fly ash, rice husk ash, etc.) or other such pozzolanic materials.

In the case of lime stabilised compressed earth products, the moisture is required for the lime-clay reactions, in addition to the lime-pozzolana reactions, if the pozzolanic materials are used along with the lime addition. The lime-clay or the lime-pozzolana reactions result in the formation of cementitious products such as silicate or aluminate hydrates, responsible for binding the inert silt and the sand particles present in the stabilised compressed earth product. The lime-clay or the lime-pozzolana reactions are slow at the ambient curing conditions, and hence, the strength development in the lime or the lime-pozzolana stabilised compressed earth products needs longer curing durations than the normally adopted 28 days curing for the cement stabilised compressed earth product.

Generally, the stabilised CEBs are not soaked in water for the purposes of moist curing. The freshly made bricks or blocks are stacked one above the other up to 5–6layers' height (i.e. up to about 1.5 m). The stack of the stabilised CEBs is covered with a hessian cloth or grass such that the direct exposure to sunlight is avoided, which can lead to quick surface drying. The stack of the stabilised CEBs receive water spray to keep the entire stack moist during the curing period.

Strength versus moist curing duration

In lime–soil mixtures, the main contributor to the strength development is lime-pozzolana reaction, though carbonation contributes slightly to the strength development. The clay is the main pozzolana source. The strength development mechanisms in lime stabilised compressed earth specimen was discussed in Chap. 3. The lime-clay reactions are slow at ambient curing conditions. The strength development in the lime stabilised earth specimens, with the moist curing duration up to six months, is depicted in Fig. 5.18. The compressive strength steadily increases with the increase in the curing duration. The rate of strength gain differs from soil to soil. It should be noted here that the compressive strength values obtained for different lime contents differ considerably as shown in Fig. 5.18. This can be attributed to the different sizes of compressed lime–soil specimens used in these studies, and also the strengths have been determined in wet, dry and partially wet conditions. There will be considerable difference in the dry and the wet compressive strengths. For few samples, there is five times increase in the strength between the 14 days and the 90-day curing period. For another sample (Saha and Ray 1967), the strength increased by two times for the curing period between 7 and 180 days. It may not be practically feasible to cure the specimens beyond 28 days in the production sites or in a factory.

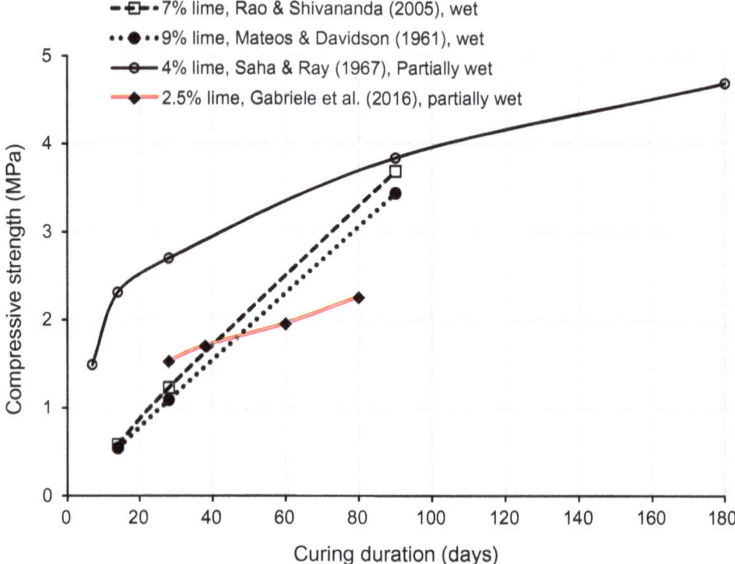

Fig. 5.18 Strength versus curing duration in lime stabilised compressed earth specimen

5.6.2 *Steam Curing*

The steam curing is very effective in accelerating the lime-clay or lime-pozzolana reactions. Generally, the lime or the lime-pozzolana stabilised compressed earth specimen need longer curing duration under the ambient temperature and moist curing conditions. The results of the lime stabilised earth discussed in the previous section showed that under ambient curing conditions the lime-clay reactions could continue for several months. Under the ambient temperature, even curing for 28 days, the lime stabilised CEB would not attain meaningful strength for the bricks or blocks. The strength development in lime stabilised CEB is sensitive to the curing temperature. Curing at higher temperature results in higher strength for the lime stabilised CEB. The strength of lime stabilised compressed earth specimen increases with the increase in curing temperature as illustrated in Fig. 5.19 (Mateos and Davidson 1961; Mateos 1964; Thompson 1966; Ranganathan and Pandian 1971; Reddy 1983; Reddy and Jagadish 1984; Bell 1996; Reddy and Lokras 1998; Reddy and Hubli 2002; Rao and Shivananda 2005; Izemmouren et al. 2015). Figure 5.19 shows the variation in the compressive strength with the curing temperature. The results shown in this figure were culled from different studies. The lime content, specimen size, density, curing duration and moisture content of the specimen during testing vary across the studies. The relationships clearly show that the compressive strength development in the lime stabilised compressed earth specimen is sensitive to the curing temperature and the strength doubles when the curing temperature increases by about 20 °C from 20 to 25 °C range.

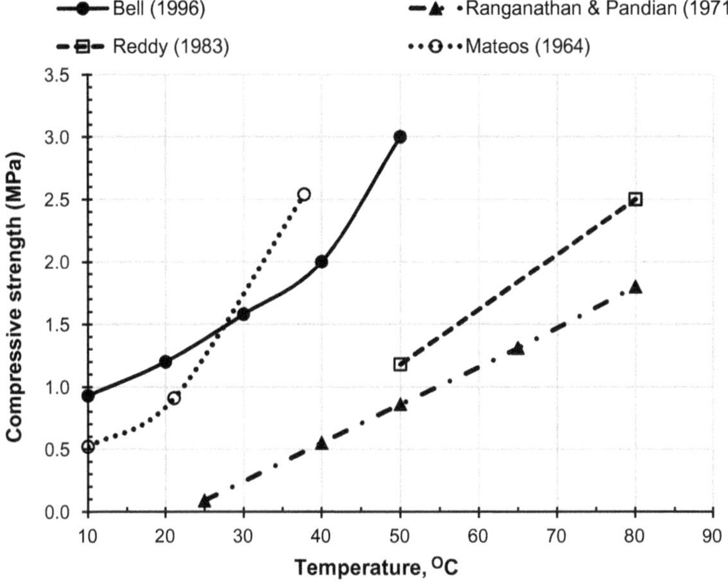

Fig. 5.19 Variation in compressive strength of lime stabilised compressed earth specimen with curing temperature

The steam curing experiments at the atmospheric pressure on stabilised compressed earth specimen can be carried out in a laboratory using a water-bath. Some of the factors affecting the characteristics of the steam-cured (at atmospheric pressure) stabilised compressed earth specimen are:

a. Type of clay mineral
b. Percentage of stabiliser (lime or lime-pozzolana)
c. Duration of curing
d. Density of the compressed earth product.

The properties of lime stabilised steam-cured CEBs was explored in greater detail by Reddy and Hubli (2002). Figure 5.20 shows the strength versus steam curing duration plots for the two types of soils containing montmorillonite and kaolinite clay minerals. The original soils were reconstituted by diluting with sand to bring down the clay content in the mix. The mix containing montmorillonite clay soil had 12.0% clay, while the soil containing kaolinite clay mineral had 14% clay. The strength results shown in the figure are wet compressive strengths of the compressed earth cube specimens (76 mm) with a dry density of 1800 kg/m^3. The stabilised compressed earth cubes were steam cured at 80 °C in a water-bath. From the plots shown in the figure, it is clear that there is a sharp increase in the strength (about threefold) with the increase in the curing period (6–24 h) irrespective of the soil type and the lime content. The strength gets saturated around 24 h of steam curing,and curing beyond 24 h is not very useful in increasing the strength. Izemmouren et al.

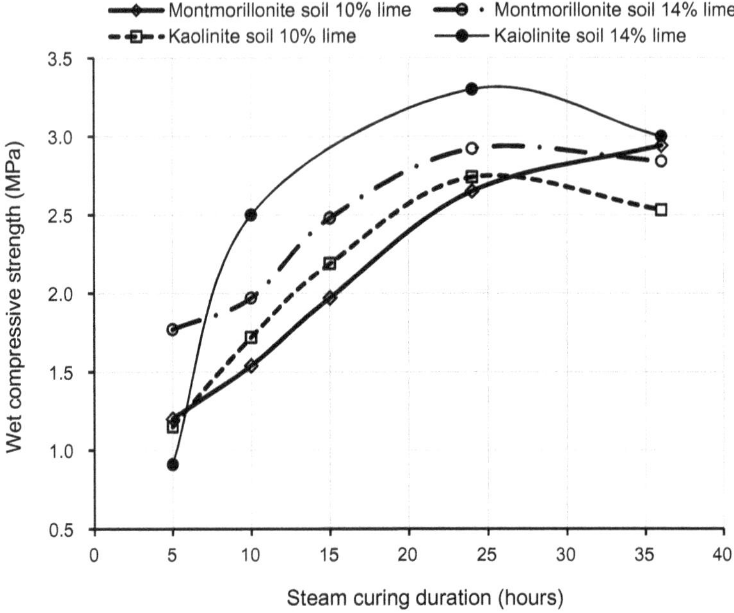

Fig. 5.20 Strength versus steam curing period for lime stabilised compressed earth specimen

(2015) noticed increase in the strength with curing duration up to 24 h of steam curing for the lime stabilised compressed earth blocks.

There will be considerable difference in the compressive strengths of 24-h steam-cured (at 80 °C) specimens and 28 days moist cured specimens at ambient temperature. Figure 5.21 shows the relationships for the wet strength and the lime content for the steam-cured and the moist-cured compressed earth cylindrical specimen. The moist curing has been performed under wet burlap for 28 days, and the steam curing is for 24 h at 80 °C. The characteristics of the two soils designated as Soil XA and Soil XB are given in Table 5.5. The clay content (~15%) of the soils is in a similar range, but the clay minerals are different. In the case of moist curing, the wet compressive strength does not vary much beyond 12% lime, whereas the strength increases with the increase in the lime content for the steam-cured specimen. For lime contents beyond 12%, the soil with montmorillonite clay (Soil XB) shows considerably higher strength when compared to the strength of the specimen using kaolinite clay soil (Soil XA). The strength of the steam-cured specimens is more than double that of moist-cured specimens.

Pozzolanic additives and the strength of steam-cured lime stabilised compressed earth

The pozzolanic materials consist of silica and alumina in reactive form, which can react with calcium hydroxide in the presence of water and develop strength. The pozzolanic materials include fly ash, silica fume, ground blast furnace slag, burnt

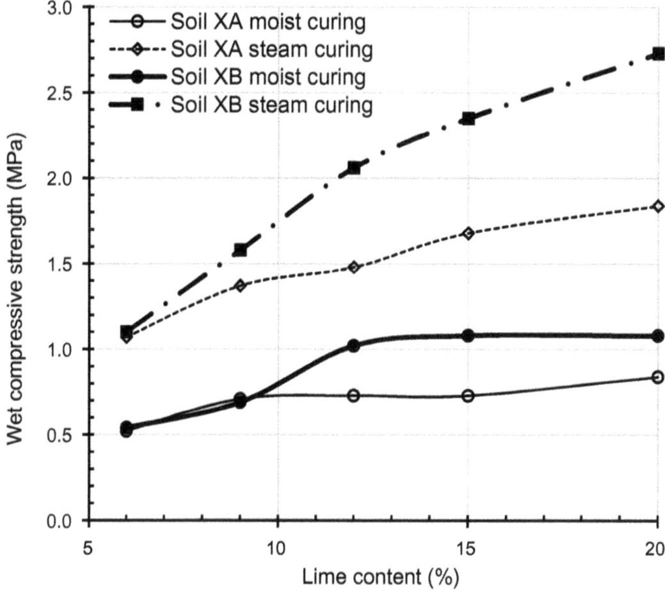

Fig. 5.21 Strength of moist-cured and steam-cured lime stabilised compressed earth specimen (38 mm diameter × 76 mm height cylindrical specimen)

Table 5.5 Properties of soils used in steam curing and wet–dry strength relationships

Property	Soil XA	Soil XB
Textural composition		
Sand (4.75–0.075 mm) (%)	75.1	79.2
Silt (0.075–0.002 mm) (%)	9.1	5.8
Clay (<0.002 mm) (%)	15.8	15.0
Atterberg's limits		
Liquid limit (%)	26.9	39.6
Plasticity Index (%)	17.5	17.2
Predominant clay mineral	Kaolinite	Kaolinite and Montmorillonite
pH	7.71	9.1

clay powder, rice husk ash, natural pozzolans such as volcanic ash. The reactivity of the pozzolanic materials depends upon the fineness and the temperature at which the parent pozzolanic materials are manufactured. Even, the natural clay minerals are also pozzolanic materials.

The fly ash has been used to improve the strength of lime stabilised compressed earth blocks. Generally, Class F-fly ash (ASTM C618-08a) is commonly used. The optimum quantity of fly ash resulting in maximum strength for the lime fly ash stabilised compressed earth depends upon the quantity and type of the clay mineral and the percentage of lime in the mix. For example, the investigations of Reddy and Hubli (2002) showed that for a soil containing 12% montmorillonite clay and with 10% lime (by mass) an optimum fly ash content of 30%. The wet compressive strength (of 76 mm cube) with and without fly ash was 7.0 and 2.65 MPa, respectively. Both the specimens were steam cured at 80 °C for 24 h. Similarly, the investigations of Izemmouren et al. (2015) showed a wet compressive strength of the steam-cured compressed earth block as 7.0 and 14.5 MPa for the cases of 10% lime and 10% lime + 30% natural pozzolana, respectively. Both the blocks were steam cured at 75 °C for 30 h. The influence of pozzolanic content of the mix on the strength of the steam-cured compressed earth specimen is shown in Fig. 5.22. The studies of Reddy (2000) were using 10% lime and fly ash as pozzolana, whereas the investigations of Izemmouren et al. (2015) were for 10% lime and natural pozzolana. In both the cases, the strength increases with increase in pozzolana content, reaches a peak and remains constant. In both the cases, the optimum pozzolana content yielding maximum strength was about 30%. This optimum value will be a function of the lime content.

The major advantages of steam curing include: (1) achieving higher strength and superior quality brick or block and (2) quicker production process. Figure 5.23 shows a small-scale steam curing unit of 1300 blocks (size: 230 × 190 × 100 mm) per day capacity. The plant consists of a horizontal steam boiler mounted in a steam chamber, where the stack of lime stabilised blocks on a tray is mounted on a rail system. The steam is generated through water boiling using biomass fuel. The stack of blocks is

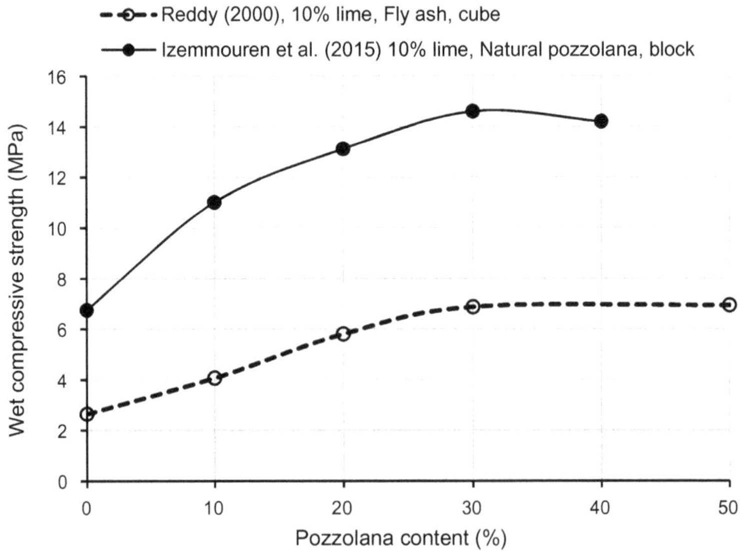

Fig. 5.22 Influence of pozzolanic materials on steam-cured stabilised compressed earth

Fig. 5.23 Small-scale (1300 blocks per day) steam curing unit

exposed to steam at 70–80 °C for 24 h. Figure 5.24 shows the temperatures attained in a typical steam curing cycle. Reddy and Lokras (1998) analysed the embodied energy in the steam-cured lime–fly ash stabilised CEB. The embodied energy in the steam-cured CEB masonry was 1395 MJ/m^3, out of which 50% was from the steam curing operation. If waste steam arising out of any industrial process is utilised, then the embodied energy in the steam-cured stabilised CEB masonry will reduce to 986 MJ/m^3. Figure 5.25 shows a load bearing masonry building built (in 1999) using steam-cured lime–fly ash stabilised CEBs in the ground floor. The wet compressive strength of these blocks was 6.8 MPa. The blocks were of size: 230 × 190 × 100 mm, and the stabiliser content was 9% lime + 2% Portland cement + 10% fly ash (the proportions were based on dry mass of soil + sand).

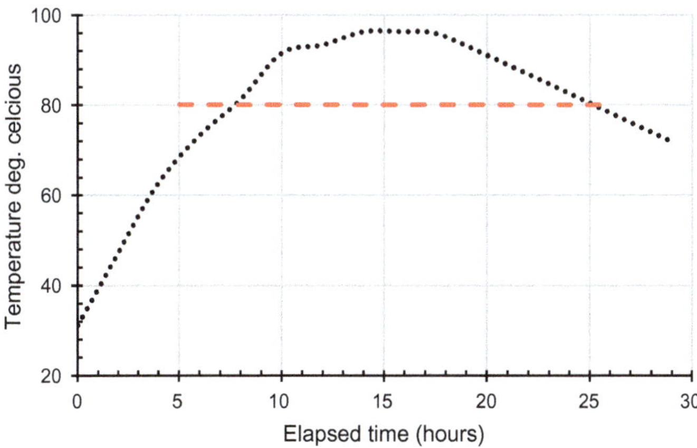

Fig. 5.24 Temperatures in a typical steam curing cycle

Fig. 5.25 Building with steam-cured lime–fly ash stabilised compressed earth block masonry in ground floor (built in 1999, picture taken in 2021)

5.7 Wet and Dry Compressive Strength of Stabilised Compressed Earth

Soil is the basic raw material for the manufacture of stabilised CEBs. The soil contains clay minerals. The residual clay minerals are present in the stabilised compressed earth products (Reddy and Latha 2014a). The residual clay minerals in the stabilised compressed earth materials soften upon encountering water. Therefore, the moisture content of the stabilised CEB at the time of testing will influence its compressive strength. There will be considerable difference in the wet and dry compressive strengths of the stabilised compressed earth materials, depending upon the extent of stabilisation and the stabiliser content. The wet compressive strength is the strength in the nearly saturated state (48–72 h of soaking in water prior to testing). Figure 5.26 displays generic representation of the variation in the compressive strength with the moisture content during testing of the unstabilised, moderately stabilised and fully stabilised compressed earth specimens. For unstabilised compressed earth specimen, there is an exponential decay in the strength with the moisture content, ultimately reaching zero strength at saturation or close to saturation. The unstabilised compressed earth products disintegrate upon soaking in water. The stabilised compressed earth material possess strength in the wet condition. For moderately stabilised compressed earth product, the strength can still decay exponentially with the moisture content, but it will not reach zero at the saturation. For

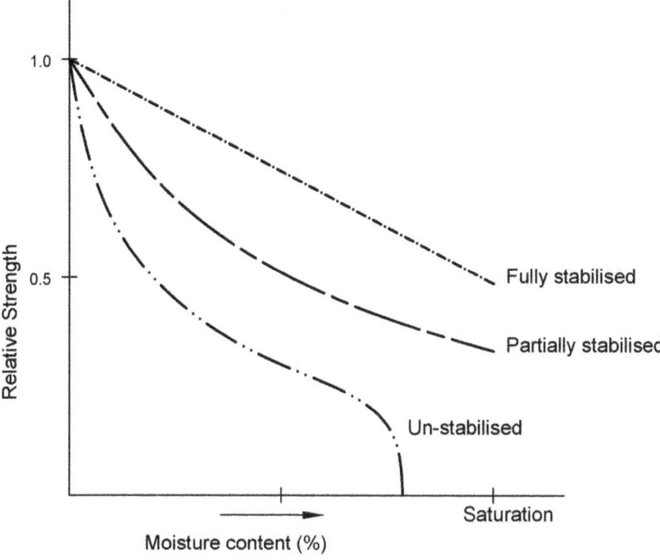

Fig. 5.26 Strength versus moisture content during testing of the compressed earth product

fully stabilised compressed earth products, the strength varies linearly with the moisture content. The wet to the dry strength ratio (strength ratio) can vary in a range of 0.2–0.7 depending upon the stabiliser content and the effectiveness of stabilisation (Ingles and Metcalf 1972; Reddy 1991; Walker and Stace 1997; Walker 2004; Reddy and Gupta 2005; Jagadish et al. 2017). The lower ratio is for the inadequately stabilised compressed earth products. For example, if the clay content in the mix is high (> 25%) and the stabiliser content is 3–4%, the wet/dry strength ratio will be low (0.2–0.3). Also, for soils with traces of montmorillonite and low cement content (no lime), the ratio will be low. The studies of Kumar (2009) showed that the compressive strength of the cement stabilised rammed earth varied linearly with the moisture content. This study was using a soil with optimum clay content and for the cement contents in the range of 7–12%. Figure 5.27 shows the plots for strength and moisture content obtained by controlled experiments on cement stabilised compressed earth cylindrical specimen. The two extreme points in these plots represent the dry and nearly saturated (wet) condition, and the two intermediate points between them represent partially saturated condition. These plots show that the relationship between the strength and the moisture content is linear.

Figure 5.28 shows a plot between the strength ratio and the clay content. The data points in the figure was pooled from several investigations, where the cement content was in the range of 2.5–12% (by mass). The plot demonstrates that the wet strength to the dry strength ratio for the cement stabilised compressed earth product decreases with the increase in the clay content. This can be attributed to the softening of the clay upon water absorption and hence not contributing to the strength in the wet state. The large scatter in the plot can be attributed to the variations in the soil

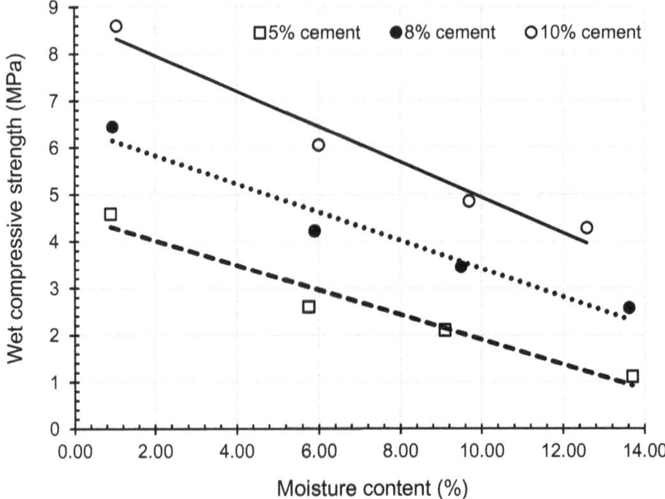

Fig. 5.27 Strength and moisture content relationships for cement stabilised compressed earth specimen

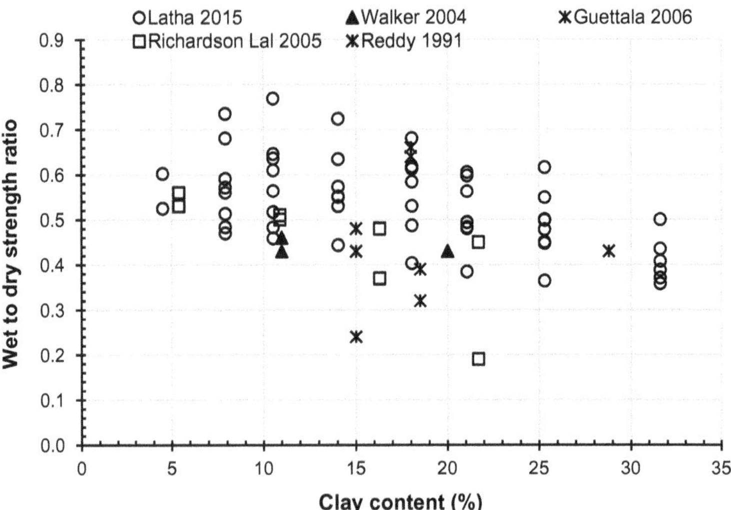

Fig. 5.28 Clay content versus wet to dry strength ratio for cement stabilised compressed earth specimen

grading (composition, mainly clay content), the specimen density and the cement content.

The plots for the strength ratio versus the lime content and the cement content are shown in Figs. 5.29 and 5.30, respectively. These plots are from the controlled

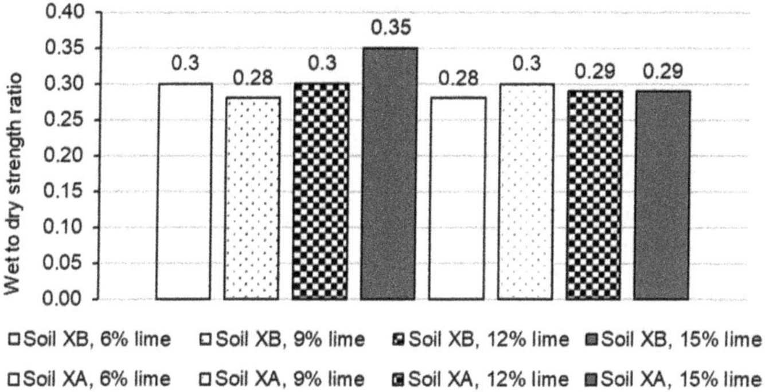

Fig. 5.29 Lime content and wet to dry strength ratio for lime stabilised compressed earth specimen

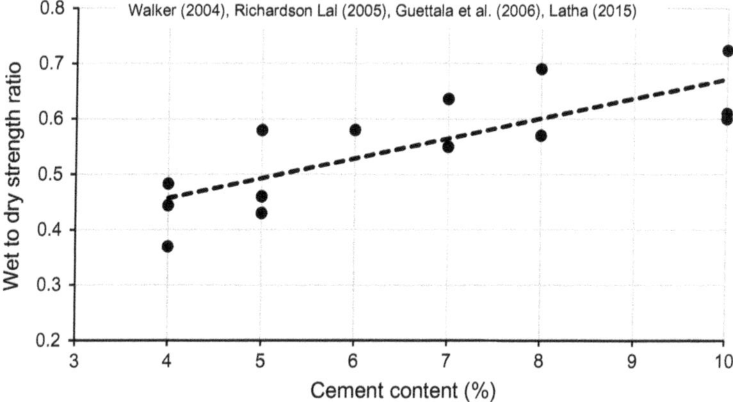

Fig. 5.30 Cement content versus wet to dry strength ratio for cement stabilised CEBs

experiments, where the clay content was controlled and was about 15%. Figure 5.29 shows a bar chart for the strength ratios as the lime content varies between 6 and 15%. The Soil XA and the Soil XB have clay content of about 15%. Soil XA has kaolinite, while Soil XB has both the kaolinite and the montmorillonite clay mineral (Table 5.5). The wet strength to the dry strength ratio is about 0.30 for the lime stabilised compressed earth specimens using lime in the range of 6–15%. Figure 5.30 shows relationship between the wet strength to the dry strength ratio and the cement content. The relationship shows linear trend line. The strength ratio increases with the increase in the cement content, and it is in the range of 0.40–0.70. The points shown in the plot represent cement stabilised CEBs using soil mixtures containing about 15% clay and dry density of about 1800 kg/m^3. The scatter in the plot can be attributed to variations in the CEB size and the soil composition.

The dry strength of the stabilised compressed earth product is attributed to bonds established by both the clay and the stabiliser. In the wet state, the contribution of clay to the strength reduces to nil. The wet strength to the dry strength ratio is sensitive to the cement content in the case of the cement stabilised CEBs. At higher cement contents (>8%) and at a clay content of <15% (optimum clay content), the difference between the wet and the dry strength reduces, attributing to higher quantity of cementitious material binding the inert sand/silt particles in the earth product. In the case of the lime stabilised compressed earth specimen, the clay gets consumed in the lime-clay reactions, and hence, the dry strength reduces accordingly. Therefore, the wet strength to the dry strength ratio is less when compared with the values of cement stabilised compressed earth product.

5.8 Long-Term Strength of Stabilised Compressed Earth Products

The stabilised compressed earth products possess residual clay minerals. The quantity of residual or retrievable clay minerals in the stabilised compressed earth product depends upon the type of the stabiliser and the stabiliser-clay ratio (Reddy and Latha 2014a; Reddy and Hubli 2002; Reddy 2002). The residual clay minerals present in the stabilised compressed earth product can undergo swelling and shrinkage due to the moisture variations in the surrounding environment and the weathering actions. The swelling and shrinkage will be cyclic in nature. The cyclic expansion and shrinkage of the clay mineral present in the stabilised compressed earth product can affect the long-term strength. The strength of the building materials should be stable during the life of the structure. The main parameter affecting the long-term strength of the stabilised compressed earth product is the stabiliser-clay ratio in the mix used for the manufacture of the earthen product. The influence of stabiliser-clay ratio on the long-term strength of the stabilised compressed earth product is displayed in Fig. 5.31. Two soils designated as the soil-A and the BC soil have been used in the controlled experiments to develop the strength and the long-term duration relationships. The properties of these two soils are given in Table 5.6. The soil-A contains predominantly kaolinite clay, whereas BC soil has expansive clay mineral such as montmorillonite. Both the cement and the lime (calcium hydroxide) have been used as stabilisers in these studies. The strength ratio shown in Fig. 5.31 has been determined in the following manner.

Strength ratio = (Long-term strength) ÷ (Compressive strength at 28 days curing period)

The stabiliser to clay ratio has been varied between 0.26 and 1.17. The compressive strength of the stabilised compressed earth specimen has been monitored up to a period of 6 years. The specimens after 28-day curing have been kept inside the laboratory and tested at the end of a 6, 12, 24, 48, 54 and 72 months. Annual variations in the ambient temperatures inside the laboratory were 16–30 °C with a relative humidity of 40–70%. The specimens were soaked in water for 48 h prior to the testing. The compressive strength corresponds to the statically compacted cube specimens

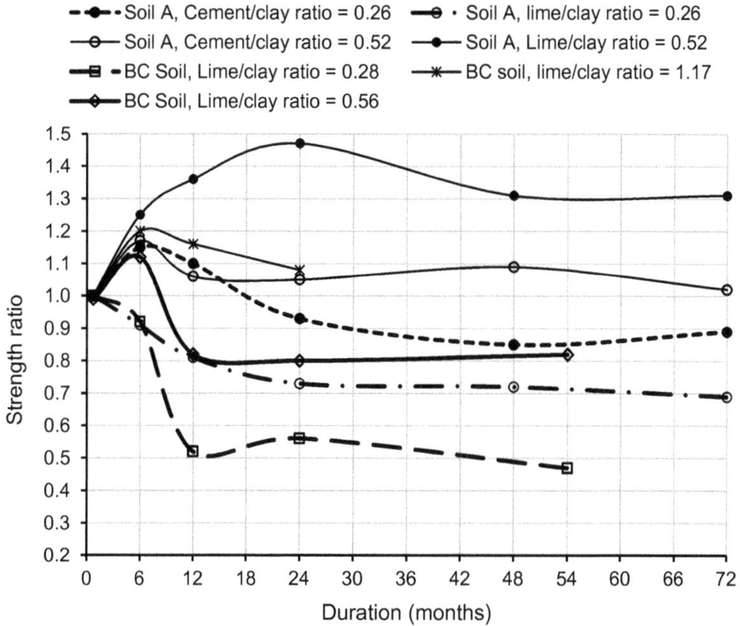

Fig. 5.31 Long-term strengths of stabilised compressed earth specimens

Table 5.6 Properties of soils used in long-term strength monitoring

Property	Soil A	BC Soil
Textural composition		
Sand (4.75–0.075 mm) (%)	48.8	35.7
Silt (0.075–0.002 mm) (%)	22.4	20.3
Clay (<0.002 mm) (%)	28.8	36.0
Atterberg's limits		
Liquid limit (%)	42.0	53.1
Plastic limit (%)	19.7	25.7
Plasticity Index (%)	22.3	27.4
USC classification	CL	CH
Predominant clay mineral	Kaolinite and traces of montmorillonite	Montmorillonite
pH	7.71	8.95
Organic matter (%)	0.72	0.67

of 76 mm size. The long-term strengths of the stabilised compressed earth specimen shown in Fig. 5.31 reveal interesting information.

Inadequate quantity of stabiliser in the mix (low stabiliser to clay ratio), i.e. earth mixes with high clay content or inadequate quantity of stabiliser, the stabilised compressed earth specimen lost strength with the age. This is true for the compressed earth specimen with either the soils having expansive or non-expansive clay minerals and irrespective of the lime or the cement as the stabiliser. The reduction in the strength with the age for the stabilised compressed earth specimen having lower stabiliser to clay ratio can be attributed to rupturing of weakly held cementitious network of bonds, caused by cyclic swell–shrink phenomenon of unstabilised clay. The residual or unstabilised clay in the compressed earth specimen undergoes swelling and shrinkage as the humidity in the surrounding region fluctuates. There is a need to understand the behaviour of stabilised CEBs, especially micro-level changes in bonds with the age. The desirable stabiliser-clay ratio shall be >0.5 for soils with non-expansive clay minerals and 0.75–1.0 for soils with the expansive clay minerals. The stabilised compressed earth specimen with sufficient quantity of stabiliser shows increase in the strength (15–40%) with the age as illustrated in Fig. 5.31. Similar observations were noticed in the studies of Reddy and Hubli (2002) and Reddy (2002) on the steam-cured stabilised compressed earth specimens using different types of soils.

The natural soils undergo swelling and shrinkage with the seasonal variations in the ambient moisture content. The magnitude of swelling caused by the clay minerals in the soils depends up on the type and the amount of the clay mineral. It follows the sequence montmorillonite > illite > kaolinite (Yong and Warkentin 1975). The free swell index tests reveal that montmorillonite has approximately 25 times higher swell potential than kaolinite (Sridharan et al. 1986). The magnitude of the shrinkage experienced by a soil is also a function of the type and the amount of clay mineral present. The montmorillonite clays shrink more than the kaolinitic clays (Yong and Warkentin 1975).

5.9 Absorption Characteristics of Stabilised Compressed Earth Products

The moisture movement in the stabilised compressed earth products is an important issue to be addressed, particularly regarding the strength of the CEB and the role it plays in the development of the bond between the brick/block and the mortar. The moisture movement in the stabilised CEBs can be examined with regards to the following.

1. Water absorption or saturated water content
2. Rate of water absorption
3. Initial rate of absorption (IRA).

5.9.1 Water Absorption or Saturated Water Content

The water absorption is a common physical property examined for the masonry units. The masonry units are two-phase materials in the dry state. The porous structure is filled with the air. When the masonry unit is exposed to water, the pores get filled with it. The amount of water a masonry unit can absorb depends upon the type of the masonry unit and its porosity/pore structure. The total amount of water absorbed by the masonry unit upon saturation is a measure of the total porosity. The standard codes of practices suggest 24-h, cold water immersion test to determine the water absorption of the masonry units. In the 24-h, cold water immersion test, the dry masonry unit (oven dried) is soaked in the water for 24 h to determine the water absorption value. Soaking dry masonry units for 24 h in the water might not lead to 100% saturation in majority of the cases. Some of the air bubbles can get trapped inside the masonry unit pores, or in other cases, the water may need more time to penetrate the pores for complete saturation of the masonry unit. IS 3495 (1992), BS EN 772-21 (2011) and many other standard codes of practices give procedures for determining the water absorption value for the masonry units such as bricks and blocks.

The water absorption value for the stabilised CEB is dependent upon the following parameters.

(a) Density of the brick/block
(b) Textural composition of the earth mix
(c) Type and quantity of the clay minerals in the mix
(d) Residual clay content in the stabilised CEB
(e) Type and quantity of the stabiliser.

5.9.1.1 Water Absorption and the Clay Content

The dry density of the stabilised CEB is an indirect measure of the porosity or the void ratio. The water absorption phenomenon in the stabilised CEBs is a bit different from those of the other masonry units such as the burnt clay bricks, the concrete blocks and the sodium silicate bricks. The stabilised CEB will have residual clay minerals, apart from the inert aggregate and silt particles. When a dry stabilised CEB is soaked in the water, the water enters the pores, the residual clay minerals in the CEB will absorb the water. Considering two types of soils (coarse grain and fine grain), controlled experiments were conducted on the water absorption values for the cement stabilised CEB by Latha (2015). Figure 5.32 shows the relationship between the water absorption values for the cement stabilised CEB as the clay content was varied over wide limits. The textural composition of the two soils is given in Table 5.2. The coarse grain and fine grain soils are designated as CG and FG in the table. The fine grain soil is a silty soil with more fines, finer than 75-micron size. The textural composition of the two types of soils was varied by reconstituting with either sand or silt fractions. Both the soils had kaolinite clay mineral. The dry density of the cement stabilised CEBs with the coarse grain soil and the fine grain soil was 1750

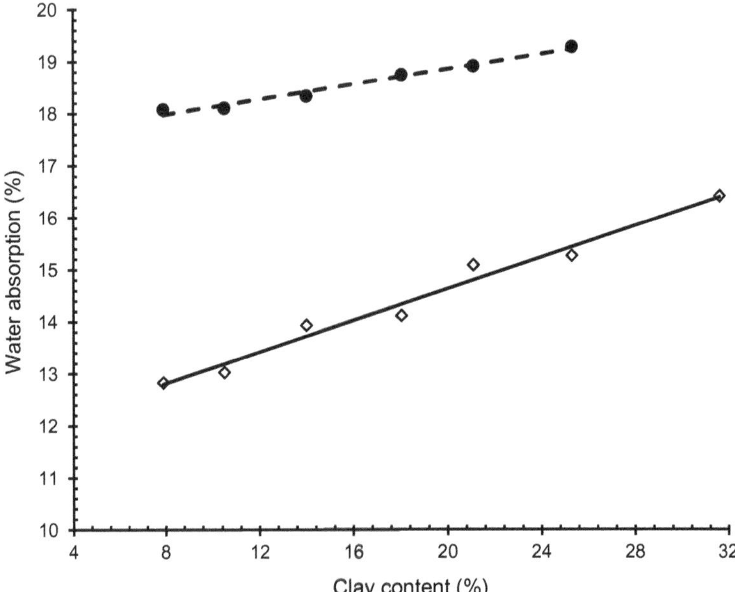

Fig. 5.32 Water absorption versus clay content of cement stabilised CEB

and 1720 kg/m^3, respectively, while the cement content was 7%, by mass. The plots in Fig. 5.32 show that the water absorption for the cement stabilised CEBs increases with the increase in the clay content irrespective of the soil type. There is a linear relationship between the water absorption and the clay content. For stabilised CEBs using coarse grain soil, the water absorption value increased from 12.8 to 16.4% as the clay content was increased from 7.9 to 31.6%. In the case of stabilised CEBs with fine grain soil, the water absorption increased from 18.1 to 19.3% as the clay content was increased from 7.9 to 25.3%. For a given clay and cement content, the stabilised CEBs with fine grain soil absorb more water than the ones from the coarse grain soil, even though the blocks have similar density and cement content. For example, at 18% clay content, the water absorption for the CEBs using the fine and the coarse grain soil was 18.8 and 14.1%, respectively. The differences in water absorption values for the bricks (having similar dry density) with the coarse and the fine grain soils can be attributed to different textural composition of the soil, and different pore structure and different void ratio for the bricks.

5.9.1.2 Water Absorption and Stabiliser Content

The water absorption value for the stabilised CEB varies with the variation in the stabiliser content. The saturated water content or the water absorption value of

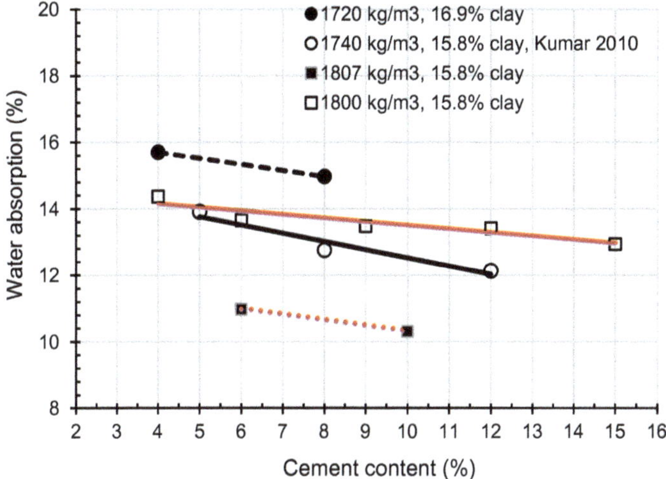

Fig. 5.33 Water absorption and stabiliser content relationships for cement stabilised compressed earth specimens

the stabilised compressed earth specimen is influenced by several parameters as discussed in the earlier sections. Controlling the density and the clay content, the water absorption values determined for the cement stabilised CEB as the cement content is varied are shown in Fig. 5.33. The clay content (kaolinite) was about 16%, and the dry density of the specimens was in the range 1720–1807 kg/m^3. The relationships are linear, and the water absorption decreases with the increase in the cement content. The reduction in the water absorption is about 10–12% as the cement content increases from 5 to 12%. At higher cement contents (>8%), part of the clay in the mix gets reacted with lime released during hydration of cement. In the lower cement content range, the quantity of lime released will be small. Also, the pore structure and the porosity of the CEB changes with the increase in the cement content.

5.9.1.3 Water Absorption and Density

The porosity or the void ratio is directly dependent upon the dry density of the stabilised compressed earth specimen as illustrated in Eq. 5.5.

$$e = (\rho_w \div \rho_d)G - 1.0 \tag{5.5}$$

where e = void ratio, ρ_w = density of water, ρ_d = dry density and G = specific gravity.

This equation shows that the void ratio and the density are inversely related. Thus, void ratio increases with the decrease in the dry density of the compressed earth product. Similarly, the water absorption and the density of the stabilised compressed

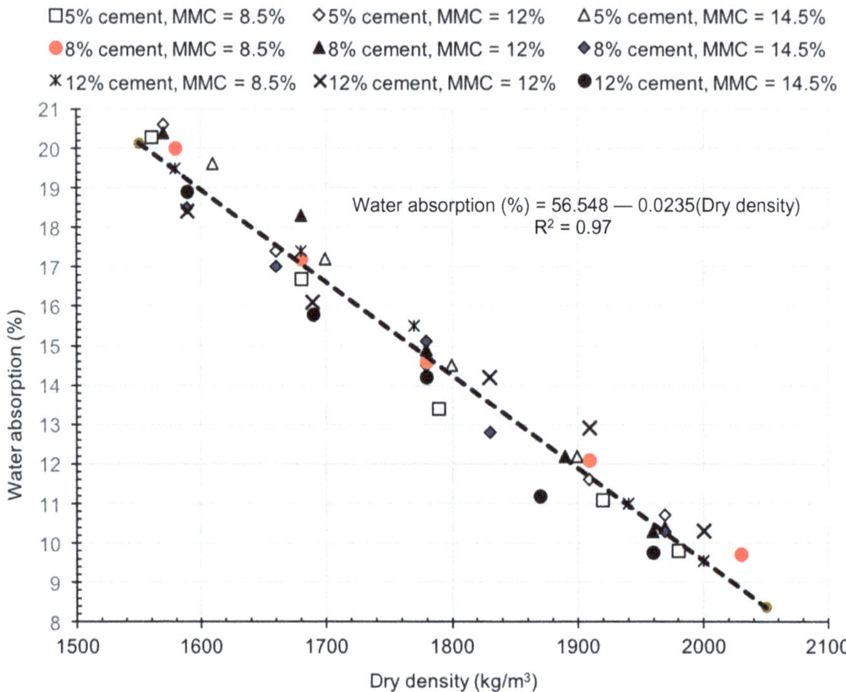

Fig. 5.34 Density and water absorption relationship for cement stabilised compressed earth specimen

earth specimen are inversely related. The Fig. 5.34 shows a plot of the water absorption and the dry density for the cement stabilised compressed earth specimens using one type of soil (having 15.8% kaolinite clay), three different cement contents (5, 8 and 12%) and three moulding moisture contents (8, 12 and 14.5%). Though there is a scatter in the points shown in the plot, the water absorption value drastically reduces with the increase in the density. A linear best fit line of the form shown in Eq. 5.6 can be obtained with a good coefficient of determination value ($R^2 = 0.97$). The linear relationship shows that for a 20% reduction in the dry density from 2000 kg/m^3, there is 125% increase in the water absorption value. The linear relationship represented by Eq. 5.6 is not a unique or generalised one, but the trend remains similar for cement stabilised compressed earth products using any other soil with clay (non-expansive clay) content in the range of 10–15% and the dry density in the range of 1500–2000 kg/m^3.

$$Water\ absorption\ (\%) = 56.548 - 0.0235\,(Dry\ density) \tag{5.6}$$

where water absorption in % (by mass), dry density in kg/m^3 (1500–2000 kg/m^3).

5.9.1.4 Water Absorption Values for Stabilised Compressed Earth Blocks

The typical water absorption values for the stabilised earth blocks are given in Table 5.7. The table gives details of block size, clay content of the mix, dry density of the block, cement content and water absorption value and the source for the data. The water absorption values of the blocks reported in the table have been determined by 24-h cold water immersion test. The results given in the table are from tests on stabilised CEBs used in several buildings in the Bangalore city, India. Generally, the soil composition (specifically the clay content of the mix) and the block density are controlled. The dry density was in the range of 1710–1835 kg/m^3, clay content ranging between 9 and 16% and the cement content in the range of 4–12%. The water absorption values for the fourteen case studies examined lie in the range of 11–16%. The Indian code for the burnt clay bricks (IS 1077-1992) specifies water absorption values in the range of 15–20% for brick compressive strengths in the range of 3.5–40 MPa.

5.9.1.5 Water Absorption Values for Lime Stabilised Compressed Earth Bricks/blocks

The water absorption values for the lime stabilised compressed earth specimen are given Table 5.8. The water absorption values have been generated using five different soils with different clay fractions. The results represent the values obtained through tests on statically compacted lime stabilised compressed earth cubes (76 mm size) having a dry density of about 1850 kg/m^3. The water absorption values for soils with kaolinite mineral is 13% when the clay content was <20% with 7.5–10% lime. When the clay fraction increases (~28%), the water absorption increases to about 16%. For the cubes using soil having montmorillonite clay, the absorption value was in the range 15–18% as the clay fraction varies from 18 to 36%. The investigations of Reddy and Jagadish (1984) showed water absorption values for lime stabilised compressed earth cubes in the range 19–25% (by mass). Here, the dry density was in the range of 1600–1700 kg/m^3 and the clay content varied between 6 and 12%, with a lime content of 5%.

The water absorption for lime stabilised compressed earth blocks is sensitive to the dry density of the product and the quantity of clay present in the mix. For a dry density of 1800 kg/m^3 and with 10–15% clay, the water absorption value will be in the range of 12–16%.

Table 5.7 Water absorption values for cement stabilised compressed earth blocks

Sl. No.	Block size (mm)	Dry density (kg/m^3)	Clay content (%)	Cement content (%)	Water absorption (%)	Source
1	305 × 143 × 100	1835	9.0	6.0	12.10	Reddy and Gupta (2005)
2	305 × 143 × 100	1835	9.0	8.0	11.20	Reddy and Gupta (2005)
3	305 × 143 × 100	1835	9.0	12.0	11.40	Reddy and Gupta (2005)
4	305 × 143 × 96	1784	10.9	4.0	15.25	Reddy et al. (2007)
5	305 × 143 × 96	1784	10.9	8.0	14.84	Reddy et al. (2007)
6	305 × 143 × 96	1784	16.3	4.0	15.70	Reddy et al. (2007)
7	305 × 143 × 96	1784	16.3	8.0	14.97	Reddy et al. 2007
8	232 × 192 × 100	1780	15.0	7.0	13.15	Ullas and Reddy (2007)
9	232 × 191 × 101	1760	– –	7.0	11.94	Ullas and Reddy (2007)
10	233 × 190 × 96	1710	18.0	8.0	14.47	Ullas and Reddy (2007)
11	231 × 190 × 102	1720	10.0	7.0	14.55	Ullas and Reddy (2007)
12	234 × 194 × 95	1800	11.0	8.0	12.27	Ullas and Reddy (2007)
13	230 × 190 × 100	1760	11.0	11.0	12.94	Ullas and Reddy (2007)
14	230 × 190 × 100	1710	11.0	8.0	14.08	Ullas and Reddy (2007)

Table 5.8 Water absorption values for lime stabilised compressed earth specimen

Sl. No.	Soil type	Clay (%)	Clay minerals	Lime (%)	Water absorption (%)
1	Red Soil-A Red Soil-A	28.8 19.2	Kaolinite Kaolinite	7.5 7.5	16.44 13.72
2	Red Soil-B	27.0	Kaolinite	7.5	16.07
3	Red Soil-C	13.9	Kaolinite	10.0	12.70
4	Red Soil-D	12.7	Kaolinite	10.0	12.80
5	BC soil BC soil BC soil	35.8 23.9 17.9	Montmorillonite Montmorillonite Montmorillonite	10.0 10.0 10.0	17.72 15.01 14.87

5.9.2 *Rate of Water Absorption in Cement Stabilised Compressed Earth Bricks/blocks*

Generally, the rate at which water is absorbed by the masonry units such as bricks or blocks are not examined for masonry construction purposes. The information on the rate of water absorption will help in determining the soaking time required for achieving the required level of saturation. The flexure bond strength of the masonry greatly depends upon the moisture content (saturation level) of the masonry unit at the time of the masonry construction. Maximum bond strength for the masonry can be achieved when partially saturated (~75% of 24-h water absorption) bricks/blocks are used for the masonry construction (Venumadhava Rao et al. 1996; Hendry 1998; Sarangapani et al. 2005; Reddy and Gupta 2006; Gumaste et al. 2007).

Figure 5.35 shows the relationships for the rate of water absorption with the soaking duration for the cement stabilised compressed earth blocks and a burnt clay brick. The relationships are for CEB having nearly similar clay content but different

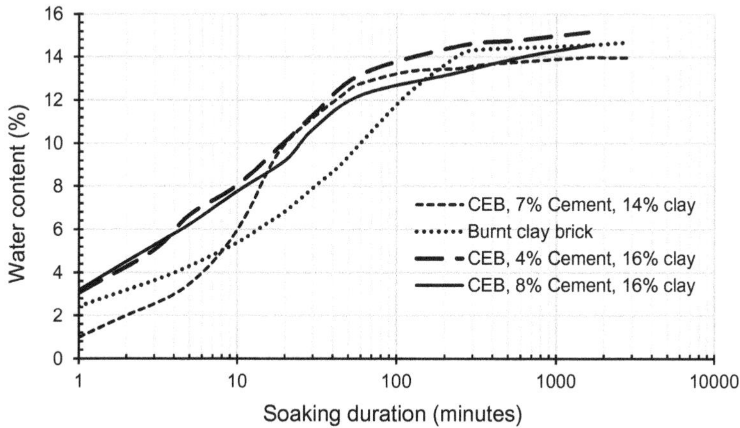

Fig. 5.35 Variation in the water content of stabilised CEB and the soaking duration

cement contents (4, 7 and 8% cement). Both the burnt clay brick and the stabilised CEB rapidly absorb water initially and reach close to 24-h absorption (near saturation) level in about 200–250 min. The masonry units saturate to about 75% of the saturation value after 48 h of soaking in water. The water content at the end of 48-h soaking is about 14.0%, 14.5% and 15.2% for 8, 7 and 4% cement blocks, respectively. The burnt clay brick absorbs 14.7% water after 48 h soaking in water. The stabilised CEBs attain 75% of 24-h soaking value, in about 20–30 min of soaking, whereas the burnt clay brick requires about 60 min. More discussions on the rate of water absorption in the stabilised CEBs can be found in Sect. 7.2.3 (Chap. 7).

5.9.3 Initial Rate of Absorption

The initial rate of absorption (IRA) is the quantity of water absorbed by the dry brick or block in one minute, when it is partially soaked in a shallow pool of water. ASTM C 67 code gives procedure to determine the initial rate of absorption for the masonry units. The IRA is generally expressed in kg/m^2/minute.

Figure 5.36 shows relationship between the IRA and the clay content of the soil mix used for the stabilised CEB production, for cement contents of 4 and 8%. The figure shows that the IRA decreases with the increase in the clay content of the mix. As the clay content of the mix increases from 5 to 20%, there is considerable decrease in the IRA. This can be mainly attributed to the reduction in the surface porosity of the cement stabilised compressed earth block with the increase in the clay content of the mix. The optimum clay content for the coarse-grained soil is about 15% to achieve maximum strength. The blocks/bricks with lower cement content

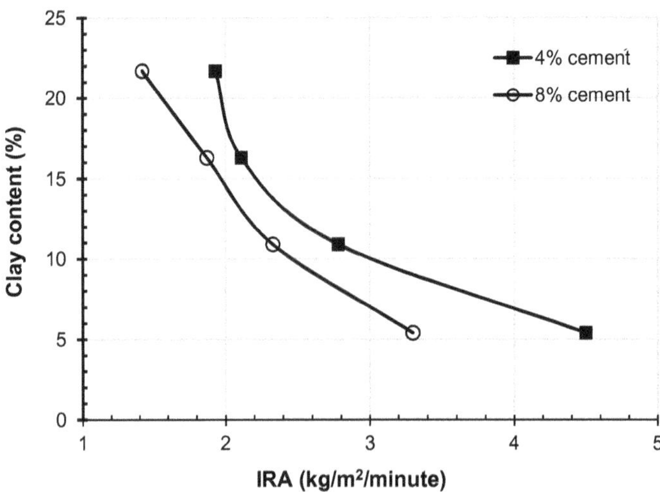

Fig. 5.36 IRA and clay content relationships for cement stabilised CEB

Table 5.9 IRA values for stabilised compressed earth blocks

Sl. No.	Block size (mm)	Wet comp. strength (MPa)	Clay content of the mix (%)	Cement content (%)	IRA (kg/m^2/min.)	References
1	305 × 143 × 96	3.14	16.3	4.0	2.11	Reddy et al. (2007)
2	305 × 143 × 96	5.73	16.3	8.0	1.87	Reddy et al. (2007)
3	305 × 143 × 100	7.19	9.0	12.0	1.60	Reddy and Gupta (2005)
4	230 × 190 × 100	6.89	14.5	7.0	0.59	Ullas (2014)
5	230 × 108 × 69	7.43	10.9	7.0	1.55	Tennant et al. (2016)
6	255 × 122 × 80	5.09	13.8	5.0	0.68	Vyas (2007)
7	255 × 122 × 80	5.09	13.8	14.0	0.24	Vyas (2007)

show higher IRA. This can be attributed to higher surface porosity for the lower cement content bricks/blocks (Reddy and Gupta 2005).

Table 5.9 gives IRA values for cement stabilised compressed earth blocks/bricks. The table gives block size, clay content, cement content and IRA values. These results show that the CEB with clay fraction in the range of 10–16% and with 4–12% cement, the IRA is in the range 0.60–2.1 kg/m^2/minute. The IRA values for the burnt clay bricks (in India) having a compressive strength in the range of 3–10 MPa are in the range of 0.8–3 kg/m^2/minute (Saranagpani et al. 2005; Ullas 2014; Gourav 2015). The results on IRA values for different types of masonry units do not show direct correlation between the IRA and the strength.

5.10 Stress–Strain Characteristics of Cement Stabilised Compressed Earth Blocks

The stress–strain characteristics (modulus, Poisson ratio, strain at peak stress and ultimate failure strains) for cement stabilised CEBs are mainly influenced by the following parameters:

(1) The clay content and composition of the soil mix
(2) Cement content
(3) Density and moisture content of the CEB.

Comprehensive investigations on the stress–strain characteristics of the cement stabilised CEBs are limited, but there are some investigations on certain aspects of stress–strain characteristics of cement stabilised CEBs (Reddy and Jagadish 1989; Reddy and Gupta 2005; Reddy et al. 2007; Reddy and Vyas 2008; Walker 2004).

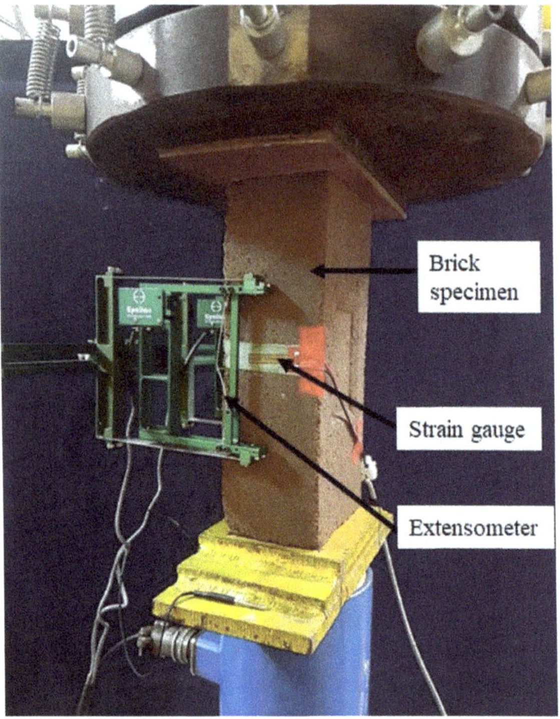

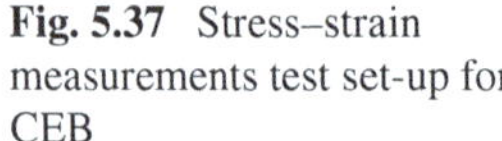
Fig. 5.37 Stress–strain measurements test set-up for CEB

An experimental set-up to ascertain stress–strain relationships for CEBs is shown in Fig. 5.37. The horizontal and vertical strain/displacement should be captured in the middle 1/3rd portion of the CEB, where the vertical compression will be uniform and free from platen friction effects.

Figure 5.38 shows typical stress–strain relationships (in wet and oven dry conditions) for the cement stabilised CEBs manufactured using optimum clay content (12–15%) and having 1800 kg/m^3 dry density with 8% cement. The initial portions of the stress–strain relationships are linear followed by non-linear portion. Beyond the peak stress, a large softening portion can be seen, especially for the dry CEB. The modulus and ultimate strain values for the cement stabilised CEBs are compiled in Table 5.10. The cement stabilised CEBs show large ultimate strain at failure in excess of 1.0% in dry condition, whereas in the wet state, the CEBs show ultimate failure strain of about 0.5–0.6%. The strain at peak stress will in the range of 0.002–0.003 for cement stabilised CEBs. The modulus of cement stabilised CEBs is sensitive to the cement content, density and the moisture content during testing. The modulus (initial tangent modulus) for CEBs (with cement content of 4–14%) is in the range of 1.5–12.0 GPa in the wet state, where the dry density of the CEBs is about 1800 kg/m^3 and are manufactured using optimum clay content (12–15%). The cement stabilised CEB modulus increases with the increase in the cement content.

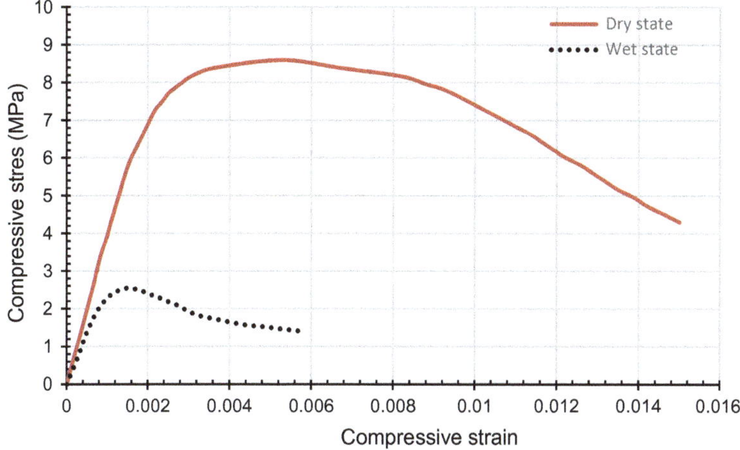

Fig. 5.38 Typical stress–strain curves for cement stabilised compressed earth block

Table 5.10 Stress–strain characteristics for cement stabilised CEBs

Cement stabilised CEB			References
Cement content (%)	Modulus (GPa)		
	Wet state	Dry state	
4	1.5–2.0	1.5–3 times more than that of the wet modulus	Gupta (2003), Richardson Lal (2005), Vyas (2007), Walker (2004), Reddy et al. (2007), Reddy and Jagadish (1989), Reddy and Gupta (2005), Reddy and Vyas (2008)
6	2.0–3.0		
8	3.5–6.0		
12	6.5–9.0		
14	10.0–12.0		

Lal (2005) explored the influence of clay content on the Poisson's ratio for the cement stabilised CEB in the wet state. Figure 5.39 shows the Poisson ratio and the clay content relationships for CEBs with 4 and 8% cement. The Poisson's ratio increases with the increase in the clay content of the CEB and can be as high as 0.26 for low cement content and high clay fraction. This can be attributed to the softening effect of clay in the wet state of the cement stabilised CEB. Generally, for blocks with 6–8% cement (and dry density >1800 kg/m^3) having optimum clay content of 15% and using coarse-grained soils, the Poisson ratio will be in the range of 0.10–0.18 in wet state (Reddy et al. 2007; Gupta 2003; Vyas 2007).

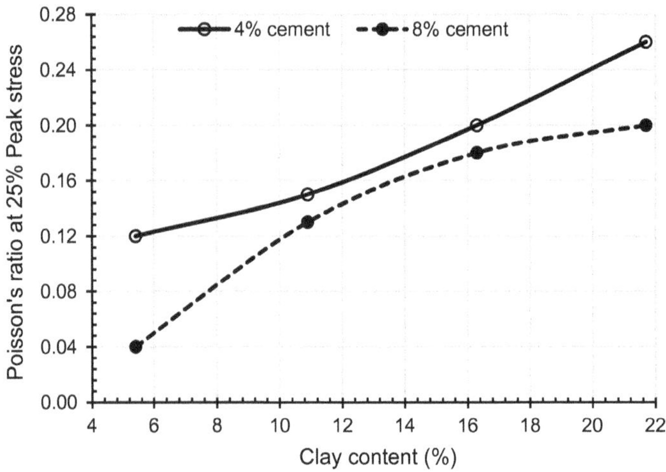

Fig. 5.39 Poisson ratio and clay content relationships

5.11 Thermal Characteristics of Stabilised Compressed Earth Blocks

The design for thermal performance and thermal comfort in the stabilised earth buildings necessitates the need for understanding the thermal properties of the stabilised CEB. The basic material property such as thermal conductivity is needed while assessing the thermal performance of the building system. The thermal conductivity of the cement stabilised CEB depends upon the following parameters.

1. Clay content of the mix used in the CEB production
2. Cement content
3. Density of CEB
4. Moisture content
5. Temperature of the surrounding environment

Investigations of Adam and Jones (1995), Akinmusuru (1994), Dondi et al. (2004), Balaji et al. (2015, 2016, 2017) and Balaji (2016) shed more light on the thermal properties of cement stabilised CEBs. Figure 5.40 (after Balaji 2016) shows the influence of clay content of the mix, cement content and density of the CEB and moisture content of CEB on the thermal conductivity of the cement stabilised CEB. The clay and cement contents and dry density affect the thermal conductivity of the CEB. The thermal conductivity increases with the increase in cement content and dry density. The conductivity increases by about 30% as the cement content goes up from 5 to 16%. The clay content is a less significant parameter, the conductivity increased by about 6% as the clay content increases from 10 to 31%. Generally, the thermal conductivity of a material increases as its density increases. About 12% increase in conductivity was noticed as the density increased from 1700 to 1900 kg/m^3. The thermal conductivity of the stabilised CEB is extremely sensitive to the moisture content. The conductivity values in wet state (nearly saturated state) are more than

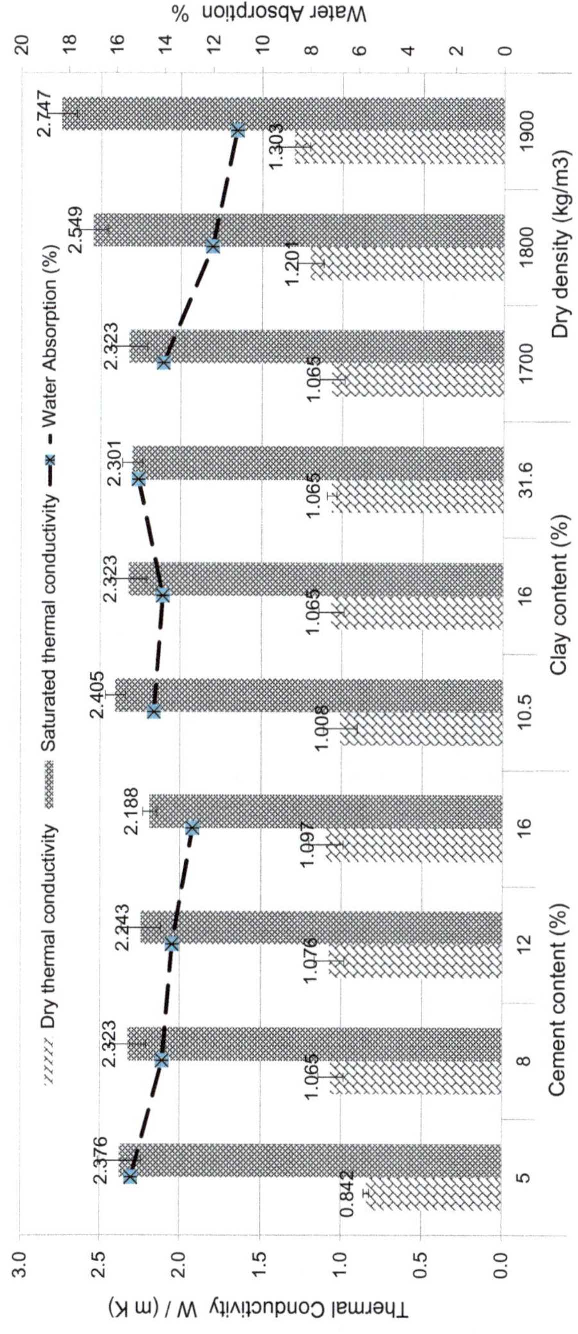

Fig. 5.40 Thermal conductivity of cement stabilised CEB (after Balaji 2016)

Table 5.11 Thermal conductivity values of walling materials

Type of material	Thermal conductivity (dry state) (W/mK)	References
Cement stabilised CEB	0.84–1.30	Balaji et al. (2017)
Fly ash brick	0.86	Balaji et al. (2016) IS 3792 (1978)
Aerated concrete block	0.17	
Burnt clay brick	0.56–0.81	
Dense concrete	1.74	
Reinforced concrete	1.58	

double that of the conductivity in the dry condition, irrespective of cement content, clay content and density. A comparison of thermal conductivity values for the cement stabilised CEB, the burnt clay brick and the concrete block is given in Table 5.11. The CEB has wide range of values due to variations in density, clay content, cement content, etc.

5.12 Durability of Stabilised Compressed Earth Blocks

The residential buildings are built to last for at least a couple of generations. The earthen building products are prone for damage due to moisture ingress. Therefore, durability of any new building material, especially any soil or earth-based technology, is of major concern. The performance of a building product when exposed to the natural environmental actions (such as sunlight, rain, floods, changes in humidity, abrasion, etc.) can be understood by performing tests for mechanical strength, dimensional stability, abrasion resistance, and resistance when exposed to cyclic wetting and drying conditions. The earthen building products possess clay minerals. The unstabilised or residual clay minerals in such products are prone for swelling and shrinkage upon exposure to cyclic wetting and drying. Therefore, swell–shrink actions of clay minerals in earthen products are of concern in the long-term durability performance of the earthen buildings and structures.

5.13 Types of Damages in Stabilised Earth Brick/block Masonry Buildings

Several factors influence the long-term performance of the stabilised earth products such as bricks/blocks and rammed earth. The deterioration/weathering of the stabilised earth products can be attributed to: (a) various factors related to material composition of the product and (b) disintegration caused by mechanical means.

Figures 5.41, 5.42, 5.43, 5.44, 5.45 and 5.46 show different types of possible damages in the stabilised earth block walls. Different types of possible damages occurring in the stabilised earth block walls can be listed as follows.

(a) Splitting cracks caused due to swelling and shrinkage of clay minerals
(b) Erosion and pitting
(c) Spalling of surface layer of the stabilised earth block
(d) Cracking and spalling
(e) Collapse of stabilised earth block wall corners after spalling is initiated
(f) Peeling off of the stiff plaster layer on stabilised earth block walls
(g) Mechanical damage to the block and the wall corners.

Fig. 5.41 Splitting cracks due to swelling and shrinkage in stabilised CEBs

Fig. 5.42 Erosion and pitting of the stabilised CEBs

Fig. 5.43 Spalling of surface layer of the stabilised CEB walls

Fig. 5.44 Cracking and spalling of the stabilised CEB wall

Fig. 5.45 Cracking and collapse of stabilised CEB wall corners after spalling was initiated

Fig. 5.46 Damage due to plaster peeling off in cement stabilised CEB walls

5.14 Deterioration Mechanisms in Stabilised Earth Products

Among the several factors pertaining to the material composition, the clay minerals present in the brick/block is a major contributing factor for the deterioration in the long-term. In the case of cement stabilised compressed earth blocks, the clay minerals are nearly intact, while in the lime stabilised earthen products, the clay minerals get consumed in lime-clay reactions (Reddy and Latha 2014a). The residual clay mineral present in the brick/block has lot of affinity for the water. On absorbing water, the clay minerals undergo swelling and upon drying the clay minerals shrink. The swell–shrink phenomenon is manifested as surface cracks (Fig. 5.41). The cyclic wetting and drying can disrupt the already established cementitious bonds between various particles causing deterioration of the stabilised earth product with age. Also, the frost action can cause snapping of already established cementitious bonds among various particles within the stabilised compressed earth block matrix, leading to the deterioration of the brick/block. Hence, the type and quantity of clay mineral present in the stabilised compressed earth product, the percentage of stabiliser, and the extent of exposure to cyclic wetting and drying can control the performance of the stabilised earth product.

After the development of cracks, erosion and pitting occurs, which is responsible for the deterioration of the stabilised compressed earth block (Fig. 5.42). The impinging rainwater droplets on the wall surface or the water from the leaking pipes can hasten the erosion and the pitting process. As the deterioration progresses with time, a layer of the block or rammed earth wall surface peels off, causing spalling type of failure (Figs. 5.43 and 5.44). The exposed wall corners are readily vulnerable for the cracking and the spalling process. Figure 5.45 shows collapsed stabilised earth block wall corners due to the excessive cracking and spalling. The stabilised earth block walls built using the poor-quality blocks (low strength, high clay soil, inadequate stabilisation, etc.) covered with stiff mortar plaster layer is also prone for the damage. The earth blocks in a wall covered with stiff/strong plaster layer can undergo swelling and shrinkage due to the moisture movement. This process leads to separation of the plaster layer from the wall surface as shown in Fig. 5.46. Over a period of time, the plaster will peel off and a part of the block surface gets attached to the plaster layer inflicting damage to the block. The exposed corners and edges of the stabilised earth block walls, especially built with the low strength blocks, are prone for mechanical damage as shown in Fig. 5.47a. Such damages can be minimised by using blocks with higher stabiliser content (i.e. stronger blocks) and by using rounded corner (bullnose) blocks as shown in Fig. 5.47b.

Fig. 5.47 Mechanical damage to corners and edges of the stabilised CEB wall and rounded corner CEB wall

5.15 Evaluation of Durability Characteristics of Stabilised Earth Products

Assessing the durability of building materials exposed to natural weathering conditions is a time-consuming process. Hence, the laboratory test methods give an idea about the durability of the building materials. There are several attempts to devise accelerated test methods for assessing the durability of earth bricks/blocks. Some of the durability assessment tests for earth bricks/blocks are:

(a) Spray erosion test
(b) Drip test
(c) Cyclic wetting and drying test
(d) Linear expansion on saturation test
(e) Freeze–thaw test.

5.15.1 *Spray Erosion Test*

The spray erosion test is an accelerated test to assess the durability of the stabilised earth block under natural weathering conditions. Investigations of Middleton (1992), Reddy and Jagadish (1987, 1995), Walker (1998, 2004), Heathcote (1995, 2002), Ogunye and Boussabaine (2002a, b) and Guettala et al. (2006) indicate several attempts to standardise the spray erosion test. Walker (1998) highlights different types of test procedures available to assess the erosion resistance of the earth blocks. Ogunye and Boussabaine (2002a) compiled information on the spray erosion tests on the stabilised compressed earth blocks carried out by many investigators. The spray erosion test procedure involves impinging a spray of water jet (at constant pressure) onto the surface of the stabilised earth block for a definite period of time. After completion of the test, the block surface is examined for pitting, surface erosion (erosion depth) and mass of the material removed by the water jet. The erosion depth and the mass loss are quantified and correlated to the durability of the stabilised CEBs. Many investigations attempted to standardise the spray erosion test to represent more closely the damage caused by the impinging rain drops on the wall surface. Figure 5.48 shows a typical spray erosion test set-up. The pressure of the spray jet, distance between the exposed block surface and the spray nozzle and the area of the exposed block surface vary across different investigations and reports/codes.

The accelerated spray erosion test can at best give some idea about the relative durability performance of the stabilised CEBs. There are attempts to correlate the erosion rate from the test with the field erosion rate due to the natural rain fall (Reddy and Jagadish 1987). Here, the standardisation becomes a problem because the nature, intensity, amount and distribution of the rain fall varies considerably from region to region. The investigation of Reddy and Jagadish (1987) showed that cement stabilised CEBs did not show any erosion or pitting even after exposure to the spray jet for 120 min.

Ogunye and Boussabaine's (2002a) compilation represents results from six different studies on the spray erosion testing of the stabilised CEBs. The investigations cited show both erosion depth and the mass loss after the erosion test. This

Fig. 5.48 Spray erosion test set-up

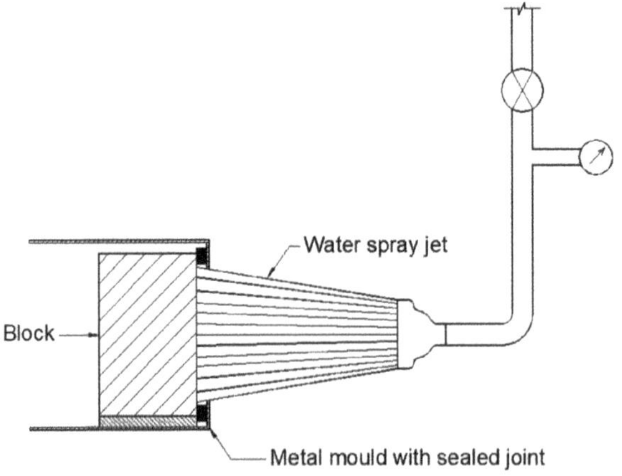

compilation clearly illustrates the lack of standardised spray erosion test procedure. Ogunye and Boussabaine (2002b) developed a rainfall test rig as an aid for the earth block weathering assessment. They tested the cement and the lime/gypsum stabilised CEBs in the new test rig. The test involved monitoring mass loss after 120 min of the erosion test. The mass loss varied between 0.4 and 1.67% for the blocks using the cement (6–10%) and the lime (6–15%), respectively.

Guettala et al. (2006) conducted durability studies on the stabilised CEBs both under laboratory and in the field conditions. Cement, lime and cement/lime stabilisers were used for the manufacture of the CEBs. The results of the laboratory spray erosion studies (for 120 min) have been compared with the erosion undergone by the sample walls weathered under the natural conditions for four years. The laboratory tests showed a spray erosion depth between 0.5 and 1.0 mm for the cement (5–8%) and the cement/lime (5–8%) stabilisation and 1.0 and 2.2 mm for the lime (5–8%) stabilisation. The study emphasises that the laboratory tests are more severe than the weathering under natural climatic conditions. Table 5.9 provides a summary of the results on the spray erosion tests from different investigations. The results clearly show that the cement/lime stabilised CEBs hardly show any erosion even after exposure for two hours in the standard spray erosion test (Table 5.12).

Table 5.12 Details of spray erosion data from different studies for stabilised CEBs

Sl. No.	Impinging distance (mm)	Water pressure (MPa)	Exposure duration (h)	Stabiliser content	Mass loss (%)	Erosion depth (mm)	References
1	175	0.07	2.0	5% cement	Negligible	0.0	Reddy and Jagadish (1987)
2	170	0.16	2.0	5% cement	–	1.0	Guettala et al. (2006)
3	170	0.16	2.0	8% cement	–	0.5	Guettala et al. (2006)
4	170	0.16	2.0	8% lime	–	2.2	Guettala et al. (2006)
5	170	0.16	2.0	12% lime	–	1.0	Guettala et al. (2006)
6	170	0.16	2.0	5% cement + 3% lime	–	1.0	Guettala et al. (2006)
7	170	0.16	2.0	8% cement + 4% lime	–	0.5	Guettala et al. (2006)
8	470	0.05	1.0	2.5–10% cement	Negligible	0.0	Walker (2004)

5.15.2 Drip Test

The drip test proposed by Yttrup et al. (1981) basically involves allowing water droplets or drip to impinge on an inclined block surface for a definite amount of time. Figure 5.49 shows schematic sketches of the drip test set-ups. The Geelong test set-up is based on the test developed by Yttrup et al. (1981). The Swinburne accelerated erosion test was developed by the Swinburne University of Technology (Weisz et al. 1995). In both these tests, after completion of the test, the block surface is monitored for the pitting damage, and depth of erosion and moisture penetration. The general reliability of these tests is unproven (HB 195-2002). Unlike the spray erosion test, the drip test gives a relative evaluation of the durability performance of different types of earth blocks.

The cement stabilised CEBs with adequate stabiliser content (cement) and manufactured using soils with optimum clay content perform exceedingly well when subjected to the spray erosion or the drip test. Evaluating the relative performance of the stabilised CEBs having optimum clay content and with the stabiliser content beyond a threshold limit (5% cement) becomes difficult. This is because the surfaces of such blocks do not show any erosion even after several hours of exposure to the water spray (Reddy and Jagadish 1987; Walker 2004). Hence, the spray erosion and the drip tests may be more suitable for the evaluation of erosion resistance of unstabilised earth blocks.

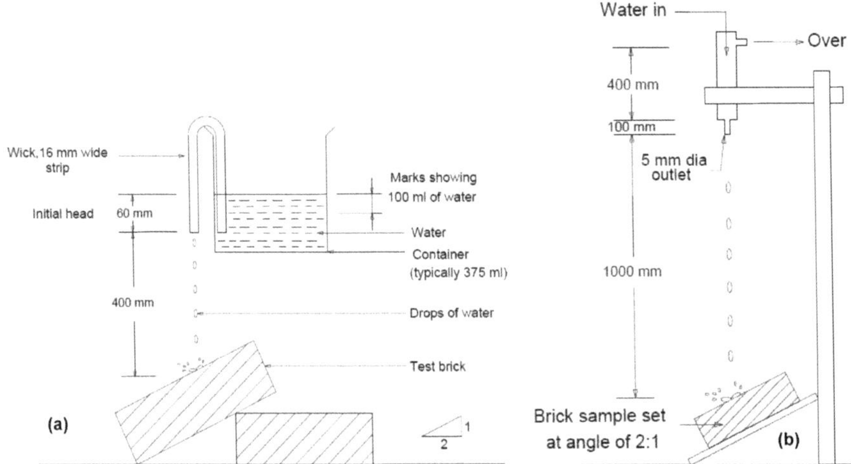

Fig. 5.49 Drip erosion test set-ups (a) Geelong drip test, (b) Swinburne drip test

5.15.3 Cyclic Wetting and Drying Test

The cyclic wetting and drying test is intended to examine the durability of the stabilised earth block. This test was originally proposed in ASTM D559 code to evaluate durability characteristics of the soil–cement used in the road construction. A modified version of the cyclic wet and dry test, more suited for the stabilised earth blocks, can be found in the revised Indian code IS 1725 (2013). This code gives test procedure and the guidelines for performing the cyclic wetting and drying test on the stabilised CEBs. The test involves monitoring the mass loss of the stabilised CEB, after exposure to 12 cycles of cyclic wetting and drying. Brief details of the cyclic wet and dry test procedure are as follows.

- Ascertain the initial dry weight of the cured stabilised earth block specimen, after drying it to constant weight in an oven at 60 ± 5 °C.
- Soak the dried specimens in potable water for a period of 5 h and then dry it in an oven at 70 ± 5 °C for 42 h. Apply two firm strokes on the surfaces of the partially dried specimen with a standard wire brush. Eighteen to twenty brush strokes are applied to cover all the faces of the specimen. This constitutes one cycle of wetting and drying.
- Repeat the above-mentioned procedure and complete 12 cycles.
- At the end of 12 cycles, the specimen is oven dried at 60 ± 5 °C till attaining constant weight and record the final dry weight.
- Ascertain the mass loss of the stabilised earth block specimen after completion of the twelve cycles of wetting and drying.

More details on the test procedure and the wire scratch brush can be found in IS 1725 (2013) code.

The test involving scratching the surface of the earth block specimen with metal wire brush, which is considered as too severe a test for evaluating the durability of the stabilised CEB (Lunt 1980). The Portland Cement Association (1956) specifies acceptable limits for mass loss ranging from 7 to 14% for various types of soils to be used in the road construction. These limits have been modified and adopted by several authorities depending upon the nature and the types of buildings and the climatic conditions. For example, Fitzmaurice (1958) recommended limits for maximum mass loss as 5% for permanent buildings and 10% for rural buildings in any type of climate. Walker and Stace (1997) and Walker (1995) attempted to correlate mass loss with clay content of the cement-soil mix used for CEB production. Spence (1975) and Spence and Cook (1983) attempted to correlate the dry density and the cement content of the cement stabilised CEBs with the mass loss. Different types of soils having different clay contents have been used in these studies. The studies show that the mass decreases with the increase in the density and the cement content. Investigations of Reddy et al. (2007), and Reddy and Latha (2014b) represent more comprehensive studies in eliciting the optimum clay content in the mix giving better durability characteristics for the cement stabilised CEBs. In these studies, the mass loss after the cyclic wet and dry test in the cement stabilised CEBs has been monitored by varying the clay content of the mix, considering the fine-grained and the coarse-grained soils.

Figure 5.50 shows plots of mass loss in cement stabilised CEB versus clay content of the mix. The soil used in these studies had the kaolinite clay mineral and the dry density of the blocks used in these studies was about 1800 kg/m^3. The plots clearly show that the mass loss is sensitive to the clay content of the soil mixture. There is an optimum clay content leading to the minimum mass loss. The optimum, clay content is about 10% for the fine-grained soils and about 15% for the coarse-grained soils. The plots show a mass loss in the range of 1–26% as the clay content of the mix was varied between 5 and 32% and the cement content between 4 and 8%. The blocks with lower cement content (4%) show higher mass loss. Figure 5.51 shows the physical condition of the cement stabilised CEBs after 12 cycles of the cyclic wet and dry test for the two cases of 4 and 8% cement. The pictures show the condition of the stabilised CEBs containing the clay fraction of 5.4, 10.9, 16.3 and 21.7%. The CEB with 16.3% clay shows the least damage in both the cases. Higher clay content blocks show more visible damage, as evidenced in the mass loss values. The stabilised CEBs used in these studies were made with a coarse-grained soil.

The tests performed on stabilised CEBs collected from different buildings with different age distributions show useful results to specify a limit for the mass loss after the twelve cycles of cyclic wetting and drying test. Based on some of the experimental studies linked to the observations on the real-time performance of the stabilised CEB buildings, a limit of 3% mass loss has been specified. This limiting number for the mass loss after the cyclic wetting and drying test has been adopted in the revised Indian code on stabilised earth blocks (IS 1725 2013).

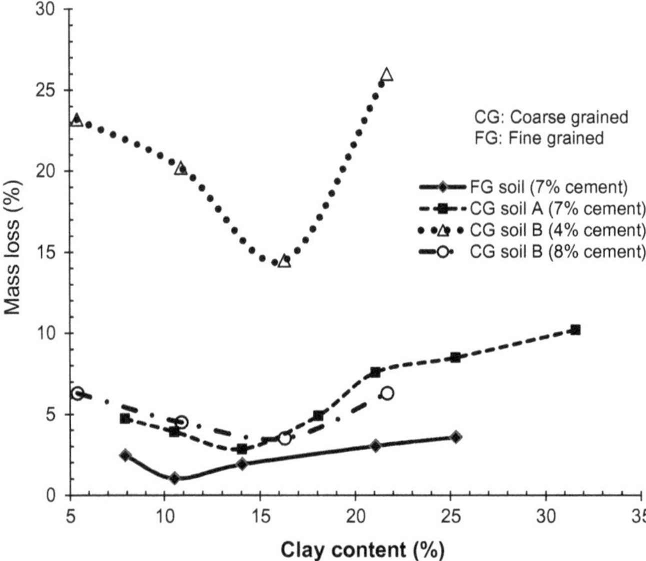

Fig. 5.50 Mass loss in cement stabilised CEB versus clay content of the soil mix

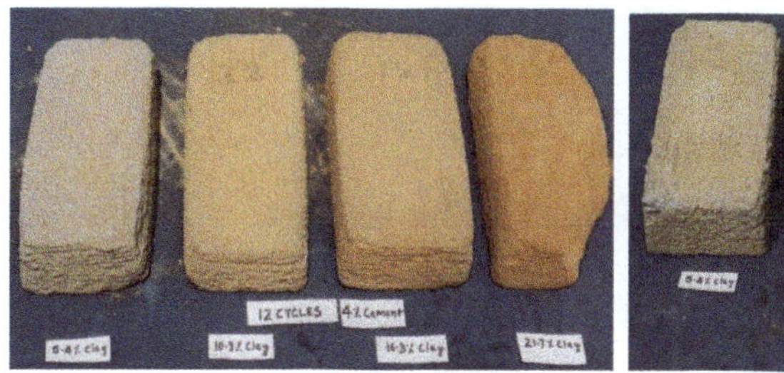

4% cement 8% cement

Fig. 5.51 Status of cement stabilised CEBs after twelve cycles of wet and dry test (Left to right—Clay: 5.4%, 10.9%, 16.3%, 21.7%) (Reddy et al. 2007, with permission from ASCE)

5.15.4 Linear Expansion on Saturation

The building materials should have satisfactory dimensional stability when exposed to normal weathering conditions. Generally, the stabilised CEB building envelopes have exposed wall surfaces. Hence, they tend to undergo wetting and drying. Even the hygroscopic moisture in the air can be a source for stabilised CEB walls to undergo wetting–drying cycles. Therefore, for satisfactory performance of the stabilised CEB walls, the blocks should be dimensionally stable. The dimensional stability of the stabilised CEB can be monitored by measuring change in the length of the CEB between the dry and the saturated conditions. This type of measurement can be designated as linear expansion on saturation.

Linear expansion on saturation (LES) for the brick or the block indicates dimension stability, when it undergoes one cycle of wetting and drying. The LES value for a stabilised CEB can be easily measured and quantified. The LES test was first proposed by Reddy (1991). The LES test has been standardised and incorporated in the Indian code on stabilised earth blocks (IS 1725 2013). An accurate length comparator is required to perform the LES test. Figure 5.52 shows a length comparator set-up for the LES test, and Fig. 5.53 shows the dimensioned sketch for the length comparator used in the LES test. The LES test procedure is as follows.

(a) The cured and air-dried block is oven dried (at 60 ± 5 °C) to attain constant weight.
(b) Measure the initial length of the oven dried block using a Vernier callipers.
(c) Position the dried block in the length comparator and then take the dial gauge reading
(d) Without disturbing the dial gauge in the length comparator, remove the block and soak it in water for 24 h
(e) After removing the block from the water, position it in the length comparator and take the dial gauge reading
(f) The difference in the dial gauge readings of the dry and wet (nearly saturated) block gives the linear expansion of the block upon soaking in water for 24 h.
(g) The linear expansion expressed in terms of the original dry length gives LES value.

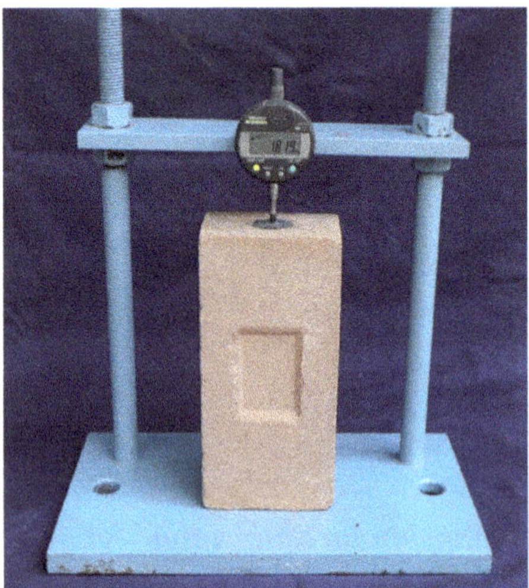

Fig. 5.52 Experimental set-up for linear expansion on saturation

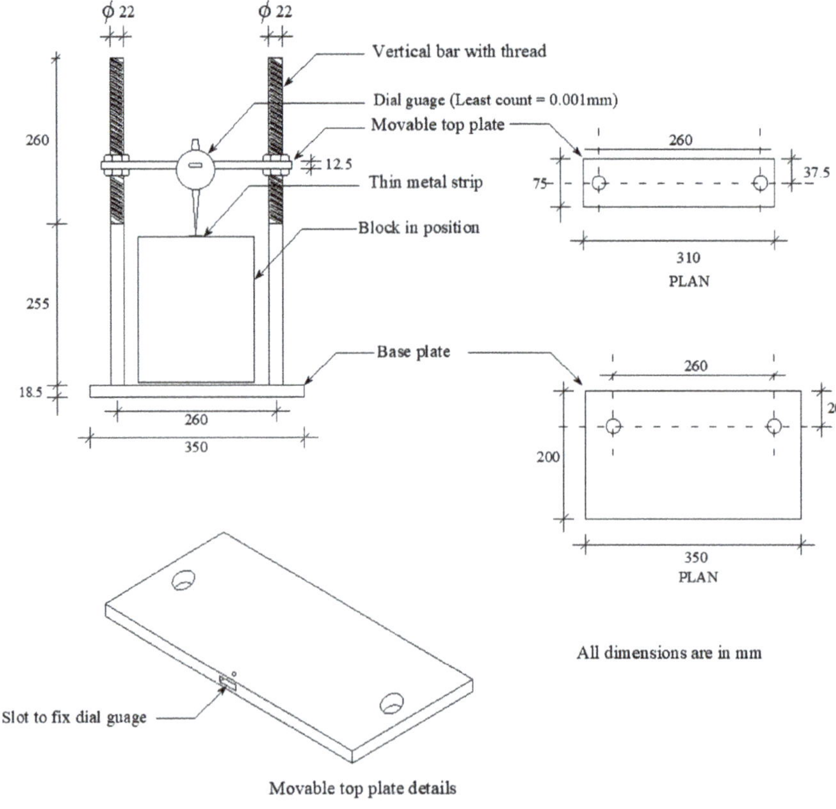

Fig. 5.53 Length comparator set-up

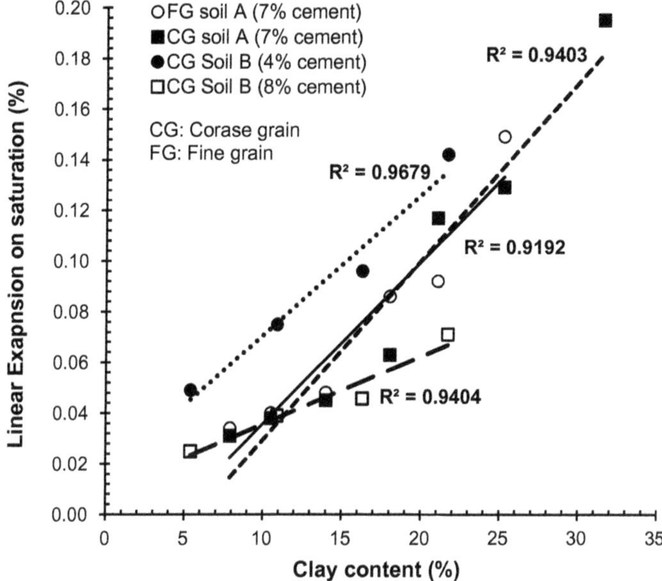

Fig. 5.54 Relationship between linear expansion on saturation and clay content of the cement stabilised CEB

The LES value of a stabilised CEB is influenced by the clay content of the soil mix, stabiliser content and the block density. Figure 5.54 shows a plot between the LES and the clay content for the CEBs using different types of soils and cement contents. The dry density of the CEBs in these studies was about 1800 kg/m^3. It is clear from these relationships that the LES and clay content are linearly related irrespective of the soil type and the cement content. The LES increases with the increase in clay content irrespective of the cement content. This can be mainly attributed to the swelling of the clay minerals after moisture absorption, more clay means more swelling and higher the value of LES.

The LES depends upon the stabiliser content. A plot of LES versus cement content of the CEB shown in Fig. 5.55. The points shown in the figure represent CEBs with 1800 kg/m^3 dry density and the clay content in the range of 11–16%. The LES of cement stabilised CEB is sensitive to the cement content. The LES decreases with the increase in the cement content. The plot (Fig. 5.55) shows LES for the cement stabilised CEBs using optimum clay content (10–15%). The LES values for the CEBs with even 4% cement are within the permissible limit of 0.10% specified in IS 1725 (2013).

A plot of LES and mass loss for a set of measurements has been shown in Fig. 5.56. A fourth-degree polynomial relationship has been shown for this data. The relationship shows nearly constant LES (~0.04%) for a mass loss of up to about 4%. Beyond 4% mass loss, LES drastically increases with increase in mass loss. Also, there were similar attempts to correlate LES and mass loss by Reddy et al. (2003).

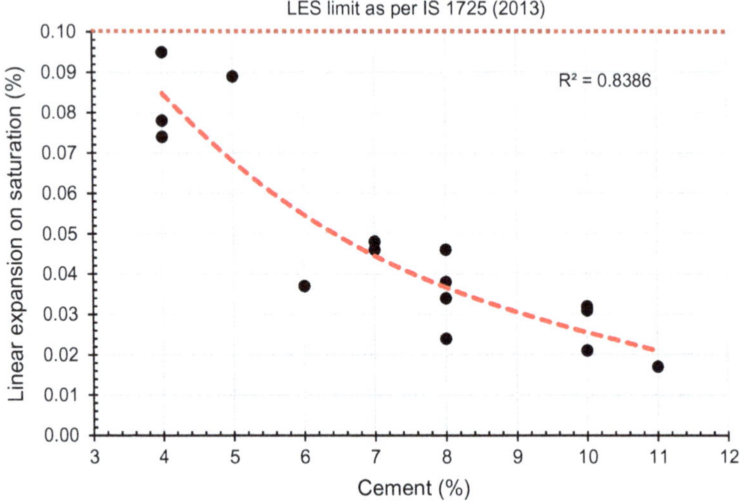

Fig. 5.55 Relationship between linear expansion on saturation and cement content of CEB

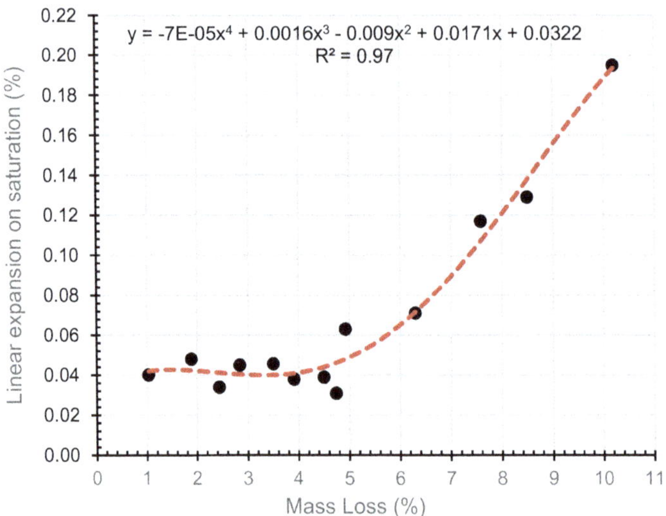

Fig. 5.56 Relationship between LES and mass loss for the cement stabilised CEB

5.15.5 Usefulness of LES and Mass Loss Values

The studies of Ullas and Reddy (2007) and Reddy et al. (2003) show performance of the existing cement stabilised CEB buildings having exposed block-work walls (without protection). These buildings were in a climate zone having an average annual rain fall of about 1000 mm spread over six months of the year. The buildings were aged between 5 and 20 years. The linear expansion on saturation values of the blocks collected from several of these constructions have been monitored. The LES values was in the range 0.006 and 0.024%. The physical performances of the stabilised CEBs in these buildings (extent of damage) has been examined and correlated with the LES values of the blocks tested in the laboratory. These results demonstrated that the stabilised CEBs with < 0.10% LES value generally led to satisfactory performance of the stabilised CEB masonry walls when exposed to natural weathering conditions over a period of several years. Another compilation by Reddy et al. (2003), for the cement stabilised CEBs collected from eight different construction sites showed LES values in the range of 0.02–0.07%.

Figures 5.57 and 5.58 show the cement stabilised CEB masonry wall surfaces exposed for 14 years (with LES of 2.32%) and 22 years (with LES of 0.03%), respectively. The wall surfaces and the blocks with higher LES value show severe damage after repeated wetting and drying caused by the natural weathering process. In contrast the blocks with low LES value (0.03%), the wall surface is intact even after 22 years of weathering. Also, the walls constructed with the blocks having <0.10% LES have performed satisfactorily in the field. Hence, keeping the linear expansion on saturation less than 0.10% and mass loss after cyclic wetting and drying test

Fig. 5.57 Damage to CEB blocks and CEB wall surface after 14 years of exposure to natural weathering (LES = 2.32%)

Fig. 5.58 Cement stabilised CEB wall surface after 22 years of exposure to natural weathering (LES = 0.03%)

<3% could assure stable long-term performance for the stabilised CEB walls. The linear expansion on saturation value can be bench marked to indicate the dimensional stability of the stabilised CEBs under natural weathering conditions. Indian code on stabilised earth blocks (IS 1725 2013) gives limits on the LES and the mass loss after durability test as 0.10% and 3%, respectively.

5.15.6 *Freeze–Thaw Test*

The performance of the stabilised CEBs when exposed to the freeze–thaw conditions in cold climates is a concern. The ASTM D560 (2016) code gives a procedure for conducting freeze–thaw tests on soil–cement mixes. This test procedure can be utilised for testing the stabilised CEBs under freeze–thaw conditions. The freezing and thawing test involves placing a sample on the absorbent medium at −23 °C for a period of 24 h in a refrigerator. Then, the sample removed from the refrigerator is allowed to thaw in a moist environment at 23 ± 2 °C for 23 h and then removed and brushed on all the faces using a standard brush with a 13 N force. The specimen is subjected to 12 freeze–thaw cycles and then oven dried to attain constant weight. The mass loss is estimated based on the initial dry weight and the weight after the freeze–thaw test.

Table 5.13 Mass loss after durability test (Guettala et al. 2006)

Type of test	Mass loss after 12 cycles (%)					
	Cement (%)		Lime (%)		Cement + lime (%)	
	5.0	8.0	8.0	12.0	5.0 + 3.0	8.0 + 4.0
Wetting and drying test	1.40	1.25	2.30	2.10	1.20	1.00
Freeze–thaw test	2.35	2.23	3.70	2.90	2.30	2.00

The freeze–thaw test is considered as a severe test for the stabilised earth products. There are limited studies in understanding the behaviour of stabilised CEBs subjected to freeze–thaw conditions. Gabriele et al. (2016) examined the effect of freeze-thaw cycles on the mechanical behaviour of lime-stabilised soil. Guettala et al. (2006) examined the mass loss after freeze–thaw tests for stabilised CEBs. Table 5.13 gives the results from the studies of Guettala et al. (2006). The results clearly indicate that.

- The mass loss after the freeze–thaw test was 60–100% more when compared to the corresponding values of mass loss after the wetting and drying test.
- The stabilised CEBs with either cement or cement-lime combinations perform better than the blocks with only lime as stabiliser.
- The mass loss after the freeze–thaw test is under 3% (except for 8% lime), which is a permissible value of mass loss for cyclic wet and dry test specified in the IS 1725 (2013) code on stabilised CEBs (Table 5.13).

5.16 Performance of Plastered Stabilised CEB Wall Surfaces

Many a times, people perceive that if a stabilised CEB wall is eroding and showing signs of failure providing a hard plaster layer (such as cement mortar or cement-lime mortar) on the surface could lead to the protection of deteriorating stabilised CEB wall. Figure 5.46 shows the failure of plastered cement stabilised CEB wall. The LES of the CEB used in this particular wall was 1.26%. The figure shows the rigid cement plaster layer getting separated from the wall surface, and in addition, a portion of the block is plucked out causing damage to the CEB wall. The main reason for the peeling off of the plaster is the differential shrinkage between the hard mortar layer and the stabilised CEB block. When the plastered CEB wall is exposed to the moisture, the block surface behind the plaster layer gets partially saturated and swells. After the moisture escapes, due to drying the swollen layer of the block shrinks and develops shrinkage cracks. A few cycles of wetting and drying lead to the separation of the plaster layer along with a layer of the block. The hard plaster layer does not swell and shrink as much as the stabilised CEB.

5.17 Summary on Durability of Stabilised CEB

Durability performance of the stabilised CEB greatly depends upon the clay mineral composition of the soil mix, the stabiliser type and the stabiliser content. The bricks/blocks with higher clay fraction in the mix are generally prone for damage due to exposure to cyclic wetting and drying cycles, especially in the exposed wall surfaces. The CEBs with optimum clay fraction and adequate stabiliser content perform exceedingly well even under exposure to cyclic wet and dry conditions. The spray erosion test and the drip test may not reveal the durability performance of the stabilised CEBs. Such tests are more suitable for evaluating the relative durability performance of unstabilised CEBs or other earthen products such as adobe blocks. Among the different accelerated weathering tests, the linear expansion on saturation (LES) test and the cyclic wet and dry tests can be relied upon for evaluating the long-term performance of stabilised CEBs. The long-term performance of exposed stabilised CEB walls and the accelerated laboratory tests revealed the limits for LES and mass loss after cyclic wet-dry test as <0.10% and 3%, respectively.

5.18 Retrofitting the Damaged Stabilised Compressed Earth Block Masonry Walls

Performance of the stabilised CEB masonry wall greatly depends upon the quality of the CEB. Different types of damages that stabilised CEB masonry walls can undergo were discussed in the earlier sections. Splitting type of surface cracks (Fig. 5.41), erosion and pitting (Fig. 5.42) and cracking and spalling (Figs. 5.43 and 5.44) are some of the typical damages. All the three types of damages can be attributed to inadequate stabilisation. Different types of repairs and retrofitting techniques are possible depending upon the extent of damage the stabilised CEB wall has suffered. The details of the repair and retrofitting techniques are discussed in the following sections.

5.18.1 Patch Plastering with Cement-Soil Mortar

Surface cracking, pitting and spalling are generally associated with excess clay fraction in the soil used for the CEB and inadequate stabilisation. The stabilised CEB with this kind of damage is associated with large amount of expansion on saturation (>0.10%). Hence, it is essential to repair the damaged block surface using a compatible mortar. Here, the compatibility is in terms of shrinkage characteristics of the mortar and the stabilised CEB. This is mainly to avoid differential shrinkage between the stabilised CEB and the patched-up mortar on the damaged surface. The cement-soil mortars having a clay fraction of about 10% and cement content of about 10% are compatible, and hence, such mortars can be used for the patch plastering of

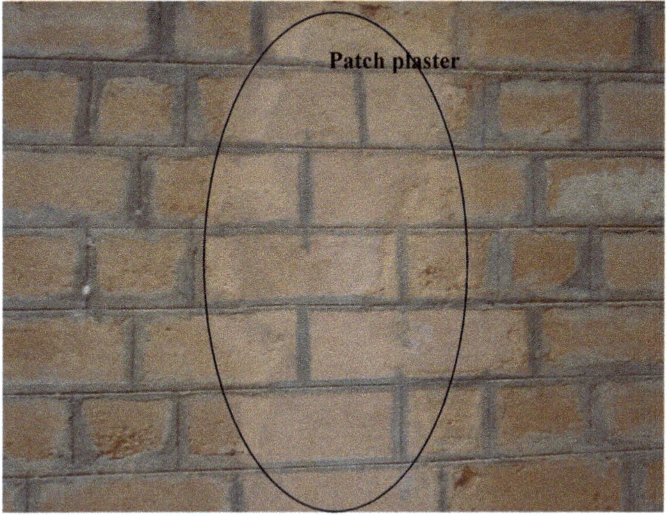

Fig. 5.59 Stabilised CEB wall with patch plaster

the damaged stabilised CEB surfaces. Figure 5.59 shows a patch plastered surface performing satisfactorily. The guidelines and the procedure for patch plastering are as follows.

1. Allow the damaged block surface to dry naturally. Remove the loose particles on the damaged surface using a brush.
2. Apply a coat of fresh cement–water (1:1, by mass) slurry on the damaged cleaned surface using a brush.
3. Apply a coat of cement-soil plaster and finish the plastered surface in flush with the neighbouring block surface.
4. Cure the plastered surface by spraying water for at least seven days.

5.18.2 Providing Additional Skirting of Stabilised CEB Veneer on the Damaged Walls

The exposed stabilised CEB walls with poor-quality stabilised CEBs are prone for damage due to splashing rainwater from the roof overhangs and from running rainwater along the height of the unprotected stabilised CEB wall. The damage could be severe leading to collapse of the wall. Such types of damaged walls can be protected by providing an additional skirting layer of thin (50–60 mm) good quality stabilised CEBs. Figure 5.60 shows the stabilised CEB veneer skirting to one of the damaged stabilised CEB wall. Precaution should be taken to ensure proper anchoring of the thin stabilised CEB masonry veneer to the main wall through bonding blocks at regular intervals.

Fig. 5.60 Stabilised CEB skirting veneer

5.18.3 Progressively Replacing the Damaged Stabilised CEB Wall

The damages to the stabilised CEB masonry walls caused by any of the several methods/processes explained in the previous sections could be very severe. The severe damages to the walls can jeopardise the safety of the building and could lead to the wall collapse. Figure 5.61 shows one such collapse. The collapsed wall (either partially or fully) can be rebuilt without disturbing the neighbouring portions or portions in the floor above. The procedure to be followed and precautions to be observed in such repair/retrofitting works are highlighted below.

1. Support the floor or roof slab with properly designed props such that the load on the wall under consideration is relieved temporarily.
2. Divide the wall into a number of sections approximately 1.0 m width. Remove one section of the wall and rebuild the portion with good quality stabilised CEBs.
3. Ensure to leave CEBs in a saw-tooth fashion in order to facilitate proper keying in of the masonry to the adjacent portion.
4. The stabilised CEB wall shall be rebuilt progressively in one metre sections and cured for a week. Figure 5.62 illustrates one such example of retrofitted stabilised CEB wall

Fig. 5.61 (Left) Collapsed stabilised CEB end wall and replaced with good quality stabilised CEB wall (Right)

Fig. 5.62 Retrofitting and rebuilding with good quality stabilised CEBs

5. Allow the wall to dry for 2–3 weeks before reloading the wall by releasing the temporary props supporting the floor or roof slab.
6. Fig. 5.63 shows a retrofitted stabilised CEB wall without disturbing the first-floor wall.

Fig. 5.63 Stabilised CEB building after retrofitting

References

ASTM D559/D559M-15 (2015) Standard test methods for wetting and drying compacted soil-cement mixtures, ASTM International, West Conshohocken, PA, 2015

ASTM D560/D560M-16 (2016) Standard test methods for freezing and thawing compacted soil-cement mixtures, ASTM International, West Conshohocken, PA, 2016

ASTM C618-08a (2008) Standard specification for coal fly ash and raw or calcined natural pozzolan for use in concrete. American Society for Testing and Materials, West Conshohocken, PA

ASTM C67 – 19, Standard test methods of sampling and testing brick and structural clay tile, American Society for Testing and Materials, West Conshohocken, PA

Adam EA, Jones PJ (1995) Thermophysical properties of stabilised soil building blocks. Build Environ 30(2):245–253

Akinmusuru JO (1994) Thermal conductivity of earth blocks. J Mater Civ Eng 6(3):341–351

Anon (1985) Lime stabilization construction manual, 8th edn. National Lime Association, Arlington, VA, USA

Anon (1990) Lime stabilization manual. British Aggregate Construction Materials Industry, London, UK

Armin R, Behzad K (2013) Stabilization of clayey soil with lime and waste stone powder. Int J Sci Res Knowl 1(12):547–556

Balaji NC, Mani M, Venkatarama Reddy BV (2015) Influence of varying mix proportions on thermal performance of the soil–cement blocks. In: Proceedings of the 2nd IBPSA Italy conference, building simulation application—2015 (BSA 2015) (Marco Baratieri FP, Corrado V and Gasparella A (eds)). BU Press, Bozen-Bolzano, Italy, p 8

Balaji NC, Mani M, Venkatarama Reddy BV (2017) Thermal conductivity studies on cement-stabilised soil blocks. Proc Inst Civ Eng Constr Mater 170(1):40–54

Balaji NC, Mani M, Venkatarama Reddy BV (2016) Dynamic thermal performance of the conventional and alterative building envelope materials. In: 4th international conference on Central Europe towards sustainable building 2016 (CESB 2016), innovations for sustainable future, 22–24 June, Prague, Czech Republic

Balaji NC (2016) Studies into thermal transmittance of conventional and alternative building materials and associated building thermal performance. PhD thesis, Department of Civil Engineering, Indian Institute of Science, Bangalore, India
Bell FG (1996) Lime stabilisation of clay minerals and soils. Eng Geol 42(1996):223–237
Bokhari AH (1976) A study of soil-cement blocks in building construction. MSc (Engg) thesis, Department of Civil Engineering, University of Engineering and Technology Lahore, Pakistan
BS EN 772-21 (2011) Methods of test for masonry Units Part 21: determination of water absorption of clay and calcium silicate masonry units by cold water absorption, British Standards Institution (BSI), London, UK
Chadda LR, Raj H (1955) Role of detrimental salts in soil stabilisation with and without cement-2: the effect of sodium carbonate. Indian Concr J 29:401–402
Clare KE, Sherwood PT (1954) The effect of organic matter on the setting of soil-cement mixtures. J Appl Chem 4 (II):625–630
Clare KE, Sherwood PT (1956) Further studies on the effect of organic matter on the setting of soil-cement mixtures, 6, 317–324
Clare KE, Cruchley AE (1957) Laboratory experiments in the stabilisation of clays with hydrated lime. Geotechnique 7:97–111
Consoli NC, Foppa D, Festugato L, Heineck KS (2007) Key parameters for strength control of artificially cemented soils. J Geotech Geoenviron Eng 133(2):197–205
Consoli NC, Dalla RA, Corte MB, Lopes LS (2011) Porosity-cement ratio controlling strength of artificially cemented clays. J Mater Civ Eng 23(8):1249–1254
Coutinho RQ, Lacerda WA (1987) Characterization and consolidation of Juturnaiba organic clays. In: Proceedings of the international symposium on geotechnical engineering of soft soils, 1, pp 17–24
Croft JB (1964) The processes involved in the lime stabilization of clay soils. In: Proceedings of 2nd conference on Australian Road Research Board, 2, pp 1169–1203
Dan Marks B, Allan Haliburton T (1972) Acceleration of lime-clay reactions with salt. J Soil Mech Found Div Proc ASCE 98(SM4):327–339
Dash SK, Hussain M (2012) Lime stabilisation of soils: reappraisal. J Mater Civil Eng 24(6):707–714. https://doi.org/10.1061/(ASCE)MT.1943-5533.0000431
Diamond S, Kinter EB (1965) Mechanisms of soil-lime stabilization: an interpretive review. Highway Research Record No 92, Highway Research Board, National Research Council, Washington, D. C., 83–102
Dogra RN, Uppal IS, Vasudeva, (1960) Strengthening the subgrades and flexible bases of roads by the use of hydrated lime. J Indian Road Congr 25:83–115
Dondi M, Mazzanti F, Principi P, Raimondo M, Zanarini G (2004) Thermal conductivity of clay bricks. J Mater Civ Eng 16(3):287
Eades JL, Grim RE (1960) Reaction of hydrated lime with pure clay minerals in soil stabilization. Bulletin 262, Highway Research Board, Washington, DC
Fitzmaurice RF (1958) Manual on stabilised soil construction for housing. U. N. Technical Assistance Programme, New York
Gabriele T, Orazi M, Orazi S (2016) Effect of freeze-thaw cycles on mechanical behaviour of lime-stabilized soil. J Mater Civil Eng 28(6). v10.1061/(ASCE)MT.1943-5533.0001509
Gidigasu MD (1976) Laterite soil engineering. Elsevier Publishing Company, Amsterdam, The Netherlands
Gourav K (2015) Studies on flexural behaviour of fly ash-lime-gypsum brick masonry. PhD thesis, Department of Civil Engineering, Indian Institute of Science, Bangalore, India
Guettala A, Abibsi A, Houari H (2006) Durability study of stabilised earth concrete under both laboratory and climatic conditions exposure. Constr Build Mater 20:119–127

Gumaste KS, Nanjunda Rao KS, Venkatarama Reddy BV, Jagadish KS (2007) Strength and elasticity of brick masonry prisms and wallettes under compression. Mater Struct 40(296):241–253
Gupta A (2003). Studies on characteristics of cement-soil mortars and soil-cement block masonry. MSc (Engg.) thesis, Department of Civil Engineering, Indian Institute of Science, Bangalore, India
HB 195 (2002) The Australian Earth building handbook by Peter Walker and Standards Australia
Heathcote KA (1995) Durability of earth wall buildings. Constr Build Mater 9(3):185–189
Heathcote KA (2002) An investigation into the erodibility of earth wall units. PhD thesis, University of Technology, Sydney Australia
Hendry AW (1998) Structural masonry. MacMillan Press Ltd., London
Herrin M, Mitchell H (1961) Lime soil mixtures. Highway Research Board, National research Council, Washington D.C., USA, Bulletin No 304
Houben H, Guillaud H (2003) Earth construction—a comprehensive guide. Intermediate Technology Publications, London UK
Ingles OG, Metcalf JB (1972) Soil stabilisation principles and practice. Butterworth's publisher, Australia
Inzemmouren O, Guettala A, Guettala S (2015) Mechanical properties and durability of lime and natural pozzolana stabilized steam-cured compressed earth block bricks. Geotech Geol Eng 33:1321–1333
IS: 3495 – 1992 (2002) Methods of test of burnt clay building bricks – Part II: determination of water absorption Bureau of Indian Standards, New Delhi, India (Reaffirmed)
IS: 1077 – 1992 (2007) Common burnt clay building bricks—specification. Bureau of Indian Standards, New Delhi, India (reaffirmed)
IS 1725 (2013) Stabilized soil blocks used in general building construction—specification (2nd revision). Bureau of Indian Standards, New Delhi, India
IS 3792 – 1978, Guide for heat insulation of non-industrial buildings. Bureau of Indian Standards, New Delhi, India
Jagadish KS, Venkatarama Reddy BV, Nanjunda Rao KS (2017) Alternative building materials and technologies New-Age International, New Delhi, India
Kenai S, Bahar R, Benazzoug, (2006) Experimental analysis of the effect of some compaction methods on mechanical properties and durability of cement stabilized soil. J Mater Sci 41:6956–6964
Kumar PP (2009) Stabilised rammed Earth for walls: materials, compressive strength and elastic properties. PhD thesis, Department of Civil Engineering, Indian Institute of Science, Bangalore, India
Lal R (2005) Characteristics of soil-cement blocks and soil-cement block masonry. MSc (Engg) thesis, Department of Civil Engineering, Indian Institute of Science, Bangalore, India
Lambe TW (1962) Soil stabilization. Foundation Engineering, Edited by Leonards GA. McGraw Hill Book Co. Inc., New York, pp 351–437
Latha MS (2015) Studies on characteristics of stabilised soil compacts for structural applications. PhD thesis, Department of Civil Engineering, Indian Institute of Science, Bangalore, India
Lunt MG (1980) Stabilized soil blocks for buildings Overseas Building Notes, No 184
Mateos M, Davidson TD (1961) Steam hardening of lime, lime and fly ash, and cement treated soil. In: Presented at 73rd session of the Iowa Academy of Sciences, Indianola, Iowa, April 14–15
Mateos M (1964) Soil-lime research at Iowa State University. J Soil Mech Found Div Proc ASCE 90(SM2):127–153
Middleton GF (revised by Schneider LM) (1992) Earth-wall construction. Bulletin 5, 4th edn. Commonwealth Scientific and Industrial Research Organization, Sydney, Australia
Mitra JN (1951) Suitability of soil for stabilised soil houses for Rangawan dam colony. Indian Concr J 15:234–238
Mitchell JK (1981) Soil improvement—state of the art report, pp 509–565. https://www.issmge.org/publications/online-library

Odell RT, Thornburn TH, McKenzie L (1960) Relationships of Atterberg limits to some other properties of Illinoi's soils. Proc Soil Sci Soc Am 24(5):297–300
Ogunye FO, Boussabaine H (2002a) Diagnosis of assessment methods for weatherability of stabilised compressed soil blocks. Constr Build Mater 16:163–172
Ogunye FO, Boussabaine H (2002b) Development of a rainfall test rig as an aid in soil block weathering assessment. Constr Build Mater 16:173–180
Olivier M, Ali M (1987) Influence of different parameters on the resistance of earth, used as a building material. In: Proceedings of international conference on mud architecture, Trivandrum, India
Oti JE, Kinuthia JM, Bai J (2009a) Compressive strength and microstructural analysis of unfired clay masonry bricks. Eng Geol 109:230–240
Oti JE, Kinuthia JM, Bai J (2009b) Engineering properties of unfired clay masonry bricks. Eng Geol 109:130–139
Oti JE, Kinuthia JM, Bai J (2008a) Developing unfired stabilised building materials in the UK. Proc ICE J Eng Sustain 161(4):211–218. https://doi.org/10.1680/ensu.2008.161.4.211
Oti JE, Kinuthia JM, Bai J (2008b) Using slag for unfired-clay masonry bricks. Proc ICE J Constr Mater 161(4):147–155. https://doi.org/10.1680/coma.2008.161.4.147
Oti JE, Kinuthia JM (2012) Stabilised unfired clay bricks for environmental and sustainable use. Appl Clay Sci 58:52–59. https://doi.org/10.1016/j.clay.2012.01.011
Oti JE, Kinuthia JM, Bai J (2010) Freeze–thaw of stabilised clay brick. Proc Inst Civil Eng Waste Resour Manag 163(3):129–135. https://doi.org/10.1680/warm.2010.163.3.129
Portland Cement Association (1956) Soil-cement construction handbook, Chicago, Illinois, USA. https://archive.org/details/buildingtechnologyheritagelibrary
Ranganatham BV, Pandian NS (1971) Strength gain in thermally cured lime stabilised clays. In: Proceedings of 4th international conference on soil mechanics, Budapest, pp 263–272
Rao SM, Shivananda P (2005) Role of curing temperature in progress of lime-soil reactions. Geotech Geol Eng 23:79–85
Reddy BVV (2000) Development and dissemination of lime based building blocks. A project report, Karnataka State Council for Science and Technology, Indian Institute of Science, Bangalore, India
Reddy BVV (2002) Long-term strength and durability of stabilised mud blocks. In: Proceedings of 3rd international conference on non-conventional materials and technologies. Construction Publishing House, 12–13 March, Hanoi, Vietnam, pp 422–431
Reddy BVV (1983) On the technology of pressed soil blocks for wall construction MSc (Engg) thesis, Department of Civil Engineering, Indian Institute of Science, Bangalore, India
Reddy BVV (1991) Studies on static soil compaction and compacted soil-cement blocks for walls. PhD thesis, Department of Civil Engineering, Indian Institute of Science, Bangalore, India
Reddy BVV, Timothy Williams and Peter Walker (2003) Durability of stabilised mud block buildings in Southern India. In: Proceedings of 9th international conference on study and conservation of earthen architecture terra 2003, 29 Nov–2 Dec, Yazd, Iran, pp 492–503
Reddy BVV, Jagadish KS (1987) Spray erosion studies on pressed soil blocks. Build Environ 22(2):135–140
Reddy BVV, Jagadish KS (1984) Pressed soil lime blocks for building construction. Mason Int 3:10–16
Reddy BVV, Kumar PP (2011) Cement stabilised rammed Earth—Part A: compaction characteristics and physical properties of compacted cement stabilised soils. Mater Struct 44(3):681–694
Reddy BVV, Gupta A (2006) Tensile bond strength of soil cement block masonry couplets using cement-soil mortars. J Mater Civ Eng 18(1):36–45
Reddy BVV, Gupta A (2005) Characteristics of soil-cement blocks using highly sandy soils. Mater Struct 38(280):651–658
Reddy BVV, Jagadish KS (1989) Properties of soil-cement block masonry. Mason Int 3(2):80–84
Reddy BVV, Jagadish KS (1995) Influence of soil composition on the strength and durability of soil-cement blocks. Indian Concr J 69(9):517–524

Reddy BVV, Lokras SS (1998) Steam-cured stabilised soil blocks for masonry construction. Energy Build 29:29–33
Reddy BVV, Hubli SR (2002) Properties of lime stabilised steam-cured blocks for masonry. Mater Struct 35:293–300
Reddy BVV, Latha MS (2014a) Retrieving clay minerals from stabilised soil compacts. Appl Clay Sci 101:362–368
Reddy BVV, Latha MS (2014b) Influence of soil grading on the characteristics of cement stabilised soil compacts. Mater Struct 47(10):1633–1645
Reddy BVV, Lal R, Nanjunda Rao KS (2007) Optimum soil grading for the soil-cement blocks. J Mater Civ Eng 19(2):139–148
Reddy BVV, Walker P (2005) Stabilised mud blocks: problems, prospects. In: Proceedings of international Earth building conference—Earthbuild 2005, 19–21 January, Sydney, Australia, pp 63–75
Reddy BVV, Uday Vyas CV (2008) Influence of shear bond strength on compressive strength and stress strain characteristics of masonry. Mater Struct 41(10):1697–1712
Saha HL, Ray P (1967) Stabilisation of the soils of deltaic region in the Sundarbans area West Bengal. J Soil Mech Found Eng 6(2):167–186
Sarangapani G, Venkatarama Reddy BV, Jagadish KS (2005) Brick-mortar bond and masonry compressive strength. J Mater Civ Eng 17(2):229–237
Sherwood PT (1993) Soil stabilization with cement and lime: state-of-the-art review. Transport Research Laboratory, Her Majesty's Stationery Office, London UK
Spence RJS (1975) Predicting the performance of soil-cement as a building material in tropical countries. Build Sci 10:155–159
Spence RJS, Cook DJ (1983) Building materials in developing countries. Wiley, Brisbane, Australia
Sridharan A, Rao SM, Satyanarayana Murthy N (1986) A rapid method to identify clay type in soils by the free swell technique. ASTM Geotech Test J 9(4):198–203
Tennant AG, Foster CD, Venkatarama Reddy BV (2016) Detailed experimental review of flexural behavior of cement stabilized soil block masonry. J Mater Civil Eng 28(6):060160004-1 to 5
Thompson MR (1966) Lime reactivity of Illinois soils. J Soil Mech Found Div ASCE 92(SM5):67–92
Ullas SN (2014) Studies on utilisation of iron ore tailings as fine aggregate in mortar and concrete. PhD thesis, Centre for Sustainable Technologies, Indian Institute of Science, Bangalore, India
Ullas SN, Reddy BVV (2007) Characteristics of soil-cement blocks from different construction sites. In: Proceedings of international symposium on earthen structures, Interline Publishers, 22–24 August, Bangalore, India, pp 141–146
Uppal IS, Kapur BP (1957) Role of detrimental salts in soil stabilisation with and without cement-3: effect of magnesium sulphate. Indian Concr J 31:228–231
Venumadhava Rao K, Venkatarama Reddy BV, Jagadish KS (1996) Flexural bond strength of masonry using various blocks and mortars. Mater Struct 29(2):119–124
Vyas UV (2007) Studies on shear bond strength—masonry compressive strength relationships and finite element model for prediction of masonry compressive strength. MSc (Engg) thesis, Department of Civil Engineering, Indian Institute of Science, Bangalore, India
Walker P (1998) Erosion testing of compressed earth blocks. In: Proceedings of 5th international masonry conference, London, U.K., 264–268
Walker P (1999) Bond characteristics of Earth block masonry. J Mater Civ Eng 11(3):249–256
Walker P (2004) Strength and erosion characteristics of Earth blocks and Earth block masonry. J Mater Civ Eng 16(5):497–506
Walker PJ (1995) Strength, durability and shrinkage characteristics of cement stabilised soil blocks. Cement Concr Compos 17:301–310
Walker P, Stace T (1997) Properties of some cement stabilized compressed earth blocks and mortars. Mater Struct 30:545–551

Weisz A, Kobe A, McManus K, Nataatmadja A (1995) Durability of mud brick—comparison of three test methods. In: Proceedings of 4th Australasian masonry conference, November, University of Technology, Sydney

Yong RN, Warkentin BP (1975) Soil properties and behaviour. Elsevier Scientific Publishing Co., New York

Yttrup PJ, Diviny K, Sottile F (1981) Development of drip test for the erodibility of mud bricks. Deakin University, Geelong, Australia

Chapter 6
Mortars for Stabilised Compressed Earth Block Masonry

6.1 Introduction

The masonry is an assemblage of the masonry units (such as brick or block) and the mortar. The mortar plays an important role in the masonry construction, by stitching the masonry units into an assemblage and preventing the moisture ingress through the joints. Also, the fresh mortar can take care of the dimensional imperfections of the masonry units while laying the masonry units over the fresh mortar bed joint. The mortar has significant influence on the flexure and the shear strength of the masonry.

The mortars are prepared at the construction site either by using the binder, fine aggregate and water or from the ready mixed mortar, which is a dry mixture consisting of the mortar ingredients. In India, generally, the mortar is prepared at the construction site using the ingredients. The mortar preparation involves intimately mixing the cementitious material, fine aggregate and water to achieve a desired consistency. The standard codes of practices give specifications for the preparation and use of the mortars. The properties of the fresh as well as the hardened mortar influence the masonry characteristics. The workability and the water retentivity are the two important properties of the fresh mortar, while the strength, the development of bond with the masonry unit and the stress–strain characteristics represent the hardened mortar characteristics.

The bond strength development in the stabilised compressed earth block (CEB) masonry depends upon the surface characteristics of the stabilised CEB. The stabilised CEB masonry using the composite mortars performs better than the CEB masonry with pure cement mortar. Instead of conventional composite mortars such as cement–lime mortar, the cement–soil mortars have been successfully used for the stabilised CEB masonry construction (Gupta 2003; Lal 2005; Walker 2004; Reddy et al. 2007; Reddy 2012; Jagadish et al. 2017; Reddy et al. 2019). The following sections reveal more information on the cement–soil mortar as well as the conventional mortars for the stabilised CEB masonry.

B. V. V. Reddy, *Compressed Earth Block & Rammed Earth Structures*,
Springer Transactions in Civil and Environmental Engineering,
https://doi.org/10.1007/978-981-16-7877-6_6

6.2 Mortar Constituents

Apart from the water, the main ingredients of the mortar are cementitious materials (called binders) and inert fine aggregates. Different types of cementitious materials used in the mortar preparation are:

(a) Portland cement
(b) Lime
(c) Lime-pozzolana cements
(d) Masonry cement

The cementitious materials impart strength to the mortar, provide workability and facilitate bond development between the mortar and the masonry unit.

Lime: The lime stone is the main source for lime. The lime stone is basically $CaCO_3$, and it may have some impurities. The burnt lime stone is called quick lime (CaO) and upon hydration it becomes the hydrated lime [$Ca(OH)_2$]. There are two types of lime: hydraulic lime and non-hydraulic lime. The hydraulic lime contains pozzolanic materials in addition to $Ca(OH)_2$. In the presence of water, the hydraulic lime reacts with the pozzolanic material, resulting in the formation of the cementitious materials, which imparts strength to the mortar. The traditional lime mortars are prepared by the addition of pozzolanic materials such as powdered burnt clay brick and ceramic tile pieces, certain types of ashes. The non-hydraulic lime, also called the fat lime, is basically $Ca(OH)_2$. The lime mortars using fat lime derive strength due to the process of carbonation. The carbonation is a very slow process, and it can extend over to several years.

Portland Cement: There are different types of Portland cements. Ordinary Portland cement (OPC), slag cement and Portland pozzolana cement (PPC) represent some of the commonly used cements for the mortar preparation. The OPC comprises of lime, alumina/silica with certain admixtures. Generally, the OPC has initial setting time of 30–45 min, and hence, pure cement mortars containing OPC alone should be consumed quickly upon mixing with the water. The strength of the mortar using the Portland cement mainly depends upon the aggregate–cement ratio and water–cement ratio.

Masonry cement: The masonry cement contains OPC (~75%), inert mineral filler such as powdered calcium carbonate and the air entraining agent (BS EN 413 and IS 3466-1988 1999). The mortars with masonry cement have better workability. The masonry cement is supposed to give a mortar that combines the desirable properties of the lime mortar and the pure cement mortar.

Lime-pozzolana (LP) cement: The LP cement is a mixture of lime and pozzolanic materials. The pozzolana reacts with $Ca(OH)_2$ at ambient temperature in the presence of water. The reaction products such as calcium silicate hydrate, calcium aluminate hydrate are responsible for the strength gain in the LP mortars. Burnt clay, fly ash, blast furnace slag and rice husk ash represent some of the pozzolanic materials. The strength of LP cement mainly depends upon the reactivity of the pozzolana. The rate of strength gain is slow in LP cement mortars cured at ambient temperatures. The 28 day cube compressive strength of LP mortars will be in the range of 1–6 MPa, depending upon the quality of pozzolana used (Yogananda 1983; Yogananda and Jagadish 1998; Jagadish 2015).

6.3 Fine Aggregate for Mortars

Both the natural river sand and the manufactured sand can be used for the mortars. Several codes of practices (IS 2116 (2002), ASTM C144-11 (2011) and BS EN 13139 (2013) and other codes) give specifications for the fine aggregate to be used in masonry mortars. The Indian code states that the sand should be clean, hard and should not contain organic matter and too much of fines such as silt and clay. The fine aggregate should be free from deleterious materials such as alkalis and salts. The grading limits specified in IS 2116 (2002) and ASTM C144-11 (2011) are shown in Fig. 6.1. The IS 2116 (2002) code gives common grading limits for both the natural and manufactured sand, whereas the ASTM C144-11 (2011) specifies different grain size ranges for the natural and the manufactured sand. Figure 6.1 shows the upper and lower bound limits for both the types of fine aggregates. The Indian code specifies maximum size of the sand particle to be 1/3 to 1/2 the joint thickness. Very large particles affect the spreading and filling of mortar in the bed joints and can lead to stress concentration in the masonry bed joint.

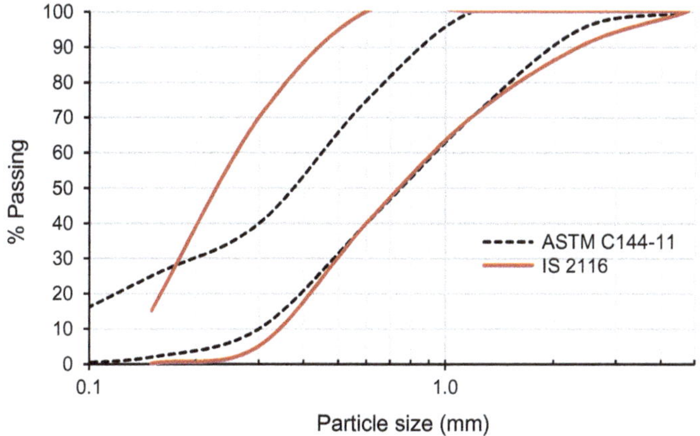

Fig. 6.1 Limiting grading curves for fine aggregates used in the mortars

6.4 Types of Mortars

The masonry mortars can be classified based on their composition and constituent materials as given Table 6.1. The Indian masonry codes (NBC 2016, IS 1905-1987 1998) specify seven grades of mortars for the masonry construction, based on the strength and the composition. The mortars are designated as H1, H2, M1, M2, M3, L1 and L2. The mortars L1 and L2 are lime-pozzolana mortars, whereas the other mortars are either cement–lime or cement–sand mortars. BS EN 1996-1-1 (2012) code specifications and requirements for the masonry mortar are given in Table 6.2. This table gives range of proportions for mortars and the compressive strength. There are guidelines for selecting the mortar based on the workability and the strength. The information provided in this table is more explicit in terms of strength, ability of the mortar to accommodate movements, workability, strength, etc. The ASTM code (ASTM C 270-10) gives specifications and requirements for mortars. This code identifies four types of mortars designated as M, S, N and O for different types of binders such as cement–lime, mortar-cement and masonry cement. For cement–lime binder, Table 6.3 gives proportions and properties for the four types of mortars. The code gives proportions and properties for mortars using other binders such as mortar-cement and masonry cement.

The composite mortars such as cement–lime mortars are the most commonly used for the masonry construction. In India, the cement–sand mortar is the most commonly used and that too using OPC binder. Such mortars possess higher strength and modulus (i.e. the mortar stiffness is high). In India, the modulus of burnt clay bricks is in the range of 500–8000 MPa as against the modulus of the commonly used cement–sand mortars in the range of 4000–10,000 MPa (Gumaste 2004; Saranagapani et al. 2005; Gumaste et al. 2007; Reddy 2012; Jagadish 2015; Jagadish et al. 2017). Another major disadvantage in using the cement–sand mortar is the development of low bond strength for the masonry, when compared with the masonry bond strength using composite mortars (Venumadhava et al. 1996, Reddy and Gupta 2005, Sarangapani et al. 2005).

The composite mortars have binders such as cement and lime. The cement–soil mortars are used for the construction of the stabilised CEB masonry. In such composite mortars, the soil replaces the lime. The systematic studies on the characteristics of cement–soil mortars can be found in the investigations of Gupta (2003), and Reddy and Gupta (2005).

Table 6.1 Mortar types

Mortar type	Main composition
Lime mortar	Lime and sand
Cement mortar	Cement and sand
Cement–lime mortar	Cement, lime and sand
Lime-pozzolana mortar	LP cement and sand
Cement–soil mortar	Cement, soil and sand

Table 6.2 Requirements for mortar (BS EN 1996-1-1:2005+A1:2012)

	Mortar designation	Compressive strength class	Type of mortar (proportion of materials by volume)				Compressive strength at 28 days (N/mm²)
			Cement:lime:sand with or without air entrainment	Cement:sand with or without air entrainment	*Masonry cement:sand	**Masonry cement:sand	
↓ Increasing ability to accommodate movement, e.g. due to settlement, temperature and moisture changes	(i)	M12	1:0–¼:3	–	–	–	12
	(ii)	M6	1:½:4–4½	1:3–4	1:2½–3½	1:3	6
	(iii)	M4	1:1:5–6	1:5–6	1:4–5	1:3½–4	4
	(iv)	M2	1:2:8–9	1:7–8	1:5½–6½	1:4½	2

* Masonry cement with inorganic filler other than lime; ** Masonry cement with lime

Table 6.3 ASTM C270-10 code specifications, proportions and properties for cement–lime binder

Mortar	Type	Proportion by volume			28 day compressive strength (MPa)	Water retentivity (%)
		Cement*	lime	Aggregate ratio		
Cement–lime	M	1	¼	Not <2¼ and not >3½ times the sum of the separate volumes of cementitious materials	17.2	75
	N	1	>¼–½		12.4	75
	S	1	>½–1¼		5.2	75
	O	1	>1¼–2½		2.4	75

* Portland cement or blended cement

6.5 Mortar Characteristics

The characteristics of the mortar should be examined in the fresh as well as in the hardened state. The fresh mortar properties include workability and water retentivity. The strength, shrinkage, development of bond with the masonry unit and the stress–strain characteristics represent some of the important properties of the hardened mortar.

6.5.1 Workability of Mortar

The workability or consistency of the fresh mortar should be such that it facilitates the mason to spread the mortar in the bed joints and fill the perpend joints easily, and the mortar adheres well to the masonry units. The mortar composition, water–cement ratio and the admixtures influence the workability of the fresh mortar. The workability of the fresh mortar can be measured and quantified. The following types of tests have been specified in different codes of practices.

1. Dropping ball test (BS 4551-2005)
2. Cone impression test (IS 2250)
3. Slump test (AS 1289)
4. Flow test (BS EN 1015-3 or ASTM C1437-07)

The dropping ball test and the cone impression tests generally do not yield consistent results. The penetration of the ball and the cone is affected when the water–binder ratio of the mortar exceeds 1.1 and especially in the case of cement–sand mortars (Gupta 2003). The flow table test specified in BS EN 1015-3 (1999) can be used to quantify the workability of the fresh mortar in terms of a flow value. Figure 6.2 shows the flow table. It has a table top plate mounted on a cam and a truncated metal cone. The flow test procedure involves vibrating the truncated cone of fresh mortar on the flow table, causing spread of the mortar. The percentage increase in the spread

Fig. 6.2 Flow table

diameter of the fresh mortar with reference to the initial diameter is expressed as a flow value for the mortar.

Figure 6.3 shows typical flow patterns on the flow table for 0, 50, 80, 100 and 150% flow. Generally, the relationship between the mortar flow and the water–binder ratio is linear. Figure 6.4 shows the flow curves for seven different types of mortar mixes such as cement–sand, cement–lime–sand and cement–soil–sand. The flow versus water–binder ratio relationships are linear and the relationships show that the mortar flow increases with the increase in the water–binder ratio irrespective of the type of the mortar. Table 6.4 gives water–binder ratio needed to achieve 100% flow for the seven different types of mortar mixes. The cement–soil mortars demand higher water–binder ratio than cement–lime mortars to achieve a similar flow value. Also, the binder rich mortars need lower water–binder ratio than the lean mortars to achieve a similar flow. The lean cement–sand (1:6, cement:sand, by volume) mortar shows poor workability and the maximum flow achievable will be about 100%. Attempt to improve the flow beyond 100% by increasing the water–binder ratio results in

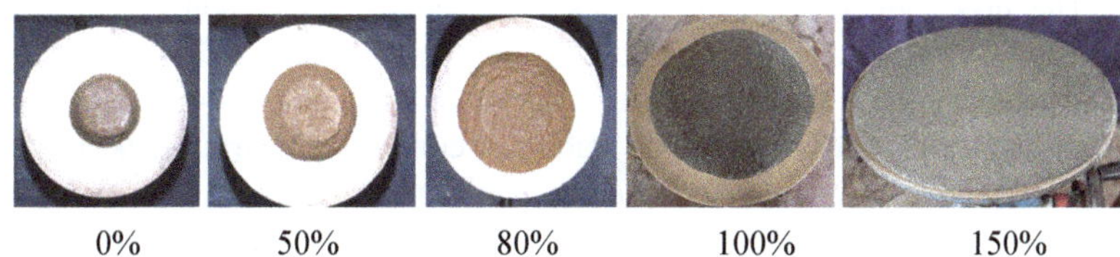

Fig. 6.3 Typical flow patterns for mortar

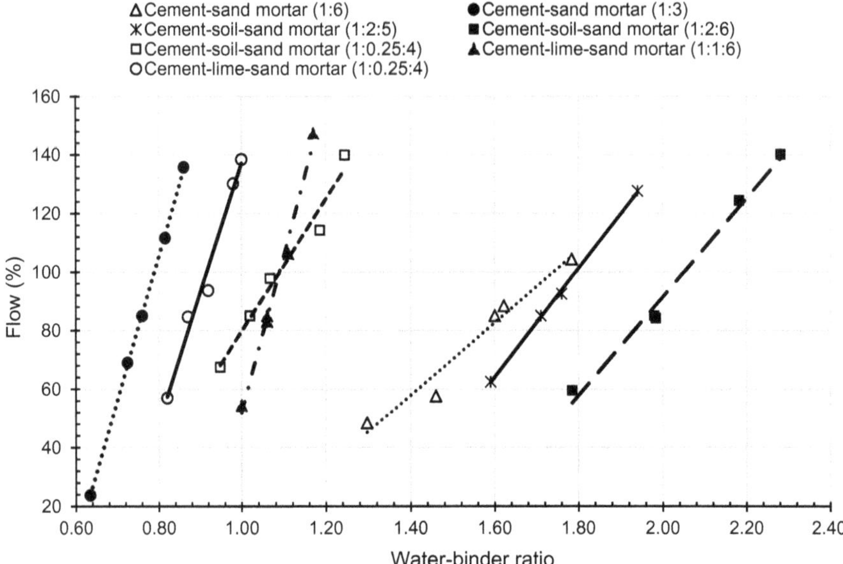

Fig. 6.4 Influence of water–binder ratio on the flow of mortar

Table 6.4 Typical water–binder ratio for mortars to achieve 100% flow

Type of mortar	Mix proportion (by volume)				Water–binder ratio to achieve 100% flow
	Cement	Lime	Soil	Sand	
Cement mortar	1	–	–	3	0.79
	1	–	–	6	1.74
Cement–lime mortar	1	0.25	–	4	0.93
	1	1	–	6	1.10
Cement–soil mortar	1	–	0.25	4	1.09
	1	–	2	5	1.80
	1	–	2	6	2.06

segregation. For the lean composite mortars such as cement–soil and cement–lime mortars, a flow as high as 150% or more is achievable, without segregation.

The flow values of fresh mortars collected from different construction sites, measured by Reddy and Gupta (2005), are given in Table 6.5. The table gives the flow values of the fresh cement–sand mortar and the cement–soil mortar used in the burnt clay brick masonry and the stabilised compressed earth block masonry. The flow value of the fresh mortars used by a mason was in the range of 86 and 119%. The flow in the vicinity of 100% seems to be favourable for the masons to construct the stabilised compressed earth block masonry walls. The masons use mortar with >100% flow for filling the perpend vertical joints.

Table 6.5 Flow values for fresh mortars collected from different construction sites (Reddy and Gupta 2005)

Construction Site	Mortar proportion (by volume)			Flow (%)	Type of masonry unit used for walls	Mortar used for
	Cement	Soil	Sand			
A	1	–	6	101	Fired clay brick	Perpend joint
A	1	–	6	86	Fired clay brick	Horizontal bed joint
B	1	2	6	115	Stabilised earth block	Horizontal bed joint
C	1	3	5	119	Stabilised earth block	Horizontal bed joint
D	1	2	6	113	Stabilised earth block	Horizontal bed joint
E	1	2	6	103	Stabilised earth block	Horizontal bed joint

6.5.2 Compressive Strength of the Mortar

The mortar composition, the water–binder ratio and the particle grading of aggregates mainly influence the compressive strength of the mortar. Generally, the mortar compressive strength is determined by testing 50 mm cube (IS 2250 1995, ASTM C109/C109M-11a 2011). The commonly used mix proportions for the construction of the stabilised compressed earth block masonry in India and elsewhere are given in Table 6.6. Also, the water–binder ratio required to achieve 100% flow and the typical cube compressive strength values are given in the Table. At 100% flow, the compressive strength of the cement–sand mortar, cement–lime mortar and cement–soil mortar (with 14–17% cement by volume) is about 5.5 MPa. With leaner mortar mixes (10% cement by volume) at 100% flow, the compressive strength of the mortar will be in the range of 3–3.5 MPa. For the mix proportions given in Table 6.6, addition of soil to the cement–sand mortar or replacing the lime with soil in the cement–lime mortar did not alter the mortar compressive strength much.

The most commonly used mortars for masonry construction using burnt clay bricks and stabilised compressed earth blocks generally have cement to aggregate (sand or sand + soil or sand + lime) ratio of 1:5–7 (by volume). Generally, such mortars with a flow value of about 100%, possess a compressive strength in the range of 4–6 MPa.

Table 6.6 Characteristics of masonry mortars (Reddy and Gupta 2005)

Mortar proportion (by volume)				Flow (%)	Water–cement ratio	Dry density (kg/m^3)	Water retentivity (%)	Compressive strength (MPa)	Drying shrinkage (%)
Cement	Soil	lime	sand						
1	0	0	6	100	1.65	1900	74	5.40	0.074
1	0	1	6	100	1.79	1900	82	5.94	0.057
1	2	0	5	100	1.70	1850	88	5.40	0.324
1	2	0	6	100	2.05	1825	87	4.05	0.351

6.5.3 *Water Retentivity*

While constructing the masonry, the fresh mortar gets sandwiched between the bricks/blocks. The water (in the fresh mortar joint) is enriched with cement hydration products and is absorbed/sucked by the masonry units. The amount of water absorbed (through suction) by the masonry unit from the fresh mortar bed joint depends upon the surface porosity and the moisture content of the brick/block at the time of construction and the ability of the mortar to retain water against masonry unit's suction. The water retentivity is defined as the ability of the mortar to retain the water against the masonry unit's suction. The mortar has cementitious materials, thus initially it requires certain amount of water for the proper hydration and the strength development. Sarangapani (1998) examined the moisture loss from the fresh mortar bed joint, when it is sandwiched between the two burnt clay bricks. The study showed that for 1:6 (cement:sand, by volume) cement–sand mortar, the water–cement ratio in the mortar reduced from 1.4 to 0.17 after one hour, when the fresh mortar bed joint was sandwiched between the dry burnt clay bricks. In the case of 1:1:6 (cement:lime:sand, by volume) cement–lime mortar, the water–cement ratio reduced to 0.62 from 1.5. The cement–lime mortar was superior to cement–sand mortar in retaining the water in the fresh mortar bed joint.

If the water loss from the mortar bed joint is large, it can lead to lower water–cement ratio and improper hydration of the cement in the fresh mortar in the bed joint, thereby affecting the mortar characteristics and the bond development. The water retentivity of the mortar depends upon many factors such as the mix proportion, water–cement ratio, type of cementitious binder, etc. The standard codes of practices such as IS 2250 (1995), ASTM C91-95 (1995) and other codes give procedure for determining the water retentivity of the mortar.

The water retentivity values for a set of lean mortars are given in Table 6.6. The results clearly indicate that composite mortars such as cement–soil and cement–lime mortars show higher water retentivity than the water retentivity of the cement–sand mortar. For example, water retentivity of 1:6 cement–sand mortar is 74%, whereas the cement–soil mortars show water retentivity in excess of 85%. The composite mortars containing soil will have clay minerals present in the mortar mix. The clay minerals can hold moisture and resist brick suction. This may be the reason for such mortars to show higher water retentivity values.

Another simple procedure can be used for estimating the water loss from the fresh mortar bed joints. Prepare a two-brick prism by sandwiching the fresh mortar bed joint between the bricks. Now, the water from the fresh mortar joint is absorbed (through suction) by the bricks. After one hour, remove the top brick and scoop out a portion of the mortar and estimate the amount of water present in the mortar. By knowing the initial water content of the mortar, the water loss from the fresh mortar bed joint can be calculated. This water loss can be minimised by using partially saturated brick at the time of construction. Sarangapani (1998) conducted such tests on mortars using both dry as well as partially saturated burnt clay bricks. The water–cement (W/C) ratio of the mortars drastically reduced when the dry brick was used. For cement–sand mortar, water–cement ratio dropped to a level of 0.15, whereas when partially saturated brick was used, W/C ratio was in the range of 0.56–0.80. The variations in the water–binder ratios of the mortars with the contact duration of cement stabilised compressed earth brick using different types of mortars (cement-mortar, cement–lime mortar and cement–soil mortar) are discussed in Chap. 7.

6.5.4 Drying Shrinkage of Mortars

The shrinkage that takes place during hardening process of the mortar can be designated as drying shrinkage. A part of the drying shrinkage is recovered upon immersion of the set mortar in the water. With time, the rate of drying and the drying shrinkage decreases. The drying shrinkage of the mortar depends upon several factors such as water–cement ratio, cement content, type of sand/grading, clay content of the soil and the curing period. The drying shrinkage of the mortar can cause shrinkage cracks within as well as at the unit–mortar interface, and it can also result in impaired bond between the masonry unit and the mortar (Baker 1979). The drying shrinkage of the mortar can be measured in the laboratory through mortar prisms in unrestrained conditions. It differs from that experienced in a masonry wall where the drying shrinkage of the mortar is influenced by the restraint offered by the masonry units, the unit absorption characteristics, the thickness of the mortar bed joint, etc. Thus, the drying shrinkage value of the mortar examined in the laboratory is more useful for the purposes of comparison rather than as the absolute shrinkage value.

The ASTM C1148-92a (1992) code gives the procedure for determining the drying shrinkage for the mortars. The drying shrinkage of mortars at 25 days of drying duration is taken as the ultimate drying shrinkage as per the ASTM code. Typical drying shrinkage values for the mortars, determined as per the ASTM code guidelines, are given in Table 6.6. The cement–lime mortar shows lower drying shrinkage when compared to the shrinkage of the cement–sand mortar and the cement–soil mortar. The cement–soil mortars show four times higher drying shrinkage, when compared to the shrinkage of the cement–lime mortar. Thus, in the case of the cement–soil mortars the drying shrinkage is very sensitive to the quantity of the clay mineral present in the mortar mix. The Gupta's (2003) studies reveal more comprehensive information on the drying shrinkage of mortars, especially cement–soil mortars. The cement–soil mortars such as 1:2:5 and 1:2:6 (cement:soil:sand, by volume) are commonly used for the construction of stabilised compressed earth block masonry walls. Even though these mortars show higher drying shrinkage in a laboratory test, there are no serious problems associated with the higher drying shrinkage in the field.

Reddy and Latha (2018) examined the influence of the mortar shrinkage on the flexure bond strength of the stabilised compressed earth brick masonry. The cement stabilised compressed earth brick and the cement–lime mortar (CLM) of proportion 1:1:6 (cement:lime:sand, by volume) were used in these investigations. The drying shrinkage of the CLM mortar was varied by varying the water–binder ratio and hence

Table 6.7 Mortar drying shrinkage and masonry flexural bond strength

Water–binder ratio	Mortar drying shrinkage (%)	Flexure bond strength of masonry (MPa)
1.55	0.06	0.28
1.58	0.07	0.37
1.63	0.09	0.43
1.65	0.10	0.50

the flow of the mortar. Table 6.7 gives the variation in the mortar drying shrinkage and the corresponding flexure bond strength of the cement stabilised compressed earth brick masonry prisms. As the drying shrinkage of the CLM mortar was varied between 0.06 and 0.10%, the masonry flexure bond strength varied between 0.28 and 0.50 MPa; i.e., 60% increase in the mortar shrinkage resulted in about 80% increase in the flexure bond strength. These results suggest that it is better to use high flow/high shrinkage value mortars, for achieving better bond strength for the cement stabilised CEB masonry.

6.5.5 Development of Bond (Adhesion) with the Brick or Block

Good bond between the mortar and the masonry unit is essential for the satisfactory masonry performance. The bond strength becomes significantly important when the masonry has to resist tensile and shear stresses. Several parameters pertaining to the bricks/blocks, the mortars and the construction practices influence the masonry bond strength. The surface characteristics of the masonry unit (pore size distribution, porosity), the moisture content of the unit at the time of construction, the absorption characteristics of the unit, the mortar composition are some of the important characteristics influencing the brick–mortar bond (Groot 1993). More detailed discussion on the bond development between the mortar and the stabilised compressed earth bricks/blocks can be found in Chap. 7.

The bond strength between the mortar and the brick or block can be measured by testing the masonry in either direct tension or flexural tension or shear. The details of the bond strength determination are discussed in Chap 7. The investigations of Venumadhava et al. (1996), Walker (1999, 2004), Sarangapani et al. (2002, 2005), Reddy and Gupta (2006) and Reddy et al. (2007), Reddy and Vyas (2008), and Reddy and Latha (2018) shed more light on the bond strength of burnt clay brick and the stabilised compressed earth block masonry. Table 6.8 gives results of the influence of the mortar type on the flexural bond strength of the cement stabilised compressed earth block masonry.

The lean cement–sand mortars lead to lower masonry bond strength. The cement–soil mortars give better bond strength for the cement stabilised compressed earth brick masonry, when compared with bond strength of the masonry using cement–sand mortars. Even the lean cement–soil mortar such as 1:1:6 (cement:soil:sand, by

Table 6.8 Flexural bond strength of cement stabilised compressed earth brick/block masonry using different types of mortars

Mortar type Cement:soil:lime:sand (by volume)	Flexural bond Strength (MPa)	Reference
1:0:0:4	0.23	Venumadhava et al. (1996), Reddy (2012)
1:0:0:6	0.10	
1:0:1:6	0.18	
1:1:0:6	0.17	
1:2:0:5	0.20	

volume) results in bond strength comparable with the bond strength using binder rich cement–sand mortar (1 cement:4 sand, by volume). The results demonstrate that characteristics of the cement–sand mortar can be easily improved by the addition of soil in smaller quantities. The cement–soil mortar instead of cement–sand mortar can be recommended for the stabilised compressed earth brick masonry.

6.5.6 *Stress–Strain Characteristics and Modulus of Mortars*

The masonry unit and the mortar in a masonry assemblage under compression experience different kinds of stresses. For the case of masonry under compression, the lateral stresses developed in the mortar and the masonry unit greatly depend upon the relative modulus (stiffness) of the mortar and the brick or the block. The modulus (stiffness) and the magnitude of displacements (strains) experienced by the mortar and the masonry unit can be quantified through the stress–strain relationships of the masonry materials. The stress–strain relationships for some of the commonly used mortars in stabilised CEB construction are shown in Fig. 6.5. The mix proportions, the mortar designation, water–binder ratio to achieve 85–100% flow and the modulus (initial tangent modulus) range are given in Table 6.9. The modulus of commonly used cement–soil and cement–lime mortars will be in the range of 4–6 GPa, whereas the cement–sand mortars will exhibit slightly higher modulus. The strain at peak stress will be in the range of 0.001–0.002. The Poisson's ratio of the lean cement–soil mortars will be in the range of 0.13–0.15.

6.6 Selection of Mortar

Several factors influence/control the mortar selection for the construction of any specific masonry type.

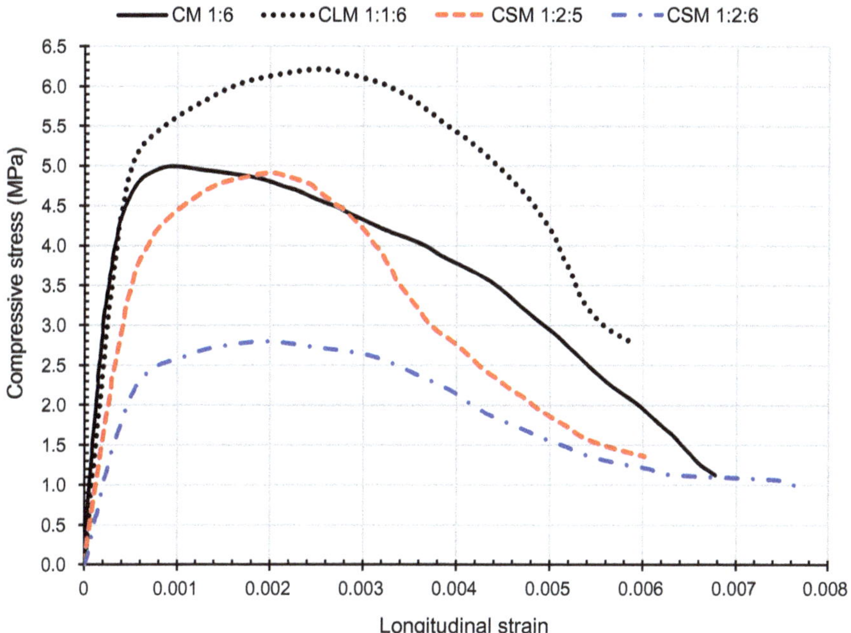

Fig. 6.5 Stress–strain curves for mortars

Table 6.9 Modulus and Poisson ratio for the mortars

Mortar type C:So:L:Sa (by volume)	Mortar designation	Water–binder ratio*	Initial tangent modulus for mortar flow 85–100% (GPa)	Poisson's ratio
1:0:0:6	CM 1:6	1.60–1.70	4.0–9.0	0.20
1:0:1:6	CLM 1:1:6	1.05–1.15	5.0–8.0	0.16
1:1/4:0:4	CSM 1:1/4:4	1.00–1.10	8.0–12.0	0.22
1:2:0:5	CSM 1:2:5	1.70–1.85	4.0–7.0	0.15
1:2:0:6	CSM 1:2:6	2.00–2.20	3.0–4.0	0.13

C: Cement; So: Soil; L: Lime; Sa: Sand; *Mortar flow: 85–100%

6.6.1 Masonry Type and the Strength of the Masonry Unit

The brick-work, stone masonry and the concrete block-work represent different types of masonry. It is preferable to select a mortar having modulus lower than that of the masonry unit for satisfactory performance of the masonry. The masonry with the lower modulus mortar will be able to accommodate movements due to settlement, temperature and moisture changes, better than the masonry with stiff mortar. The stone has very high modulus and will absorb very little water from the fresh mortar bed joint, and hence, for the stone masonry used in superstructure, richer mortars are preferable to develop satisfactory bond strength.

6.6.2 Masonry Application

The masonry is used for varieties of applications such as foundations, superstructure (walls), roofing panels, retaining structures, masonry shell structures (domes, vaults, etc.). Generally, for applications such as foundations, it is preferable to use lean and low modulus mortars which can accommodate movements due to settlements, etc. The superstructure is subjected to both the gravity and the lateral loads. Some of the walls may have to be designed to resist tensile stresses. Similarly, roof panels and masonry shells are subjected to tensile stresses. In such cases, the mortar has to be carefully selected to give maximum tensile resistance for the masonry.

6.6.3 Load Carrying Capacity of the Masonry

Based on the masonry strength required to resist the gravity loads, masonry unit strength and mortar strength combination gives the basic or limiting compressive stress. For example, the standard codes on masonry design give guidelines to choose basic or limiting compressive strength of the masonry based on a combination of the mortar strength and the masonry unit strength. There will be design procedures and data tables relating the masonry unit strength, geometric parameters of the masonry unit and the mortar type/strength to compressive strength of the masonry.

6.6.4 Moisture Penetration and Frost Resistance

Impervious mortar is essential to keep away the water entry into inside of the buildings and preventing leakage of water retaining masonry structures. Generally, the composite mortars are more impervious than the cement–sand mortars. The mortars using masonry cement and the plasticiser additives have better resistant to frost attack.

6.7 Mortars for Stabilised Compressed Earth Block Masonry

The stabilised compressed earth blocks are used for the construction of load bearing masonry in the walls, shell structures such as vaults and domes, water storage tanks and other structures. The shell structures demand higher bond strength. The cement–soil mortars have been used for stabilised compressed earth block masonry. For the loadbearing masonry walls of low-rise (up to four storeys) structures, cement–soil mortars of proportions 1:2:5 and 1:1:6 (cement:soil:sand, by volume) can be used. For the construction of unreinforced masonry vaults and domes, richer mortar mixes such as 1:½:4 or 1:¼:3 (cement:soil:sand, by volume) are recommended.

References

AS-1289 (1993) Methods of sampling and testing mortar for masonry construction. Standards Australia, Sydney

ASTM C 91–95 (1995) Standard specification for masonry cement. American Society for Testing Materials, West Conshohocken, USA

ASTM C109/C109M-11a (2011) Standard test method for compressive strength of hydraulic cement mortars (using 2-in or 50 mm cube specimens). American Society for Testing Materials, West Conshohocken, USA

ASTM C1148–92a (1992) Standard test methods for measuring the drying shrinkage of masonry mortar. American Society for Testing Materials, West Conshohocken, USA

ASTM C1437-07 (2007) Standard test method for flow of hydraulic cement mortar. American Society for Testing Materials, West Conshohocken, USA

ASTM C144–11 (2011) Standard specification for aggregate for masonry mortar. American Society for Testing Materials, West Conshohocken, USA

ASTM C270-10 (2010) Specifications and proportions and properties for cement–lime binder. American Society for Testing Materials, West Conshohocken, USA

Baker LR (1979) Some factors affecting the bond strength of brickwork. In: Proc. 5th international brick masonry conference, Washington, II-9, pp 84–89

BS 4551:2005+A2:2013 (2013) Mortar. Methods of test mortar and screed and plasters. Chemical analysis and physical testing. British Standards Institution, UK

BS EN 13139 (2013) Aggregates for mortar. British Standards Institution (BSI), London, UK

BS EN 1996-1-1:2005+A1:2012 (2012) Eurocode 6—design of masonry structures, part 1—1. British Standards Institution (BSI), London, UK

BS EN 413-1 Masonry cement—part 1: specification. British Standards Institution (BSI), London, UK

BS EN:1015-3 (1999) Methods of tests for mortar for masonry mortar—part 3: determination of consistence of fresh mortar (by flow table). British Standards Institution (BSI), London, UK

Groot C (1993) Effects of water on mortar-brick bond. PhD thesis, Faculty of Civil Engineering, Delft University of Technology, Netherlands

Gumaste KS (2004) Studies on strength and elasticity of brick masonry walls. PhD thesis, Department of Civil Engineering, Indian Institute of Science, Bangalore, India

Gumaste KS, Nanjunda Rao KS, Venkatarama Reddy BV, Jagadish KS (2007) Strength and elasticity of brick masonry prisms and wallettes under compression. Mater Struct 40(296):241–253

Gupta A (2003) Studies on characteristics of cement-soil mortars and soil-cement block masonry. MSc (Engg.) thesis, Dept. of Civil Engineering, Indian Institute of Science, Bangalore, India

IS 1905:1987 (reaffirmed 1998) Code of practice for structural use of unreinforced masonry. Bureau of Indian Standards, New Delhi, India

IS 2116:1980 (reaffirmed 2002) Specification for sand for masonry mortars. Bureau of Indian Standards, New Delhi, India

IS 2250-1981 (reaffirmed 1995) Code of practice for preparation and use of masonry mortars. Bureau of Indian Standards, New Delhi, India

IS 3466:1988 (reaffirmed 1999) Specification for masonry cement. Bureau of Indian Standards, New Delhi, India
Jagadish KS (2015) Structural masonry. I. K. International Publishing House, New Delhi, India
Jagadish KS, Venkatarama Reddy BV, Nanjunda Rao KS (2017) Alternative building materials and technologies. New-Age International, ISBN: 978-93-859-2387-6
Lal R (2005) Characteristics of soil-cement blocks and soil-cement block masonry. MSc (Engg.) thesis, Dept. of Civil Engineering, Indian Institute of Science, Bangalore, India
NBC (2016) National building code of India 2016, group 2, sections 1 and 4. Bureau of Indian Standards, New Delhi, India
Reddy BVV, Gupta A (2005) Characteristics of cement–soil mortars. Mater Struct 38(280):639–650
Reddy BVV, Gupta A (2006) Tensile bond strength of soil–cement block masonry couplets using cement–soil mortars. J Mater Civ Eng 18(1):36–45
Reddy BVV, Latha MS (2018) Mortar shrinkage and flexure bond strength of stabilised soil brick masonry. J Mater Civ Eng 30(5):05018002
Reddy BVV, Vyas CVU (2008) Influence of shear bond strength on compressive strength and stress strain characteristics of masonry. Mater Struct 41(10):1697–1712
Reddy BVV, Lal R, Nanjunda Rao KS (2007) Enhancing bond strength and characteristics of soil–cement block masonry. J Mater Civ Eng 19(2):164–172
Reddy BVV (2012) Stabilised soil blocks for structural masonry in earth construction. In: Hall MR, Lindsay R, Krayenhoff M (eds) Modern earth buildings: materials, engineering, construction and applications. Woodhead Publishing Ltd., UK, pp 324–363
Reddy BVV, Nikhil V, Nikhilash M (2019) Moisture transport in cement stabilised soil brick–mortar interface. In: Reddy BVV, Mani M, Walker P (eds) Earthen dwellings and structures: current status in their adoption. Springer Transactions in Civil and Environmental Engineering, Singapore, pp 27–37
Sarangapani G (1998) Studies on the strength of brick masonry. PhD thesis, Department of Civil Engineering, Indian Institute of Science, Bangalore, India
Sarangapani G, Venkatarama Reddy BV, Groot CJWP (2002) Water loss from fresh mortars and bond strength development in low strength masonry. Mason Int 15(2):42–47
Sarangapani G, Venkatarama Reddy BV, Jagadish KS (2005) Brick-mortar bond and masonry compressive strength. J Mater Civ Eng 17(2):229–237
Venumadhava RK, Venkatarama Reddy BV, Jagadish KS (1996) Flexural bond strength of masonry using various blocks and mortars. Mater Struct 29:119–124
Walker P (1999) Bond characteristics of earth block masonry. J Mater Civ Eng 11(3):249–256
Walker P (2004) Strength and erosion characteristics of earth blocks and earth block masonry. J Mater Civ Eng 16(5):497–506
Yogananda MR (1983) Studies on surkhi and rice husk ash pozzolana. MSc (Engg.) thesis, Dept. of Civil Engineering, Indian Institute of Science, Bangalore, India
Yogananda MR, Jagadish KS (1998) Pozzolanic properties of rice husk ash, burnt clay and red mud. Build Environ 23(4):304–308

Chapter 7
Stabilised Compressed Earth Block Masonry

7.1 Introduction

The bonded masonry units and the mortar constitute the masonry. The characteristics of the masonry units and the mortar influence the masonry behaviour. The bond between the masonry unit and the mortar is crucial in imparting strength to the masonry, especially to resist flexure and shear stresses. Evaluating and understanding the characteristics of the masonry unit and the mortar is not enough to assess the strength of the masonry in compression, shear and flexure. The behaviour of the masonry needs to be examined especially when the new type of masonry unit such as the stabilised compressed earth brick/block (CEB) is used. The stabilised CEB has distinctly different characteristics, when compared to the characteristics of the conventional masonry units such as burnt clay bricks, concrete blocks, stone, etc. The distinctive characteristics of the stabilised CEB are:

(a) The strength and the modulus are sensitive to the moisture content of the stabilised CEB. For example, the wet-to-dry strength ratio in compression is in the range 0.40–0.70.
(b) The stabilised CEBs have higher straining capacity, especially in the dry state. The strain at failure can reach a value of 1.5% or more.
(c) The strength and the modulus of the stabilised CEB greatly depend upon the soil composition, block density, stabiliser content and the moisture content.

In view of the above-mentioned distinct characteristics, the characteristics of the stabilised CEB masonry especially using stabilised earth mortars needs a careful examination. Various aspects of stabilised CEB masonry including the design are discussed in the following sections.

B. V. V. Reddy, *Compressed Earth Block & Rammed Earth Structures*,
Springer Transactions in Civil and Environmental Engineering,
https://doi.org/10.1007/978-981-16-7877-6_7

7.2 Bond Strength in Stabilised Compressed Earth Block Masonry

The adhesion between the masonry unit and the mortar can be quantified through measurement of the bond strength. The bond development is influenced by several parameters pertaining to the stabilised CEB and the mortar. The important parameters are:

(1) The stabilised CEB's surface texture and the microstructure
(2) The rate of water absorption and the moisture content of the stabilised CEB at the time of the construction
(3) Mortar composition and its workability or the flow characteristics

7.2.1 *Mechanism of Bond Development*

The bond development between the masonry unit and the mortar can be attributed to the mechanical interlock of cement hydration products into the brick/block pores or chemical bond or both. During the masonry construction, the bricks or the blocks are placed over the fresh mortar bed joint. The fresh mortar comprises of a binder (cement/lime, etc.), fine aggregates, water and may be some admixtures/additives. The binder (especially cement) hydrates and results in the formation of the cementitious materials such as silicate/aluminate hydrates. Therefore, the fresh mortar is rich with water-containing cementitious products. The dry or the partially saturated brick absorbs the water (through suction, Fig. 7.1), rich with cementitious products. This movement of the liquid results in depositing cementitious products into the tiny brick pores. Also, the cementitious products at the brick/block–mortar interface grow into brick/block pores. After the mortar sets, the hydration products harden. Thus, the bond between the brick/block and the mortar is established through mechanical interlocking of the binder hydration products into the brick pores (Fig. 7.2). The investigations of Garndet et al. (1972), Barnes et al. (1978), Lawrence and Cao

Fig. 7.1 Moisture suction by the brick/block from the fresh mortar bed joint

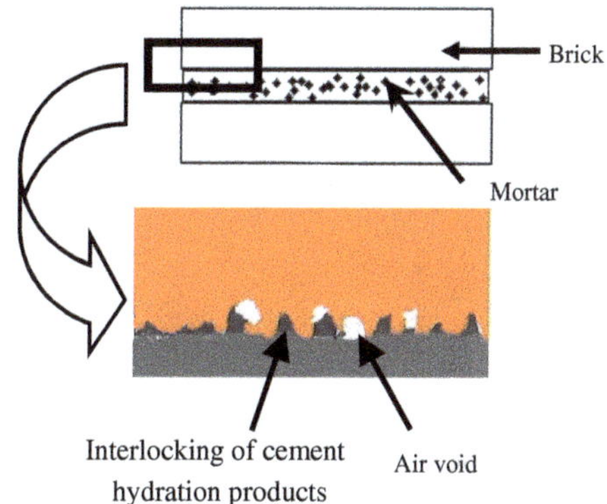

Fig. 7.2 Schematic sketch showing the mechanical interlock of cementitious products into the brick pores

(1987, 1988), and Gourav and Reddy (2018) shed more information on the bond development mechanism at the masonry unit–mortar interface.

Another type of mechanism for bond development at the brick–mortar interface is due to the chemical reaction between the pozzolanic materials on the brick/block surface and the free calcium hydroxide in the mortar and/or released during cement hydration. This type of bonding is termed as chemical bond. Majority of the masonry units do not possess minerals/pozzolanic materials on the bed surface. Masonry units such as the fly ash bricks contain unreacted fly ash particles, which readily react with the calcium hydroxide from the mortar establishing chemical bond (Gourav 2015; Gourav and Reddy 2018).

7.2.2 Surface Texture of Cement Stabilised Compressed Earth Blocks

The texture of the masonry unit's surface influences the development of bond between the mortar and the brick/block. Rough textured surface of the masonry unit gives better bond strength than the smooth textured surface (Groot 1993; McBurney et al. 1946; Thornton 1953; Kampf 1963). The texture of the brick/block surface can be better understood through a microstructure analysis. The microstructure of the surface can be examined through scanning electron microscopy (SEM) imaging. Typical SEM images for the cement stabilised CEB surfaces are displayed in Fig. 7.3 for 6, 8 and 10% cement contents. These SEM images reveal the surface texture, pore size and pore structure. The surface pore size and the surface porosity are sensitive to the cement content and the density of the CEB. Figure 7.3 shows that the pores are of irregular shape and the pore sizes vary considerably across the surface. The number of pores and the pore sizes decrease with the increase in cement content of the CEB. Reddy and Gupta (2005) monitored the surface pore sizes of cement stabilised CEB manufactured using a sandy soil (clay content of 9%) employing a manual graphical method. The study showed a decrease in the surface porosity with

Fig. 7.3 SEM image of the cement stabilised compressed earth block surface

the increase in the cement content, mainly attributing to more cementitious material coating the soil/sand particles.

Figure 7.4 shows the linear relationship between the surface porosity of the CEB (with 8% cement) and the dry density. The surface porosity reduces with the increase

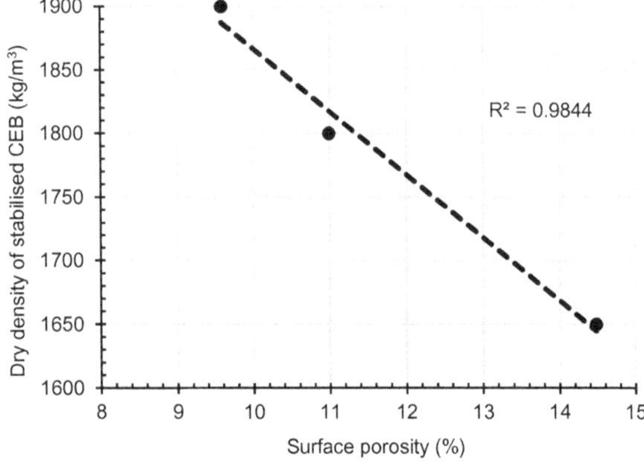

Fig. 7.4 Relationship between the surface porosity and the dry density of the cement stabilised CEB

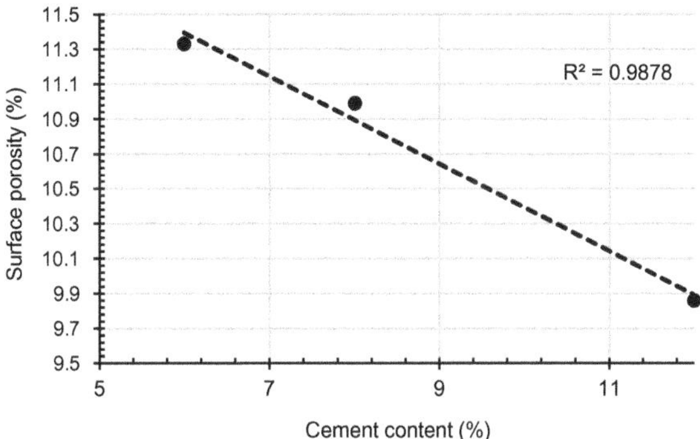

Fig. 7.5 Variation in the surface porosity with the cement content of the CEB

in the CEB density. The figure shows the surface porosity of the CEB reducing from 14.5 to 9.5%, as the density was increased from 1650 to 1900 kg/m^3. Figure 7.5 shows the variation in the surface porosity with the cement content of the CEB, where the dry density of the CEB was 1800 kg/m^3. The surface porosity decreased (by ~13%) from 11.33 to 9.9% as the cement content increased by from 6 to 12%. The surface porosity of the cement stabilised CEB ranges between 10 and 20% depending upon the density, cement content and the particle size distribution of the soil (Reddy and Gupta 2005; Latha 2015).

The surface pore sizes of the cement stabilised CEB vary from few microns and up to 300 microns. The studies of Garndet (1973) and Garndet et al. (1972) reveal that the bond between the brick and the mortar is due to the anchorage of needle like entringite crystals (having diameter ~ 0.05 μm) into the brick pores. Therefore, the pore radii < 0.05 μm results in lower bond strength. Microstructure studies on the brick surface showed that coarser pores give better bond strength (Garndet et al. 1972).

7.2.3 Rate of Water Absorption in Stabilised CEB

The rate at which the block absorbs water can influence the bond strength development. This is mainly due to the movement of cementitious materials from the fresh mortar bed joint into the brick/block pores during the masonry construction. The percentage of clay minerals present in the soil mix used for the stabilised CEB, the quantity of stabiliser (cement/lime), block density, porosity and the pore size distribution of the CEB control the rate of water absorption (Walker and Stace 1997; Reddy and Gupta 2005; Reddy and Latha 2014).

Figure 7.6 shows actual water absorption rate for the cement stabilised CEBs as the soaking duration in the water was varied for a set CEBs with 7–8% cement made using the coarse-grained and the fine-grained soils. The CEBs of size 230 × 108 ×

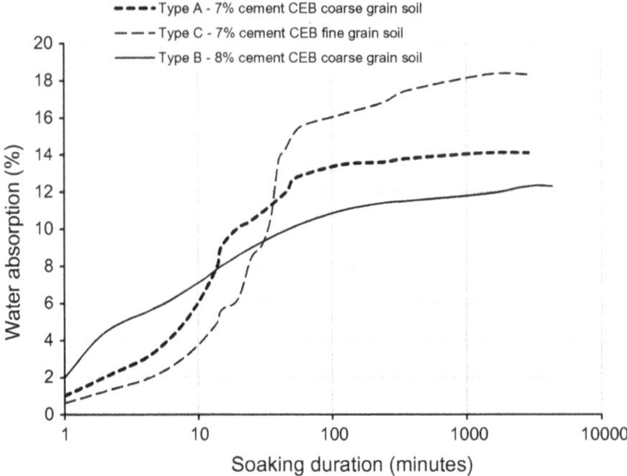

Fig. 7.6 Soaking time and water absorption relationships for CEBs with 8% cement

75 mm were used in these studies. The water absorption values at different soaking durations and the degree of saturation for the three CEBs (Fig. 7.6) are given in Table 7.1. In the case of the CEBs using coarse-grained soils, soaking of dry CEBs in the water results in rapid absorption of water in the first few minutes and the rate of water absorption is highest in the initial few minutes. The CEBs attain about 70% of 24 h water absorption value in 20 min of soaking. For the case of CEBs with fine-grained soil, the absorption rate is slow in the initial soaking period (<10 min) and raises sharply between 10 and 60 min of soaking. In one hour of soaking, the CEBs absorb 85–90% of 24 h water absorption value. For both the CEBs using the two types of soils (coarse-grained and fine-grained soils), beyond 24 h of soaking hardly any change in the water absorption value.

The CEBs do not get completely saturated even after 72 h of soaking in the cold water in the ambient room temperature of about 20–25 °C. The CEB with coarse-grained soil attains 67–71% saturation after 48 h of soaking, while the one with fine-grained soils it goes up to 88% saturation. The reason for not getting fully saturated even after 48 or 72 h of soaking in the cold water could be due to entrapped air in the pores inside the CEBs.

A generalised water absorption–soaking duration curve for cement stabilised CEBs is depicted in Fig. 7.7 considering variations in soil composition, density, cement content, etc. The CEB requires 20–30 min to attain 75% of 24 h absorption value and 200–400 min to near 24 h absorption value. The block density, soil type and grading, and the cement content have significant influence on the rate of water absorption. The soaking duration required to attain 75 and 100% of 24 h absorption value decreases with increase in the cement content and the density of the blocks (Reddy and Gupta 2005).

Table 7.1 Water absorption and degree of saturation for cement stabilised CEBs (Fig. 7.6)

CEB type	Block details				Soaking duration											
	Dry density (kg/m^3)	Cement (%)	Clay (%)	Soil type	10 min		20 min		60 min		24 h		48 h		72 h	
					M. C. (%)	S_r (%)	M. C. (%)	S_r (%)	M. C. (%)	S_r (%)	M. C. (%)	S_r (%)	M. C. (%)	S_r (%)	M. C. (%)	S_r (%)
A	1750	7.0	14.0	CG	6.04	30.45	10.07	50.78	12.90	65.06	14.10	71.11	14.10	71.11	–	–
B	1800	8.0	15.8	CG	7.10	38.92	8.43	46.21	10.30	56.47	11.90	65.24	12.30	67.43	12.61	69.13
C	1720	7.0	10.5	FG	3.75	18.00	6.25	30.01	15.57	74.77	18.33	88.02	18.33	88.02	–	–

A and B: CEB with coarse-grained soils; C: CEB with fine-grained soil; M.C.: Moisture content by mass; S_r: Degree of saturation

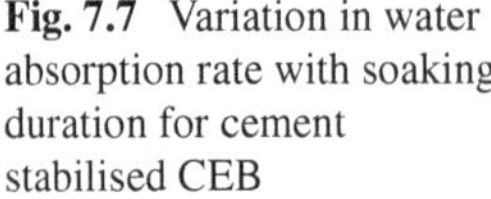

Fig. 7.7 Variation in water absorption rate with soaking duration for cement stabilised CEB

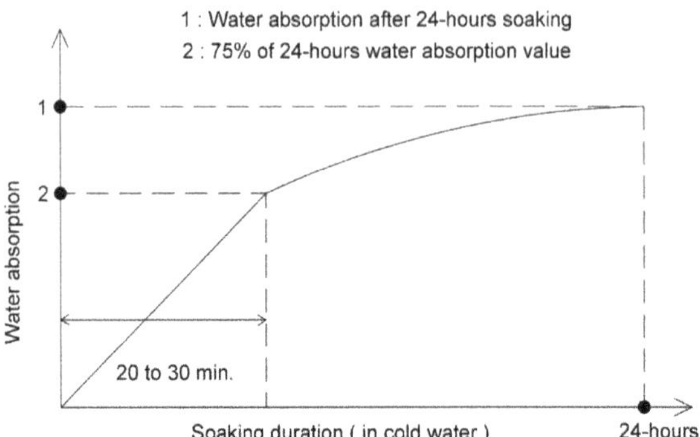

7.2.4 Assessing Bond Strength

The bond strength between the stabilised CEB and the mortar joint can be determined through a flexure or direct tension or shear bond test. Figure 7.8 shows the schematic representation of these three types of tests. Among these, the most commonly used test to evaluate the bond strength is the flexure bond test. Standard procedure for assessing the flexure bond strength of a masonry prism can be found in ASTM C1072-10 (2010). This type of test is called bond-wrench test. The Fig. 7.9 shows a schematic representation of ASTM bond-wrench test set-up and a simplified bond-wrench test set-up (called modified bond-wrench test). The testing procedure involves mounting a prism and fixing it at the base. The horizontal load is applied at the top of the prism causing the cantilevered masonry prism to bend. The lateral load causing the failure of the prism is recorded to estimate the bending moment at failure and the flexure strength of the masonry prism. Here, the flexural tension developed is perpendicular to the bed joint. In the modified bond-wrench test, only one joint in the prism fails and one flexure strength value can be obtained in each test, whereas in the bond-wrench test suggested in ASTM C1072-10 (2010) all the joints in the prism can be tested, yielding more bond strength values per prism. However, in this type of test, the wrench mechanism grips the uppermost brick at discrete points, thus applying concentrated moments on a 230 mm long brick. Consequently, the stress

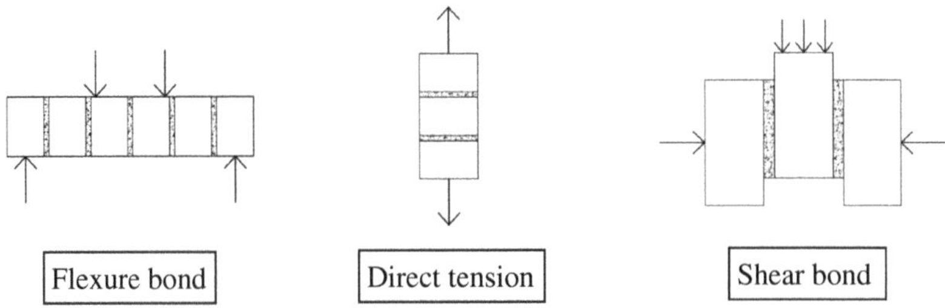

Fig. 7.8 Bond strength evaluation methods

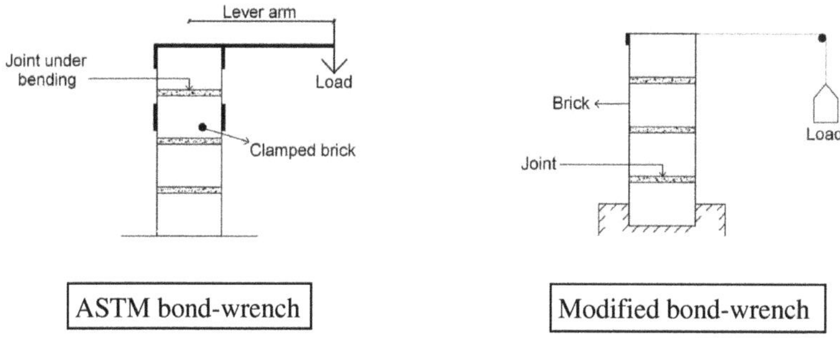

Fig. 7.9 Bond-wrench test set-up

in a joint, just 80 mm below the point of application of the moment, is not likely to be uniformly distributed over the length of the masonry unit. Therefore, this type of test while indicating the bond strength may not give good measure of the bond stress at failure. These types of problems are addressed in the modified bond-wrench test.

7.2.5 Optimum Moisture Content in the Stabilised CEB for Bond Development

The stabilised CEB absorbs water from the fresh mortar bed joint during the initial couple of hours of masonry construction. The water transport from the mortar to the CEB (during initial phase of masonry construction) is essential for the development of the bond at the block–mortar interface. Excess quantity of water lost (rapidly) from the fresh mortar bed can impair the setting of the cement present in the bed joint and can affect the bond development.

The moisture loss (with time) from the fresh mortar bed joint can be easily monitored through a simple experiment on the masonry couplets, where a mortar bed joint is sandwiched between the two CEBs. Figs. 7.10, 7.11 and Fig. 7.12 show plots for residual water/binder ratio in the mortars with the duration of contact with the CEBs for the three types of mortars (cement–sand, cement–lime and cement–soil mortars). The cement stabilised CEB manufactured with a soil containing 10.90% clay and 8% cement (by mass) having a dry density of 1850 kg/m^3 has been used in these experiments. The CEB had a wet compressive strength of 9.09 MPa (standard deviation 0.83 MPa). The oven-dried, partially saturated (9.3%) and nearly saturated (12.61%) bricks were used in the experiments.

The dry and partially saturated stabilised CEBs absorb the moisture from the fresh mortar bed joint thus affecting the water–binder ratio of the mortar. The plots (Figs. 7.10, 7.11 and 7.12) clearly show that the water–binder ratio of the mortar decreases sharply in the initial one-hour duration of contact with the dry and partially saturated CEBs. Beyond one hour of contact duration, the water–binder

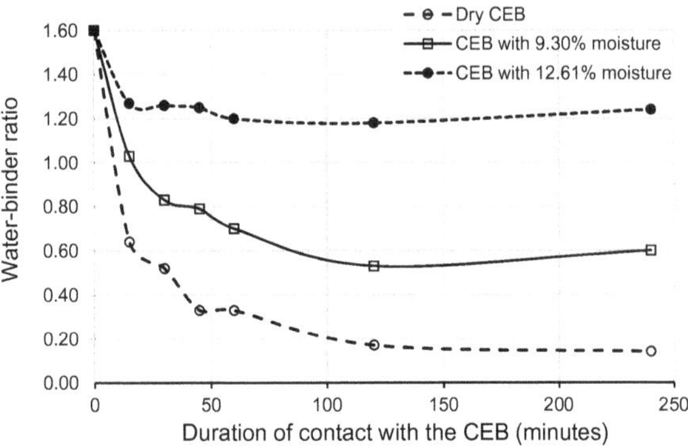

Fig. 7.10 Water–binder ratio versus duration of contact with the fresh mortar for CM mortar

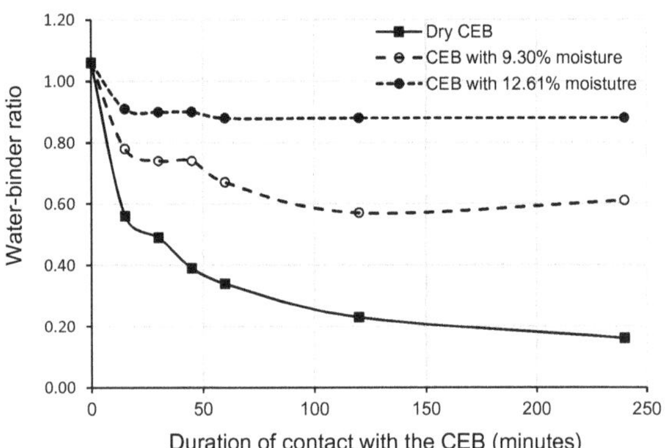

Fig. 7.11 Water–binder ratio versus duration of contact with the fresh mortar for CLM mortar

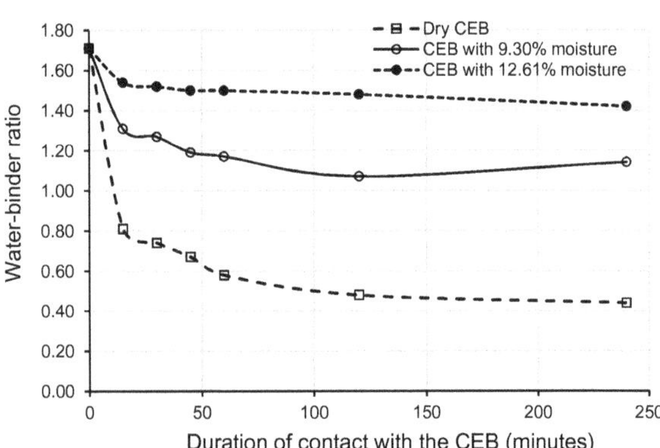

Fig. 7.12 Water–binder ratio versus duration of contact with the fresh mortar for CSM mortar

ratio remains nearly constant irrespective of the initial brick moisture content and the mortar type. Table 7.2 gives residual water–binder ratios after one and two hours of casting the masonry couplet for different moisture contents in the CEBs (i.e. dry, partially saturated and nearly saturated conditions).

Table 7.2 Water loss from fresh mortar bed joint

Mortar proportion (by volume)					Mortar flow (%)	Water binder ratio						
Cement	Lime	soil	sand	Designation		Initial	After 1.0 h contact with CEB			After 2.0 h contact with CEB		
							Dry	*Partially saturated	**72 h soaked	Dry	*Partially saturated	**72 h soaked
1	–	–	6	CM	85	1.60	0.33	0.70	1.20	0.17	0.53	1.18
1	1	–	6	CLM	85	1.06	0.34	0.67	0.88	0.23	0.57	0.88
1	–	2	5	CSM	85	1.71	0.58	1.17	1.50	0.48	1.07	1.48

* Partially saturated, moisture content: = 9.30%; ** 72 h soaked, moisture content = 12.61%

For example, after two-hour contact with the dry CEB, the water–binder ratios of the mortars were 0.17, 0.23, 0.48 for CM, CLM and CSM mortar, respectively. The corresponding initial water–binder ratios were 1.60, 1.64 and 1.71, respectively. The reduction in the water–binder ratio (after two-hour contact duration) was maximum for the CM mortar (~90%) followed by CLM and CSM mortars (75%). Use of partially saturated CEB's with 9.3% moisture resulted in 65, 45 and 37% reduction in the water–binder ratio after two-hours of contact. These results demonstrate that the use of partially saturated CEBs during construction result in sufficient residual moisture in the fresh mortar bed joint, for proper hydration of the cement, facilitating the bond development. The minimum water–cement ratio required for proper hydration of the Portland cement is 0.38 (Neville 1995).

Moisture content of the stabilised earth brick at the time of masonry construction has significant influence on the interfacial bond strength between the masonry unit and the mortar. The investigations of Sinha (1967), Venu madhava Rao et al. (1996), Sarangapani et al. (2005), and Reddy and Gupta (2006) showed that the maximum bond strength for the masonry can be achieved while using partially saturated bricks/blocks having 75% of 24 h water absorption value during the masonry construction. Figure 7.13 shows a plot of moisture content of the cement stabilised CEB (during masonry construction) and the masonry bond strength using different types of mortars. The CEB used in these experiments had a wet compressive strength of 9.09 MPa (standard deviation 0.83 MPa) and contained 8% cement by mass. The water content of the brick was 11.9% after 24 h soaking and 12.61% after 72 h soaking in cold water. The pore structure of the block surface is shown in Fig. 7.14. The pore sizes were in the range of 2–315 μm with a surface porosity of 9.67%. The maximum pore size was 315 μm.

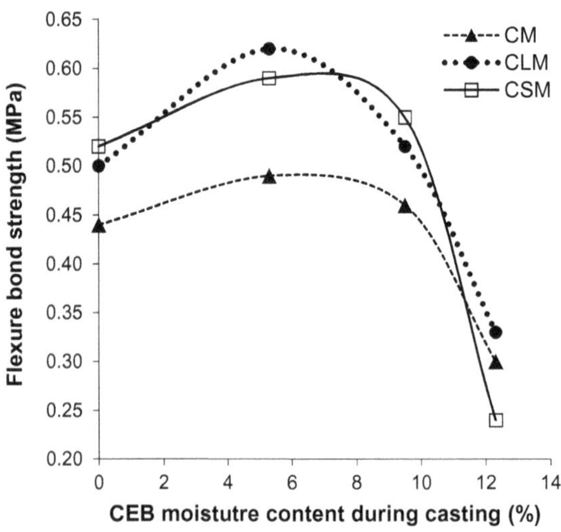

Fig. 7.13 CEB moisture content during construction versus masonry bond strength

Fig. 7.14 SEM image showing the surface pore structure of the cement stabilised CEB

The plots in Fig. 7.13 show that masonry the bond strength increases with the increase in moisture content (during casting) of the stabilised CEB, which reaches a peak value (when moisture content is close to 60–75% of 24 h water absorption value) and drops drastically at the nearly saturated water content, irrespective of the mortar type. There is a difference in the bond strength between the peak and the extreme values, especially on the saturated side. For example, the bond strength at 60–75% of 24 h water absorption value is 25 and 50–60% higher than the bond strength at dry and nearly saturated conditions, respectively, for the cement–lime and the cement–soil mortars. The main reason for this type of behaviour can be attributed to the drastic lowering of water–binder ratio due to excessive water suction by the dry brick from the fresh mortar bed joint. Lowering the water–binder ratio below a critical value will impair the cement setting and the bond development. Whereas, when the brick/block is in nearly saturated state at the time of construction, there will be hardly any movement of moisture from the water-rich mortar bed joint to the CEB and therefore affecting the bond development due to the lack of interlocking of hydration products into the block pores.

Moisture movement from the fresh mortar bed joint into the brick/block pores is essential for developing mechanical interlock of the cement hydration products into the brick/block pores. Use of partially saturated CEB's will help in achieving the maximum bond strength without impairing the setting of the cement in the fresh mortar bed joint.

7.3 Flexure Strength of Stabilised CEB Masonry

Masonry realises bending stresses, when subjected to lateral loads or eccentric loads. Masonry walls have bed joints and perpend joints, and hence, masonry flexure strength in the two orthogonal directions will be different. Figure 7.15 shows schematic representation of the masonry wall undergoing bending in the two orthogonal directions. Masonry under flexure can be recognised with reference to the direction of flexural tension developed: (1) tension developed perpendicular to the bed joints and (2) tension developed parallel to the bed joints. The flexure strength of the masonry when tension developed is parallel to the bed joints is about 2–3 times the flexure strength of the masonry when tension developed is perpendicular to the bed joints (Hendry 1981). The flexural stresses in the unreinforced masonry are resisted by the interfacial bond strength between the masonry unit and the mortar.

7.3.1 Flexure Bond Strength

The flexure bond strength of the masonry prisms measured using a bond-wrench set-up represents the case of tension developed perpendicular to the bed joints. The flexure bond strength of the stabilised CEB masonry can be found in the investigations of Venu Madhava Rao et al. (1996), Walker (1999), and Latha and Reddy (2018), Reddy et al. (2019). Table 7.3 gives the results of some flexure bond strength values for the cement stabilised CEB masonry prisms using cement–sand, cement–lime–sand and cement–soil–sand mortars. The masonry specimens have been tested in the wet condition (except the results by Walker 1999), and the stabilised CEB has been manufactured using soils with a clay content in the range of 10–15%. Cement–sand mortar shows low flexure bond strength, whereas the cement–lime and the cement–soil mortars show higher bond strength (0.17–0.52 MPa).

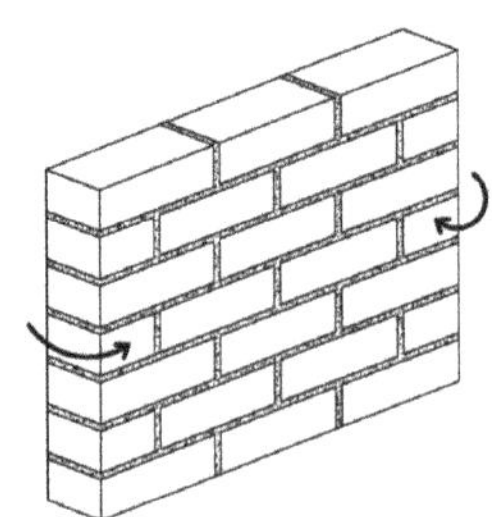

Bending parallel to bed joints

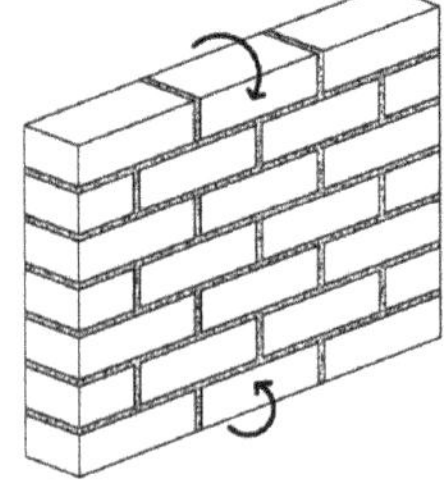
Bending perpendicular to bed joints

Fig. 7.15 Bending of masonry wall panel in the two orthogonal directions

Table 7.3 Flexure bond strength of the cement stabilised CEB masonry prisms

Mortar proportion (by volume)				Cement stabilised CEB brick/block		Flexure bond strength (MPa) (wet state)	Reference
C	L	So	Sa	Cement (%)	Wet strength (MPa)		
1	–	–	6	6.0	4.60	0.10	Venu Madhava Rao et al. (1996)
1	–	1	6	6.0	4.60	0.17	
1	1	–	6	7.0	5.32	0.44	Latha and Reddy (2018)
1	–	2	8	10.0	4.26	0.31*	Walker (1999)
1	–	2	5	8.0	9.09	0.52	Reddy et al. (2019)

* Partially dry
C: Portland cement, L: lime, So: Soil, Sa: Sand

Table 7.4 Cement–soil mortars for stabilised CEB masonry

Mortar proportion (by volume)			Mortar flow (%)	Used for
Cement	Soil	Sand		
1	1	6	85–95	• Bed and perpend joints in masonry walls • Plastering/rendering
1	2	6		
1	2	5		
1	¼	3	80–90	• Bed and perpend joints in masonry shells (vaults/domes) and arches • Ceiling plasters/pointing
1	½	4		

The soil/earth-based mortars such as cement–soil mortars show considerable improvement in the bond strength when compared to the bond strength using conventional mortars such as the cement–sand mortar. In the case of composite mortars such as the cement–lime mortar, replacing lime with soil results in cement–soil mortar showing similar or better bond strength for the CEB masonry. Typical proportions for the cement–soil mortars suitable for stabilised CEB masonry are listed in Table 7.4 along with their suitability for different types of the CEB masonry applications. The soil used in the cement–soil mortars should contain non-expansive clay minerals such as kaolinite.

7.3.2 Flexure Strength of Stabilised CEB Masonry Walls

The flexural strength of the masonry walls can be assessed by testing the masonry wallettes. Figure 7.15 shows schematic diagrams of single wythe masonry wallettes undergoing lateral bending in the two orthogonal directions. In one case, the tension developed is parallel to the bed joints (TPBJ), and in the other case the tension developed is normal to the bed joints (TNBJ). The flexural strength of the masonry is dependent upon the pre-compression. The load-bearing masonry walls in a structure experience vertical pre-compression due to gravity loads. Therefore, understanding the lateral strength of the walls with reference to the vertical pre-compression will be useful.

The investigations of Tennant et al. (2016) reveal the flexural strength of the cement stabilised CEB masonry walls in the two orthogonal directions subjected to lateral bending. The flexure strength of CEB masonry as the vertical pre-compression was varied, for the two cases of TPBJ and TNBJ has been examined. Figure 7.16 shows the test set-up for testing the masonry walls under lateral bending with pre-compression. The orthogonal flexure strength ratio (TNBJ/TPBJ) was found to be in the range of 0.50–0.60. The flexure strength when tension developed is parallel to the bed joints is more than double that of the flexure strength when tension developed is normal to the bed joints. The CEB having wet compressive strength of 7.43 MPa and 1:2:5 (cement:soil:sand, by volume) cement–soil mortar have been used in these investigations.

The pre-compression in the lower floor walls (of 2–3 storey load bearing masonry buildings) could be as much as 0.50 MPa. The pre-compression enhances the flexure strength of the masonry. The flexure strength of CEB masonry and pre-compression are linearly related as shown in Fig. 7.17. The linear regression fit for twelve experimental points is shown in the figure. These results have been generated using single wythe (107 mm thick) wallettes built using cement–soil mortar. The flexure strength increases with the increase in the pre-compression.

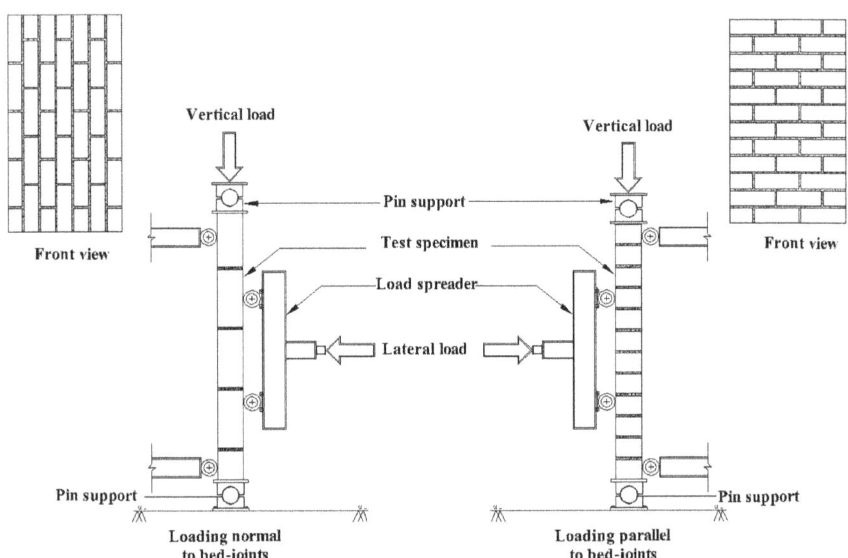

Fig. 7.16 Lateral loading of CEB masonry wallettes

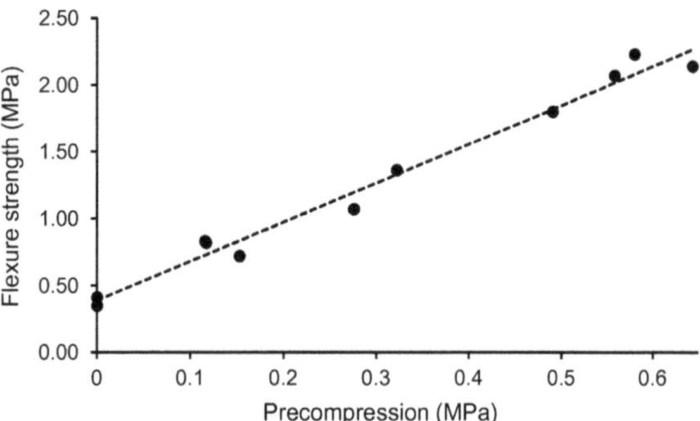

Fig. 7.17 Flexure strength of CEB masonry wall panel (tension normal to bed joints)

7.4 Shear Strength of Stabilised Compressed Earth Block Masonry

The masonry walls in a structure experience in-plane shear stresses, especially under lateral loading due to wind and seismic effects. Such stresses are resisted by the bond strength between the masonry unit and the mortar. In-plane shear strength of the stabilised CEB masonry can be experimentally determined either through a raking test with some pre-compression or through diagonal tension test. Figures 7.18 and 7.19 show such test set-ups.

Shear bond strength of the stabilised CEB masonry couplets was examined by Reddy and Vyas (2008). Two types of CEB and two types of cement–lime mortars were used in these studies. The CEB with 5% cement (by mass) designated as CEB5 had a wet compressive strength of 5.09 MPa, and the CEB with 14% cement (by mass) designated as CEB14 had a wet compressive strength of 11.46 MPa. Two types of cement–lime mortars of proportion 1:1:6 (cement:lime:sand, by volume) designated as CLM1 and 1:0.5:4 (cement:lime:sand, by volume) designated as CLM2 have been used. The shear bond strength (in wet condition) was 0.12, 0.15 and 0.22 MPa for CEB5–CLM1, CEB5–CLM2 and CEB14–CLM1 brick–mortar combinations,

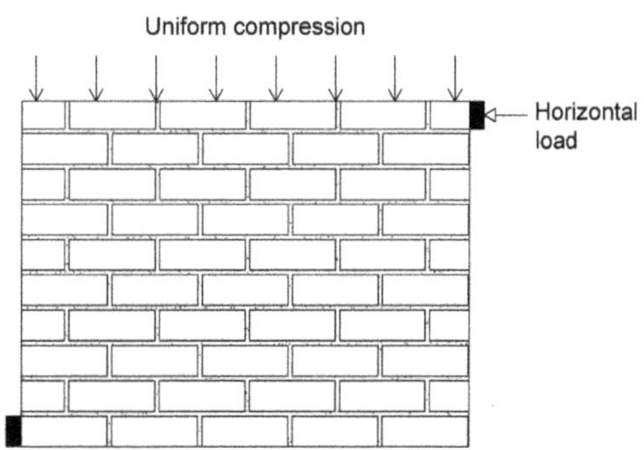

Fig. 7.18 Masonry wallette under in-plane shear test

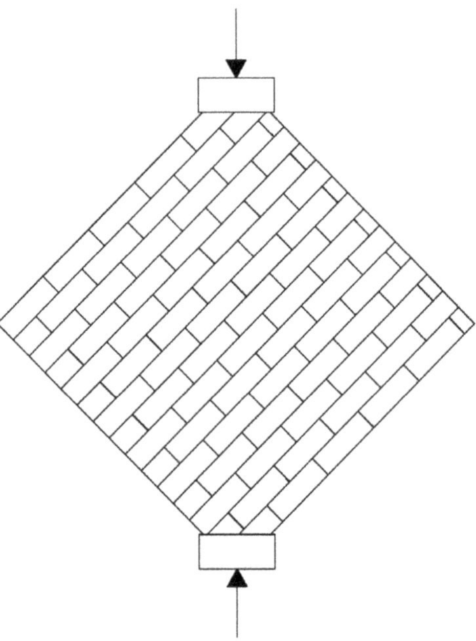

Fig. 7.19 Diagonal tension test

respectively. Richer cement–lime mortar and CEB with higher cement content has resulted in higher shear bond strength.

The shear bond strength is influenced by the pre-compression in the direction normal to the bed joints experiencing shear displacement. Vyas (2006) conducted tests on masonry couplets using CEB5 and cement–lime mortar (CLM1). Figure 7.20 shows the relationship between shear bond strength and pre-compression for CEB couplet. It is a bi-linear relationship. Initially, portion of the bi-linear relationship has much steeper slope than the slope of the later one. The shear bond strength increased from 0.12 to 1.10 MPa as the pre-compression was increased from 0.0 to 1.0 MPa. There will be upper limit for the pre-compression value depending upon the masonry compressive strength. In the absence of data on the shear strength of

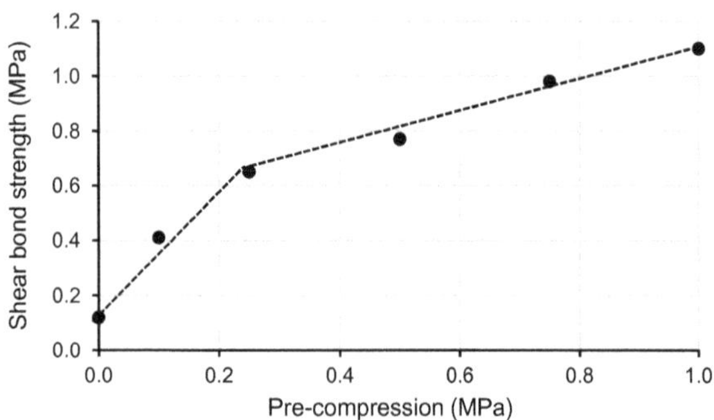

Fig. 7.20 Shear bond strength versus pre-compression for cement stabilised CEB couplet (after Reddy and Vyas 2008)

the CEB masonry walls, the shear bond strength of the masonry couplet gives some information for assessing the shear strength of the CEB masonry walls.

7.5 Compressive Strength of Stabilised Compressed Earth Block Masonry

Masonry consists of two materials: the masonry unit and the mortar. Interfacial bond between the masonry unit and the mortar is essential for the masonry to behave as one structural entity. Figure 7.21 shows a masonry prism and a masonry wallette. When the masonry specimen is subjected to vertical compression, the masonry materials experience stresses and displacements different from those experienced by the masonry unit and the mortar when tested individually in a testing machine. This can be mainly attributed to the differing deformation characteristics of the two materials under compression. Therefore, the strength of the masonry will be different from that of the masonry unit and the mortar.

7.5.1 Compressive Strength of the Masonry Unit and the Mortar

The compressive strength of the masonry unit such as brick or block can be determined following a standard test procedure outlined in any of the standard codes of practices. Typical testing procedure involves levelling the two opposite bed faces of the masonry unit, positioning the masonry unit between the two steel platens of the testing machine, applying the compressive load at a specified rate till the specimen fails. Typical failure pattern of the masonry unit tested for compressive strength following standard test procedure is shown in Fig. 7.22. The failure is controlled by

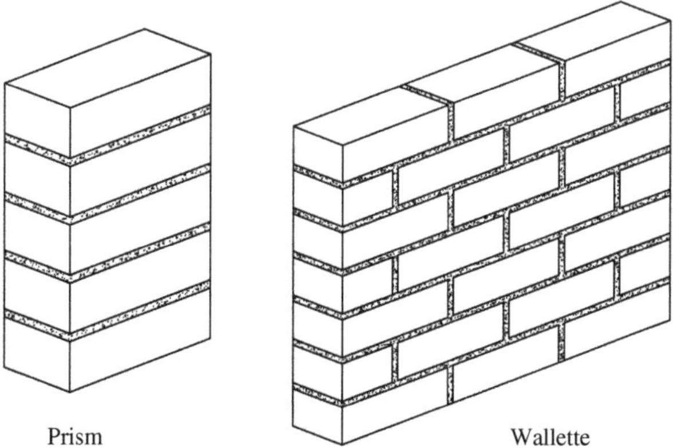

Fig. 7.21 Masonry prism and wallette

Fig. 7.22 Masonry unit failure pattern during standard compression test

the restrained lateral deformation of the masonry unit. Figure 7.23 shows the forces acting on the masonry unit under compression in a machine set-up. The masonry unit experiences self-equilibrating horizontal frictional shear at the platen and the brick/block interface. Hence, the masonry unit experiences diagonal shear failure due to the restrained deformation close to the steel platen as shown in the Fig. 7.22. As the masonry unit becomes thinner, the horizontal frictional shear dominates the failure. The restrained compressive strength of the masonry unit therefore depends upon the height-to-thickness ratio of the masonry unit. Figure 7.24 shows the restrained strength variation with the height of the masonry unit representing cement stabilised CEB. The compressive strength decreases with the increase in the masonry unit height. The relationship between the compressive strength and the masonry unit's height-to-width ratio is a second-degree polynomial of the form:

$$\text{Wet compressive strength (MPa)} = 1.43X^2 - 4.52X + 7.58 \tag{7.1}$$

where X = Height to width ratio of the masonry unit.

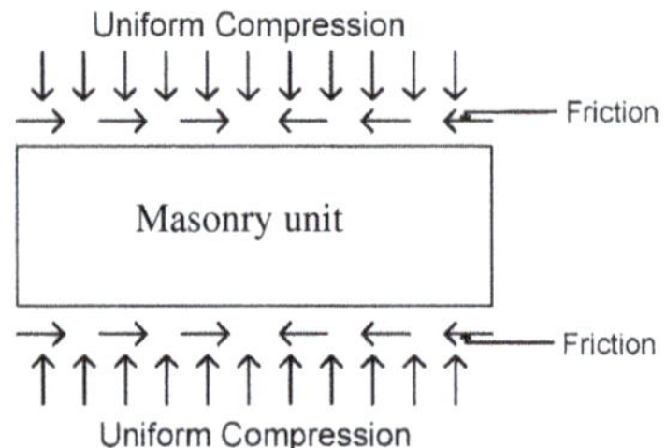

Fig. 7.23 Forces acting on the masonry unit during compression test

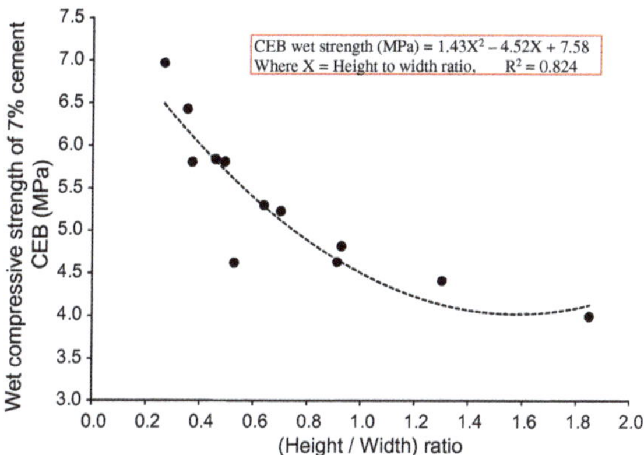

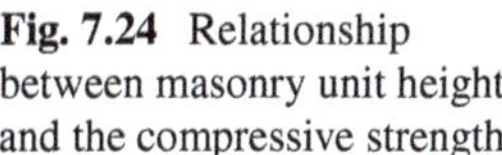
Fig. 7.24 Relationship between masonry unit height and the compressive strength

The compressive strength of the CEB (7% cement by mass and 1750 kg/m^3 dry density) as the block height was varied from 50 to 100 mm is given in Table 7.5 (Latha 2015). The compressive strength increases by 50% as the block height was reduced from 100 to 50 mm.

The compressive strength of the masonry unit drastically reduces if the horizontal friction at the interface between the machine platen and the masonry unit surface is removed. The friction can be removed by the use of Teflon sheet, waxing or special brush platens. The masonry unit tested without horizontal friction undergoes free lateral expansion under uniform compression due to Poisson effect. The free lateral deformation induces horizontal tension in the masonry unit, ultimately leading to the development of the vertical splitting cracks (Fig. 7.25). This kind of strength is designated as the unrestrained compressive strength. The unrestrained compressive strength of the masonry unit such as burnt clay brick will be about 50% of the restrained compressive strength tested following the standard test procedure (Lenczner 1972).

Generally, the mortar compressive strength is determined through the testing of the cube specimens. The cube will be sandwiched between the two steel platens of the testing machine. Like the masonry unit testing, the mortar cube will also experience platen friction, and therefore, it is the restrained compressive strength.

Table 7.5 Variation in the compressive strength of cement stabilised CEB with the height of the block (Latha 2015)

CEB dimensions (mm)			Wet compressive strength (MPa)
Length	Width	Height	
230	190	50	6.97
230	190	70	5.81
230	190	100	4.18

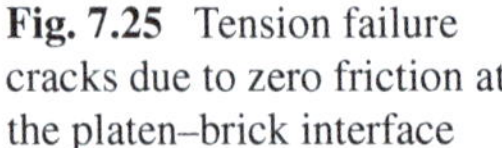

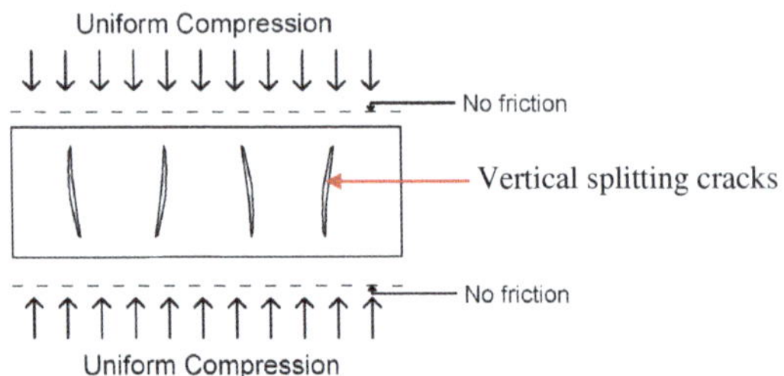

Fig. 7.25 Tension failure cracks due to zero friction at the platen–brick interface

The compressive strength of the masonry unit and the mortar determined following the standard testing procedures is basically restrained compressive strength. The failure pattern of the masonry unit or the mortar cube under such compressive loading conditions resembles typical diagonal shear failure. The nature of stresses induced in the masonry unit and the mortar bed joint in a masonry wall under compression will be different from those experienced by the individual masonry unit or the mortar cube while testing in a machine for assessing the compressive strength.

7.5.2 Masonry Under Compression

The masonry consists of the masonry unit and the mortar joints, with a good bond between them. When the masonry is subjected to concentric compression the masonry unit and the mortar bed joints experience different types of stresses. This is attributed to the different stiffness and deformation characteristics of the masonry unit and the mortar. The nature of stresses developed in the masonry unit and the mortar can be easily understood by analysing a stack-bonded masonry prism. The nature of stresses developed will depend upon the relative stiffness of the masonry unit and the mortar. The modulus and ultimate strain values for the commonly used typical mortars and the modulus values for the cement stabilised CEBs are compiled in Table 6.9 (Chap. 6) and Table 5.10 (Chap. 5), respectively. The modulus of the different types of mortars used for the masonry is in the range of 3.0–12.0 GPa. The modulus of the cement stabilised CEB is sensitive to the cement content and the moisture content at the time of testing. The modulus for CEB is in the range of 1.5–12.0 GPa in the wet state, and the modulus of the dry cement stabilised CEB will be 1.5–3 times the wet CEB modulus, where the dry density of the CEB's is about 1800 kg/m^3 and is manufactured using optimum clay content (12–15%). Based on the modulus values of the mortars and the CEBs, there can be two categories of CEB masonry: (1) $E_b/E_m > 1$ and (2) $E_b/E_m < 1$. E_b and E_m are the modulus of the CEB and the mortar, respectively. The nature of stresses developed in the stabilised CEB, and the mortar are depicted in Fig. 7.26 for the two cases. In a masonry prism under uniform compression, the CEB can experience bi-axial tension or bi-axial compression depending upon the E_b/E_m ratio and similarly the mortar. When the material (brick or mortar) is under bi-axial

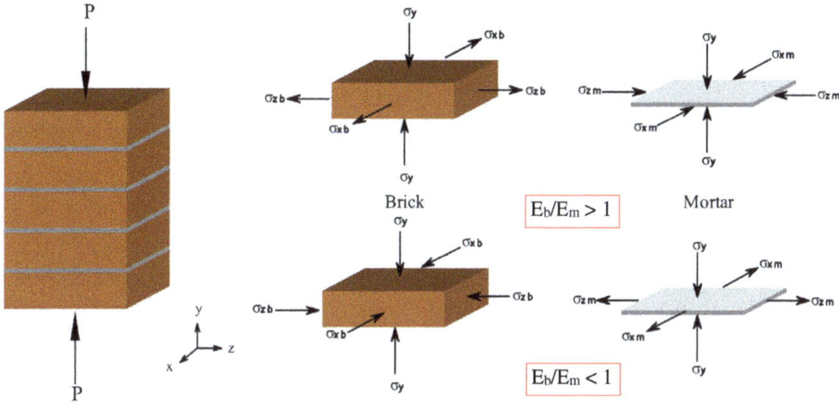

Fig. 7.26 Nature of stresses developed in the masonry unit and the mortar, when the masonry is subjected to axial compression

tension, the masonry prism can develop vertical splitting cracks as shown in Fig. 7.27. The compressive strength of the masonry unit or the mortar under the influence of bi-axial tensile stresses can be much different from the compressive strength determined in a standard test under the platen restraint. Therefore, the compressive strength of the brick or the mortar joint in the masonry is much different from that of the compressive strength of the individual brick or the mortar cube, determined in a testing machine

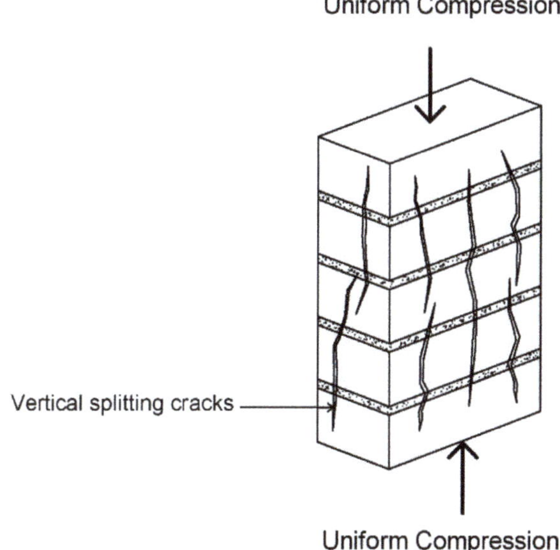

Fig. 7.27 Vertical splitting cracks in a masonry prism under compression

adopting standard test procedure. The compressive strength of the CEB masonry will be about 25–40% of the compressive strength of the CEB.

7.5.3 Factors Influencing the Compressive Strength of Stabilised CEB Masonry Walls

Use of any new type of masonry unit such as stabilised CEB for the construction of structural masonry walls of the buildings or other structures demands understanding of the masonry wall strength and its behaviour under gravity loading. The compressive strength of the CEB masonry wall is influenced by the characteristics of the stabilised CEB, the mortar, type of bonding, physical dimensions of the wall and the nature of fixity at the ends. The major factors affecting the strength of the stabilised CEB masonry wall are listed below.

(a) Characteristics of stabilised CEB

1. Compressive strength
2. Height of the CEB unit
3. Moisture content
4. Pore structure of the bedding faces
5. Stress–strain characteristics

(b) Characteristics of the mortar

1. Mortar composition
2. Workability of the fresh mortar
3. Compressive strength
4. Thickness of the mortar bed joint
5. Stress–strain characteristics

(c) Type of masonry

1. Type of masonry wall and bonding detail
2. Bed joint orientation with reference to loading direction

(d) Wall geometry and end conditions

1. Slenderness ratio of the masonry wall
2. Load eccentricity
3. Nature of slab-wall connectivity or end conditions

7.5.4 Compressive Strength of Stabilised CEB Masonry Elements

The basic compressive strength or the characteristic compressive strength of the masonry is generally assessed by testing a masonry prism or a wallette. ASTM C1314

(2016), BS EN 1052-1 (1999), NBC (2016) and other standard codes on masonry give procedure for assessing the compressive strength of the masonry. ASTM and Indian codes suggest testing stack-bonded masonry prisms and normalising the prism compressive strength by multiplying with correction factors to account for height-to-width ratio of the prism. The Eurocode (EN 1052-1) suggests testing a masonry wallette.

7.5.4.1 Stabilised CEB Masonry Prism Strength

The stack-bonded masonry prisms are used in the laboratory testing to assess the compressive strength of the CEB masonry. Investigations of Shrinivasa Rao (1993), Shrinivasa Rao et al. (1995), Reddy (1991), Reddy and Jagadish (1989), Gupta (2004), Walker (2004), Lal (2005), Vyas (2006), Reddy et al. (2007), Reddy and Gupta (2008), Reddy and Vyas (2008), Latha (2015) and Tennant et al. (2016) generated information on the cement stabilised CEB masonry prism strength. These investigations used cement stabilised CEB with different types of mortars combinations. The size and shape as well as the cement content of the CEBs used in these investigations varied considerably. Considering the studies focused on cement stabilised CEB using optimum clay content in the 10–15% range, Table 7.6 gives the compressive strength of cement stabilised CEB masonry prisms collated from different investigations. The brick/block size and the masonry prism sizes are different, and hence, the prism strength has been normalised using correction factors from ASTM C1314 (2016) code.

Table 7.7 gives the height-to-width ratio correction factors for masonry prism compressive strength extracted from ASTM C1314 (2016) code. Similarly, the CEB compressive strength should be normalised to account for the shape of the masonry unit (height and width). Some of the standard codes on masonry and masonry unit testing give correction factors based on the height-to-width ratio, which is called shape modification factor, while some standard codes (BS EN 772-1) give shape modification factors based on height and width of the masonry unit. Table 7.8 gives the shape factors extracted from BS EN 772-1 code. The normalised compressive strength of the stabilised CEB using the shape factors given in Table 7.8 were used in Fig. 7.28. A plot of normalised prism compressive strength versus the normalised stabilised CEB brick or block compressive strength is shown in Fig. 7.28. The prism compressive strength increases with the increase in the stabilised CEB strength. The prism strength-to-block strength ratio is in the range 0.45–0.60 with some exceptions in few cases. For the burnt clay brick masonry, the range is 0.25–0.35 (Hendry 1981, Sarangapani et al. 2002, Gumaste et al. 2007). The higher value of prism strength-to-block strength ratio for the CEB masonry can be mainly attributed to the larger masonry unit height (~100 mm) when compared to the height of the burnt clay brick (~70 mm).

The mortar strength is poorly related to masonry prism compressive strength as illustrated in Fig. 7.29. An empirical expression of the form given below was derived from the data given in Table 7.6 using multivariate regression analysis.

Table 7.6 Compressive strength of cement stabilised CEB masonry prisms

Sl. No.	Stabilised CEB			Mortar					Prism					Reference
	Size (mm)	C (%)	Wet strength (MPa)	Proportion (by volume)				Strength (MPa)	H/W ratio	Wet compressive strength (MPa)				
				C	L	So	Sa			Exptl	Normalised (1)	Predicted* (2)	Ratio (2) ÷ (1)	
1	305 × 145 × 101	7	3.49	1	–	–	6	6.07	2.21	1.44	1.464	3.040	2.08	Srinivasa Rao (1993)
2	305 × 143 × 100	6	3.13	1	–	–	6	5.40	3.05	1.23	1.321	2.607	1.97	Gupta (2004)
	305 × 143 × 100	12	7.19	1	–	–	6	5.40	3.05	4.84	5.198	9.092	1.75	
3	305 × 143 × 100	6	3.13	1	1	–	6	5.94	3.05	1.33	1.428	2.586	1.81	
	305 × 143 × 100	8	5.63	1	1	–	6	5.94	3.05	2.37	2.545	6.246	2.45	
	305 × 143 × 100	12	7.19	1	1	–	6	5.94	3.05	5.27	5.660	9.019	1.59	
4	305 × 143 × 100	6	3.13	1	–	2	5	2.70	3.05	1.37	1.471	2.763	1.88	
	305 × 143 × 100	8	5.63	1	–	2	5	2.70	3.05	2.50	2.685	6.674	2.49	
	305 × 143 × 100	12	7.19	1	–	2	5	2.70	3.05	5.25	5.639	9.637	1.71	

(continued)

Table 7.6 (continued)

Sl. No.	Stabilised CEB			Mortar					Prism					Reference
	Size (mm)	C (%)	Wet strength (MPa)	Proportion (by volume)				Strength (MPa)	H/W ratio	Wet compressive strength (MPa)				
				C	L	So	Sa			Exptl	Normalised (1)	Predicted* (2)	Ratio (2) ÷ (1)	
5	305 × 143 × 96	4	3.14	1	–	2	5	3.45	3.69	1.20	1.350	2.629	1.95	Lal (2005)
	305 × 143 × 96	4	2.77	1	–	2	5	3.45	3.69	1.02	1.148	2.179	1.90	
	305 × 143 × 96	8	5.73	1	–	2	5	3.45	3.69	2.38	2.678	6.490	2.42	
	305 × 143 × 96	8	4.99	1	–	2	5	3.45	3.69	2.13	2.400	5.273	2.20	
6	255 × 122 × 80	5	5.09	1	1	–	6	3.42	3.61	2.30	2.573	5.344	2.08	Vyas (2006)
	255 × 122 × 80	5	5.09	1	½	–	4	9.40	3.61	2.50	2.797	4.909	1.76	
7	230 × 190 × 100	7	4.62	1	1	–	6	4.12	2.84	3.25	3.446	4.161	1.21	Latha (2015)
	230 × 190 × 100	7	4.62	1	–	2	5	3.47	2.84	3.72	3.945	4.222	1.07	
	230 × 190 × 100	7	5.23	1	1	–	6	4.12	3.01	3.58	3.834	5.766	1.50	
	305 × 143 × 100	7	5.23	1	–	2	5	3.47	3.01	4.09	4.380	5.850	1.34	

(continued)

Table 7.6 (continued)

Sl. No.	Stabilised CEB			Mortar					Prism					Reference
	Size (mm)	C (%)	Wet strength (MPa)	Proportion (by volume)				Strength (MPa)	H/W ratio	Wet compressive strength (MPa)				
				C	L	So	Sa			Exptl	Normalised (1)	Predicted* (2)	Ratio (2) ÷ (1)	
8	230 × 108 × 69	7	7.43	1	–	2	5	6.80	3.57	4.74	5.288	8.903	1.68	Tennant et al. (2016)
9	230 × 109 × 71	8	7.32	1	–	–	6	5.00	3.81	4.40	4.993	8.767	1.76	New data
	230 × 109 × 71	8	7.32	1	1	–	6	6.11	3.81	4.17	4.732	8.620	1.82	
	230 × 109 × 71	8	7.32	1	–	2	5	4.90	3.81	3.73	4.233	8.781	2.07	

C: Portland cement, L: Hydrated lime, So: Soil, Sa: Sand, H: Prism height, W: Prism width, *predicted by Eq. (7.2)

Table 7.7 Height-to-thickness correction factors for masonry prism compressive strength (ASTM C1314 (2016))

H_p/T_p	1.30	1.50	2.00	2.50	3.00	4.00	5.00
Correction factor	0.75	0.86	1.00	1.04	1.07	1.15	1.22

H_p: Masonry prism height; T_p: Least lateral dimension of the prism

Table 7.8 Shape factor *d* to allow for the tested dimensions of the specimens after surface preparation (BS EN 772-1)

Height (mm)	Width (mm)			
	100	150	200	≥250
65	0.85	0.75	0.70	0.65
100	1.00	0.90	0.80	0.75
150	1.20	1.10	1.00	0.95
200	1.35	1.25	1.15	1.10

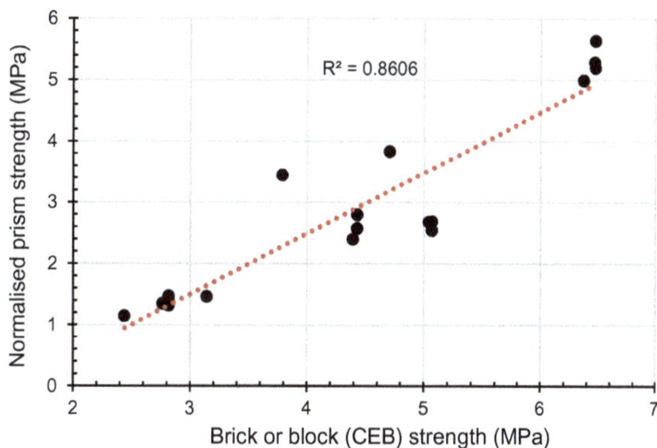

Fig. 7.28 Prism strength versus brick or block strength

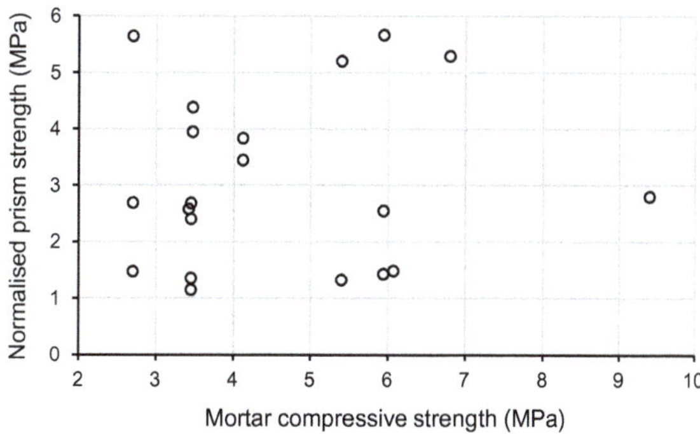

Fig. 7.29 Mortar compressive strength versus prism compressive strength

$$f_p = kf_b^k f_m^l = 0.634(f_b)^{1.502}(f_m)^{-0.084} \quad (7.2)$$

where f_p: Normalised masonry prism wet compressive strength in MPa
f_b: Normalised CEB wet compressive strength in MPa
f_m: Mortar cube compressive strength in MPa.

The values predicted by Eq. (7.2) are given in Table 7.6. The ratio between the predicted and normalised prism strength varies in the range of 1.07–2.45. Equation (7.2) predicts the masonry prism strength on the much higher side. Such an empirical formula should be used with caution, as they are derived from a specific set of experimental data. Generalising such formulae is difficult. The Eurocode (BS EN 1996-1-1:2005) gives an equation to predict the characteristic compressive strength of the masonry based on the masonry unit and the mortar type/strength. The equation is as follows.

$$f_k = K(f_b)^{0.7}(f_m)^{0.3} \quad (7.3)$$

where f_k: Characteristic compressive strength of the masonry in MPa
f_b: Normalised mean compressive strength of the masonry unit in MPa
f_m: Compressive strength of the mortar in MPa.

Considering solid CEB masonry unit and using general purpose mortar, the value of $K = 0.55$ as given Table 3.3 of BS EN 1996-1-1:2005.

7.5.4.2 Basic Compressive Stress and Characteristic Compressive Strength of CEB Masonry

Based on the masonry prism or wallette strength one can derive the basic compressive stress and the characteristic compressive strength of the masonry. In the working stress design approach for masonry, the basic compressive stress is used to assess the limiting compressive stresses in the masonry walls. Masonry codes based on the working stress design approach such as Indian codes (NBC 2016; IS 1905) recommend obtaining the basic compressive stress for the masonry as follows.

Basic compressive stress = (0.25) × (Normalised masonry prism compressive strength).

The method of assessing the characteristic compressive strength of masonry using Eurocodes is given in Sect. 8.3 of Chap. 8.

7.5.5 Influence of Joint Thickness on the Stabilised CEB Masonry Compressive Strength

The relative thickness of the bed joint and the masonry unit influence the masonry compressive strength significantly. Generally, the masonry compressive strength decreases as the joint thickness-to-masonry unit height ratio increases. For a given masonry unit height, increase in the bed joint thickness leads to reduction in the masonry compressive strength. This type of masonry behaviour is true when the masonry unit modulus is more than that of the mortar modulus. The state of stress in the masonry unit and the mortar for this type of masonry is illustrated in Fig. 7.26. In this type of masonry under compression, the masonry unit is under bi-axial tension in the two lateral directions. The force equilibrium equation is as follows.

$$(\sigma_{bt})t_b = (\sigma_{mc})t_j \tag{7.4}$$

where σ_{bt} = Lateral tension in the masonry unit
σ_{mc} = Lateral compression in the mortar
t_b = Thickness of masonry unit
t_j = Thickness of the bed joint.

It is clear for the force equilibrium Eq. (7.4) that the lateral tensile stress in the masonry unit increases as the bed joint thickness increases. For a given masonry unit with limited tensile strength, the masonry prism compressive strength decreases as the bed joint thickness increases.

In the case of the stabilised CEB manufactured using optimum clay content in the soil, the modulus of the block depends upon the stabiliser content and the density. Two situations arise here, the stabilised compressed earth block modulus is greater than the mortar modulus and vice versa. The state of stress in the CEB and the mortar joint for these two cases is illustrated in Fig. 7.26. The stabilised CEB masonry compressive strength and the bed joint thickness relationships for these two cases will be different. There are few focused studies (Shrinivasa Rao 1993; Gupta 2004; Lal 2005; Vyas 2006) in understanding the variation in the masonry prism compressive strength as the joint thickness was varied for the above-mentioned two cases for stabilised CEB masonry. The information culled from these studies is displayed in Fig. 7.30. Figure 7.30 shows variation in the CEB masonry prism compressive strength with the joint thickness-to-block height, ratio. The CEBs used in these investigations had optimum clay content in the mix used for the block production. The block dry density was about 1800 kg/m^3, and the strength of the blocks and prisms were generated in wet (48 h soaking in water prior to testing) condition.

Figure 7.30 shows the two trend lines. For the case of $E_b > E_m$, the masonry prism compressive strength decreases as the ratio of joint thickness to block height increases. The behaviour is similar to what was observed for masonry with burnt clay bricks (Francis et al. 1971; Totaro 1994). The reasons for the strength reduction for thicker bed joint thickness was discussed earlier. The other trend line (Fig. 7.30)

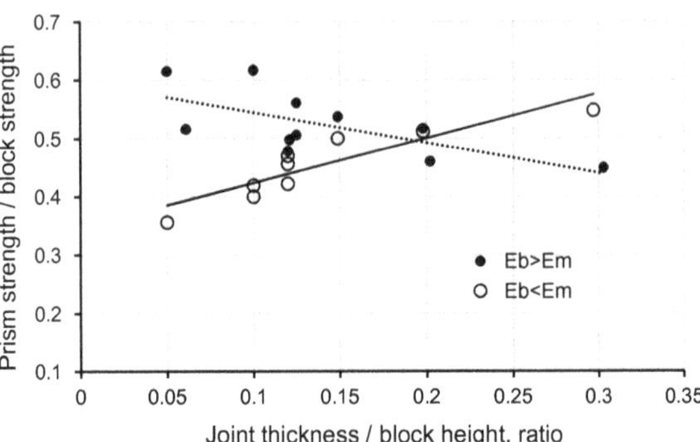

Fig. 7.30 Influence of joint thickness-to-block height ratio on masonry compressive strength

for the case of $E_b < E_m$, shows increase in masonry prism compressive strength as the joint thickness-to-block height ratio increases. In this case, the mortar will be under bi-axial tension and compression, and the masonry unit is under tri-axial compression as illustrated in Fig. 7.26. The equilibrium equation for the forces in the bed joint and the block or brick is as follows.

$$(\sigma_{bc})t_b = (\sigma_{mt})t_j \tag{7.5}$$

where σ_{bc} = Lateral compression in the masonry unit
σ_{mt} = Lateral tension in the mortar
t_b = Thickness of the masonry unit
t_j = Thickness of the bed joint.

The force equilibrium Eq. (7.5) clearly illustrates that the lateral tensile stress in the bed joint decreases as the bed joint thickness increases. For a given mortar type with limited tensile strength, the masonry prism compressive strength increases as the bed joint thickness increases.

Generally, the masonry prism under uniform compression fails by developing vertical splitting cracks (Fig. 7.27). The vertical splitting cracks will be in a direction perpendicular to the direction of lateral tensile stress developed either in the masonry unit or in the mortar joint, depending upon their modular ratio. For the case of $E_b > E_m$, the vertical splitting crack first develops in the masonry unit and then extends into the mortar joint and to the whole prism height. It is vice versa for the case of $E_b < E_m$. The arguments put forward here are based on elastic analysis. The masonry materials will enter nonlinear phase as they approach failure stage. Also, there will be Poisson's ratio influence on the nature and magnitude of stresses developed in the masonry materials in a masonry prism under compression. The investigations of Reddy et al. (2007) provide more information on the modulus and Poisson's ratio effect on the stresses developed in the masonry materials in a masonry prism under uniform compression.

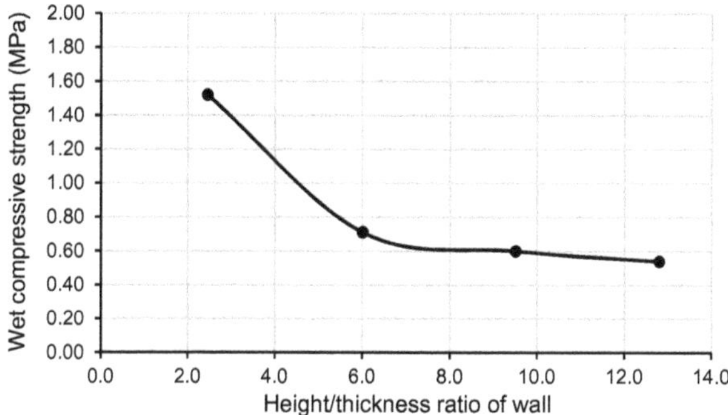

Fig. 7.31 Effect of slenderness ratio on the compressive strength of the cement stabilised CEB masonry wall

7.5.6 Compressive Strength of Stabilised CEB Masonry Walls

Irrespective of the strength of the brick/block and the mortar, assessing the load-carrying capacity of the full-scale wall in a structure is essential. The major factors affecting the compressive strength of the masonry wall are:

(1) Strength of the masonry unit and the mortar, and their geometrical dimensions
(2) Elastic properties of the masonry materials
(3) Slenderness ratio of the wall
(4) Load eccentricity
(5) Boundary conditions
(6) Bonding pattern

The compressive strength of the CEB masonry prisms and the wallettes was discussed in the earlier sections. The compressive strength of the CEB masonry wall will be much lower than that of the CEB masonry prism or the wallette. Figure 7.31 shows a plot of CEB wall strength (in wet condition) and the slenderness ratio subjected to the concentric loads (Reddy 1991). The cement (5%) stabilised CEB used in generating the plot had a wet compressive strength of 2.51 MPa, and the cement-mortar had a compressive strength of 3.38 MPa. The wall strength decreases with increase in the slenderness ratio as depicted in Fig. 7.31. Since the slenderness ratio of the walls tested was small (≤ 12), the ultimate failure was due to the material failure as depicted in Fig. 7.32. The slender masonry walls (height-to-thickness ratio > 25) fail due to buckling. There were partial restraints at the ends of the CEB walls tested in these experiments. The bottom portion of the wall was resting on the floor, and at the top end, there was a spreader concrete beam. The stress or capacity reduction factors given in the masonry codes to account for slenderness and eccentricity of loading can be used for the design of CEB masonry.

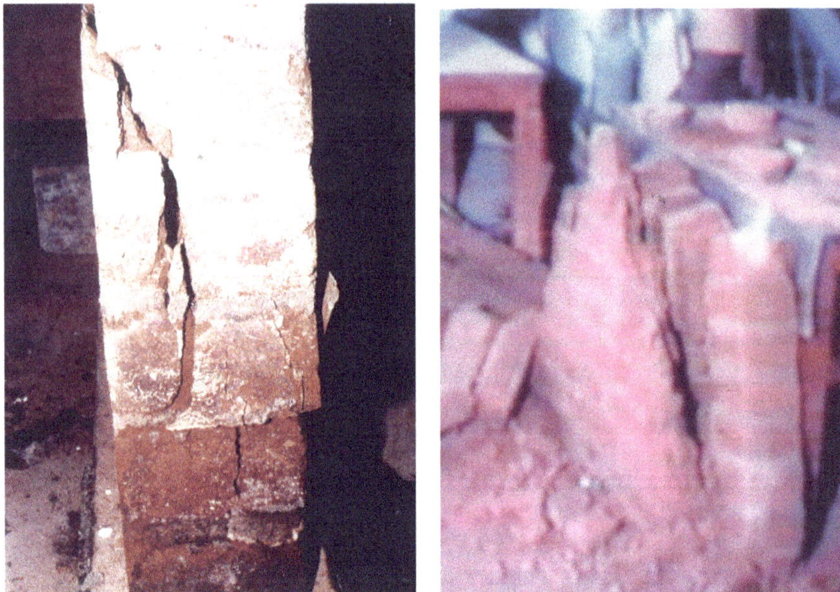

Fig. 7.32 Material failure (shear failure) of the masonry wall (height/thickness ratio of wall: 12)

7.6 Stress–Strain Characteristics of Mortars, Stabilised CEB and CEB Masonry

The stress–strain characteristics of the stabilised CEB and the mortars were discussed in Sect. 5.10 (Chap. 5) and Sect. 6.5.6 (Chap. 6), respectively. Typical stress–strain curves (in wet state) for the mortar, cement stabilised CEB and CEB masonry are shown in Fig. 7.33. The stress–strain curves show that the strain at peak stress for the masonry is higher than the strain at peak stress for the CEB and the mortar. Also, the

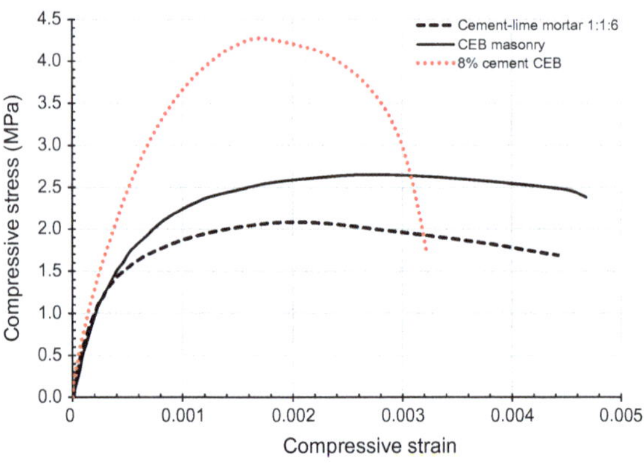

Fig. 7.33 Typical stress–strain relationships for cement stabilised CEB, cement–lime mortar and their masonry (in wet condition)

Table 7.9 Stress–strain characteristics for CEB, mortar and masonry (in wet condition)

Details of CEB		Mortar details			Masonry details		Reference
Modulus (MPa)	Strain at peak stress	Composition (by volume) C: So: L: Sa	Modulus* (MPa)	Strain at peak stress	Modulus (MPa)	Strain at peak stress	
6100	0.0033	1: 0: 0: 6	4500	0.0040	4200	0.0044	Gupta (2004)
6100	0.0033	1: 0: 1: 6	4800	0.0026	4500	0.0052	
6100	0.0033	1: 2: 0: 5	4300	0.0023	5900	0.0046	
8000	0.0017	1: 2: 0: 5	6600	0.0019	7200	0.0029	Lal (2005)
6700	0.0016	1: 0: 1: 6	6500	0.0020	5300	0.0027	Vyas (2006)
14,500	0.0014	1: 0: 1: 6	6500	0.0020	13,100	0.0025	Vyas (2006)

* Modulus: Initial tangent modulus, C: Portland cement, L: lime, So: Soil, Sa: Sand

post-peak response for the masonry shows large flattening portion with higher strains at failure than the post-peak behaviour of the CEB. The stress–strain characteristics of the stabilised CEB, mortars and their masonry are given in Table 7.9. Generally, the modulus of cement stabilised CEB masonry lies in between the modulus of CEB and the mortar. The strain at the peak stress for the masonry is much larger than the strain at peak stress for the mortar and the CEB.

References

ASTM C1072-10 (2010) Standard test methods for measurement of masonry flexural bond strength, ASTM Standards, Conshohocken, PA, USA

ASTM C1314-16 (2016) Standard test method for compressive strength of masonry prisms, ASTM Standards, Conshohocken, PA, USA

Barnes BD, Diamond S, Dolch WL (1978) The contact zone between Portland cement paste and glass aggregate surfaces. Cem Concr Res 8(2):233–243

BS EN 1052-1 (1999) Methods of test for masonry: determination of compressive strength. The British Standards Institution, UK

BS EN 1996-1-1:2005+A1:2012 (2012) Eurocode 6: design of masonry structures, part 1—1: general rules for reinforced and unreinforced masonry structures. The British Standards Institution, UK

BS EN 772-1:2011+A1:2015 (2015) Methods of test for masonry units part 1: determination of compressive strength. The British Standards Institution, UK

Francis AJ, Horman CB, Jerrems LE (1971) The effect of joint thickness and other factors on compressive strength of brickwork. In: Proc. 2nd international brick masonry conference, Stoke-on-Trent, pp 31–37

Garndet J, Javelas R, Perrin B, Theonoz B (1972) Role de l'ettringite dans la liaison de type mecanique entre la terre cuite et la pate de ciment-portland. Revue Terre Cuite 48:21–28 (in Greek)

Garndet J (1973) Physikalich-chemische mechanism der Haftung zwischen ziegel and zement. In: Proc. 3rd Int. Brick masonry conference, Essen, pp 217–221
Gourav K (2015) Studies on flexural behaviour of fly ash-lime-gypsum brick masonry. PhD thesis, Department of Civil Engineering, Indian Institute of Science, Bangalore, India
Gourav K, Reddy BVV (2018) Bond development in burnt clay and FaL-G brick masonry. J Mater Civ Eng 30(9):04018202, 1–10. https://doi.org/10.1061/(ASCE)MT.1943-5533.0002412
Groot C (1993) Effects of water on mortar-brick bond. PhD thesis, Delft Univ. of Technology, The Netherlands
Gumaste KS, Nanjunda Rao KS, Venkatarama Reddy BV, Jagadish KS (2007) Strength and elasticity of brick masonry prisms and wallettes under compression. Mater Struct 40(296):241–253
Gupta A (2004) Studies on characteristics of cement-soil mortars and soil-cement block masonry. MSc (Engg.) thesis, Dept. of Civil Engineering, Indian Institute of Science, Bangalore, India
Hendry AW (1981) Structural brickwork. Macmillan, London, UK
IS 1905 (1987) Code of practice for structural use of unreinforced masonry. Bureau of Indian Standards, New Delhi, India
Kampf L (1963) Factors affecting bond of mortar to brick. In: Proceedings of symposium on masonry testing, ASTM International
Lal R (2005) Characteristics of soil-cement blocks and soil-cement block masonry. MSc (Engg.) thesis, Dept. of Civil Engineering, Indian Institute of Science, Bangalore, India
Latha MS (2015) Studies on characteristics of stabilised soil compacts for structural applications. PhD thesis, Department of Civil Engineering, Indian Institute of Science, Bangalore, India
Latha MS, Reddy BVV (2018) Mortar shrinkage and flexure bond strength of stabilised soil brick masonry. J Mater Civ Eng 30(5):05018002
Lawrence SJ, Cao HT (1987) An experimental study of the interface between brick and mortar. In: International proceedings of the fourth North American masonry conference, Dublin, pp 194–204
Lawrence SJ, Cao HT (1988) Microstructure of the interface between brick and mortar. In: International proceedings of the eighth international brick/block masonry conference, Dublin, Ireland, pp 194–204
Lenczner D (1972) Elements of loadbearing brickwork. Pergamon Press, Oxford, UK
McBurney JW, Copeland MA, Brink RC (1946) Permeability of brick-mortar assemblages. In: Proceedings—American society for testing and materials, ASTM, W Conshohocken, PA, pp 1333–1354
NBC (2016) National building code of India 2016, group 2, sections 1 & 4. Bureau of Indian Standards, New Delhi, India
Neville AM (1995) Properties of concrete, 4th edn. Pearson, Essex, UK
Reddy BVV (1991) Studies on static soil compaction and compacted soil-cement blocks for walls. PhD thesis, Dept. of Civil Engineering, Indian Institute of Science, Bangalore, India
Reddy BVV, Gupta A (2005) Characteristics of soil-cement blocks using highly sandy soils. Mater Struct 38(280):651–658
Reddy BVV, Gupta A (2006) Strength and elastic properties of stabilised mud block masonry using cement soil mortars. J Mater Civ Eng 18(3):472–476
Reddy BVV, Gupta A (2008) Influence of sand grading on the characteristics of mortars and soil-cement block masonry. Constr Build Mater 22(8):1614–1623
Reddy BVV, Jagadish KS (1989) Properties of soil-cement block masonry. Mason Int 3(2):80–84
Reddy BVV, Latha MS (2014) Influence of soil grading on the characteristics of cement stabilised soil compacts. Mater Struct 47(10):1633–1645
Reddy BVV, Vyas CVU (2008) Influence of shear bond strength on compressive strength and stress strain characteristics of masonry. Mater Struct 41(10):1697–1712
Reddy BVV, Lal R, Nanjunda Rao KS (2007) Enhancing bond strength and characteristics of soil-cement block masonry. J Mater Civ Eng 19(2):164–172
Reddy BVV, Nikhil V, Nikhilash M (2019) Moisture transport in cement stabilised soil brick-mortar interface. In: Venkatarama Reddy BV, Mani M, Walker P (eds) Earthen dwellings and structures:

current status in their adoption. Springer Transactions in Civil & Environmental Engineering, Singapore, pp 27–37

Sarangapani G, Venkatarama Reddy BV, Jagadish KS (2002) Structural characteristics of bricks, mortars and masonry. J Struct Eng 29(2):101–107

Sarangapani G, Venkatarama Reddy BV, Jagadish KS (2005) Brick-mortar bond and masonry compressive strength. J Mater Civ Eng 17(2):229–237

Shrinivasa Rao S (1993) Studies on soil-cement blocks and block masonry. MSc (Engg.) thesis, Dept. of Civil Engineering, Indian Institute of Science, Bangalore, India

Shrinivasa Rao S, Venkatarama Reddy BV, Jagadish KS (1995) Strength characteristics of soil-cement block masonry. Indian Concr J 69(2):127–131

Sinha BP (1967) Model studies related to load bearing brickwork. PhD thesis, Univ. of Edinburgh, Edinburgh, UK

Tennant AG, Foster CD, Venkatarama Reddy BV (2016) Detailed experimental review of flexural behavior of cement stabilized soil block masonry. J Mater Civ Eng 28(6):060160004-1–060160004-5. https://doi.org/10.1061/(ASCE)MT.1943-5533.0001548

Thornton JC (1953) Relation between bond and the surface physics of masonry units. J Am Ceram Soc 36(4):105–120

Totaro N (1994) A hybrid elastic theory for evaluation of compressive strength of brick masonry. In: Proc., 10th Int. brick and block masonry Conf., pp 1443–1451

Vyas VCU (2006) Studies on shear bond strength—masonry compressive strength relationships and finite element model for prediction of masonry compressive strength. MSc (Engg.) thesis, Dept. of Civil Engineering, Indian Institute of Science, Bangalore, India

Venu Madhava Rao K, Venkatarama Reddy BV, Jagadish KS (1996) Flexural bond strength of masonry using various blocks and mortars. Mater Struct 29:119–124

Walker PJ (1999) Bond characteristics of earth block masonry. J Mater Civ Eng 11(3):249–256

Walker PJ (2004) Strength and erosion characteristics of earth blocks, and earth block masonry. J Mater Civ Eng 16(5):497–506

Walker P, Stace T (1997) Properties of some cement stabilised compressed earth blocks and mortars. Mater Struct 30:545–551

Chapter 8
Design of Stabilised Compressed Earth Block Masonry

8.1 Design for Gravity Loads

Standard codes (Indian, British, Eurocode, Australian standards, and many other codes) are used for the design of masonry structures. In the absence of dedicated masonry design code on the stabilised compressed earth block (CEB) masonry, the existing masonry design codes can be made use of, for the design of cement stabilised CEB masonry.

Masonry consists of two different materials: the masonry unit and the mortar. Generally, the two materials have different strength and deformation characteristics. Therefore, the nature of stresses experienced by the two materials (in a masonry assembly subjected to uniform compression) will be different as discussed in the Sect. 7.5.2 (Chap. 7). Under compression, the failure pattern of a masonry prism/wallette or wall will be much different when compared with the masonry unit's failure. The slenderness and load eccentricity influence the compressive strength of the masonry walls.

The design of load bearing wall or a column for gravity loads finally reduces to determination of the compressive strength of the masonry to support the design loads. The masonry design leads to assessing the compressive strength of the masonry unit required for a specific type of mortar and the wall thickness chosen. The basic principle of the masonry wall design can be expressed as follows.

Design loading $\leq$ *Design load resistance*

The design for the vertical loading is assessed from the known applied loading. The design resistance is a function of the basic or characteristic compressive strength of the masonry, the slenderness of the wall and the load eccentricity. The design can be carried out following either the working stress or the limit state design principles. The Indian code NBC (2016) (Part 6, Group 2) gives the design procedure for the unreinforced masonry, which is based on the working stress design principles. Majority of the masonry design codes are based on the limit state design principles.

B. V. V. Reddy, *Compressed Earth Block & Rammed Earth Structures*,
Springer Transactions in Civil and Environmental Engineering,
https://doi.org/10.1007/978-981-16-7877-6_8

The design procedure outlined in the NBC (2016) code (working stress method) and the Euro code-6 (BS EN 1996-1-1:2005+A1:2012) (limit state method) is discussed in the following sections.

8.2 Design Procedure for Gravity Loading as Per NBC (2016) Code

Some of the basic definitions mentioned in the two codes are highlighted in the Table 8.1. The basic principle followed in NBC (2016) code is as follows.

The compressive stress developed at the base of the wall or column due to the total design gravity loads $\leq$ (The permissible compressive stress)

$$\begin{aligned}\text{The permissible compressive stress} =& (\text{Basic compressive stress})\\ &\times (\text{Stress reduction factors})\end{aligned} \tag{8.1}$$

8.2.1 Stress Reduction Factors

There are three types of stress reduction factors to account for the slenderness of the wall and the load eccentricity (k_s), area reduction factor (k_a) and shape modification factor (k_p). The NBC (2016) code gives procedure to assess the values of the stress reduction factors.

Brief procedure for the design of a wall under vertical gravity loading is as follows.

1. For a chosen wall thickness, estimate the total load realised on the wall and the compressive stress developed at the base of the wall.
2. Determine the slenderness ratio and load eccentricity for the wall based on the guidelines given in the codes.
3. Based on the physical wall dimensions, end conditions and the geometry of the masonry unit chosen, assess the three stress reduction factors (k_s, k_a and k_p).
4. Calculate the permissible compressive stress and determine the *basic compressive stress* using the Eq. (8.1).
5. Referring to the Table 9 in NBC (2016) (Part 6, Group 2, Sect. 4), select the brick or block strength and the mortar type for a given value of the basic compressive stress.

Design examples in the subsequent sections illustrate the design procedure based on NBC (2016) code.

Table 8.1 Terminology used in the NBC 2016 and Eurocode 6

Terminology	NBC 2016 (Part 6 Sect. 4)	Eurocode-6
1. Masonry unit		
(a) Compressive strength	Mean compressive strength of a specified number of masonry units as per relevant Indian Standard	Mean compressive strength of a specified number of masonry units (see EN 771–1 to EN 771-6)
(b) Normalised compressive strength	Compressive strength of masonry units having height to width ratio > 0.75 is normalised using shape modification factors	Compressive strength of masonry units converted to the air-dried compressive strength of an equivalent 100 mm wide & 100 mm high, masonry unit (see EN 771–1 to EN 771-6)
2. Mortar joints		
Bed joint	Horizontal mortar joint upon which masonry units are laid	Mortar layer between the bed faces of the masonry units
Perpend joint (head joint or cross joint)	Vertical joint normal to the face of the wall	Mortar joint perpendicular to the bed joint and to the face of the wall
Longitudinal joint or Wall joint	Vertical joint parallel to the face of the wall	Vertical mortar joint within the thickness of a wall, parallel to the face of the wall
3. Masonry		
Characteristic strength of masonry	Not relevant in working stress design procedure	Value of the strength of masonry having a prescribed probability of 5% of not being attained in a hypothetically unlimited test series
Compressive strength of masonry	Normalised masonry prism strength	Strength of the masonry in compression without the effects of platen restraint, slenderness or eccentricity of loading
Basic compressive stress	Crushing strength of masonry with slenderness ratio less than 6, zero eccentricity and masonry unit height to width ratio ≤ 0.75 or 0.25(normalised masonry prism compressive strength)	Not relevant in limit state design procedure

8.3 Design Procedure for Gravity Loading as Per Eurocode-6 (BS EN 1996-1-1:2005+A1:2012)

The code is based on the limit state design principles. The unreinforced masonry walls subjected to mainly vertical gravity loading, the resistance of the masonry walls shall be based on the wall geometry, load eccentricity and material properties.

At the ultimate limit state

$$N_{\text{Ed}} \leq N_{\text{Rd}} \tag{8.2}$$

N_{Ed} = Design value of the vertical gravity load on the wall
N_{Rd} = Design value of the vertical resistance of the wall

$$N_{\text{Rd}} = \Phi t f_{\text{d}} \tag{8.3}$$

where,
Φ = Capacity reduction factor accounting for slenderness and load eccentricity
t = Wall thickness
f_{d} = Design compressive strength of the masonry which is based on the characteristic compressive strength of the masonry.

Reduction factor Φ can be assessed based on the guidelines given in BS EN 1996-1-1.

The design value of the loads (N_{Ed}) can be obtained by assessing actions (which include loads and imposed deformations) based on BS EN 1991 (Eurocode 1) or other relevant sources. A distinction must be made between the actions which are permanent, or which vary with time or change position or extent. Assess the ultimate loading actions based on the combinations of actions and the partial safety factors.

The partial safety factors for the masonry units and mortars are given in Table 8.2 (BS EN 1996-1-1). Refer to the BS EN 998-2 and BS EN 1996-2 for the details on designed and prescribed mortar definitions. The partial safety factors are applied to the characteristic material strengths to obtain the design strengths.

The characteristic compressive strength of the masonry should be determined from either

(1) tests on masonry specimens (tested as per EN 1052-1) or from the existing database. The test results can be expressed in a tabular form or in terms of the following equation.

$$f_{\text{k}} = K(f_{\text{b}})^{\alpha}(f_{\text{m}})^{\beta} \tag{8.4}$$

Table 8.2 Partial safety factors for the masonry units (γ_{M}) (BS EN 1996-1-1)

Masonry unit	Material partial safety factor (γ_{M})				
	Class				
	1	2	3	4	5
Units of category I, designed mortar	1.5	1.7	2.0	2.2	2.5
Units of category I, prescribed mortar	1.7	2.0	2.2	2.5	2.7
Units of category II, any mortar	2.0	2.2	2.5	2.7	3.0

Refer to BS EN 998-2 and BS EN 1996-2 for designed and prescribed mortars

where
f_k is the characteristic compressive strength of the masonry, in MPa
K, α, β are constants
f_b is the normalised mean compressive strength of the masonry unit, in MPa
f_m is the mortar compressive strength, in MPa

or

(2) from the following relationship for the masonry with general purpose mortar or light weight mortar

$$f_k = K(f_b)^{0.7}(f_m)^{0.3} \tag{8.5}$$

The value of K can be obtained from the Tables given in the code. For example, for the masonry units having holes ≤25% falling under Group 1, for which the value of $K = 0.55$ when general purpose mortar is used. For the mortar joint parallel to the face of the wall, through all or any part of the length of the wall and using general purpose mortar, K is multiplied by 0.8 [3.6.1.2(6), BS EN 1996-1-1]. The normalised compressive strength (f_b) can be obtained from the mean compressive strength of the masonry unit by multiplying with a shape factor to account for masonry unit size (especially height and width) given in BS EN 772-1. The design procedures using the limit state concepts are illustrated in the following sections.

8.4 Stabilised CEB Masonry Design Using Working Stress Method

8.4.1 Example 1—Design of a Dormitory

The masonry design calculations are based on the working stress design principles of NBC (2016) code of India. The plan and sections of the four storey load bearing cement stabilised CEB building are shown in Figs. 8.1 and 8.2. It is a dormitory building, located close to Bangalore city, India. The roof and the floor slabs are continuous reinforced concrete in-situ construction. The design of the two critical walls in the building indicted as Wall-A and Wall-B has been considered.

Loadings
Roof
The slab thickness: 150 mm including the weatherproof course and ceiling finishes
The bulk density of the reinforced concrete slab and finishes: 25 kN/m^3
Dead load = 0.15 m × 25 kN/m^3 = 3.75 kN/m^2
Live load or imposed load = 1.5 kN/m^2 (Clause 3.4.1, NBC 2016, Group 2, Part 2, Sect. 1)

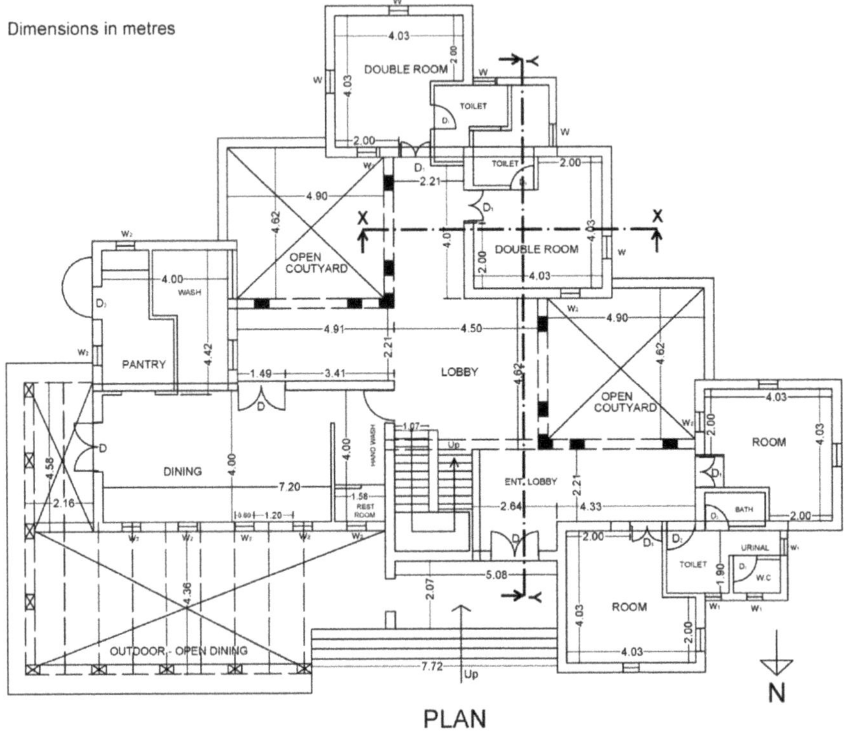

Fig. 8.1 Plan of the dormitory building

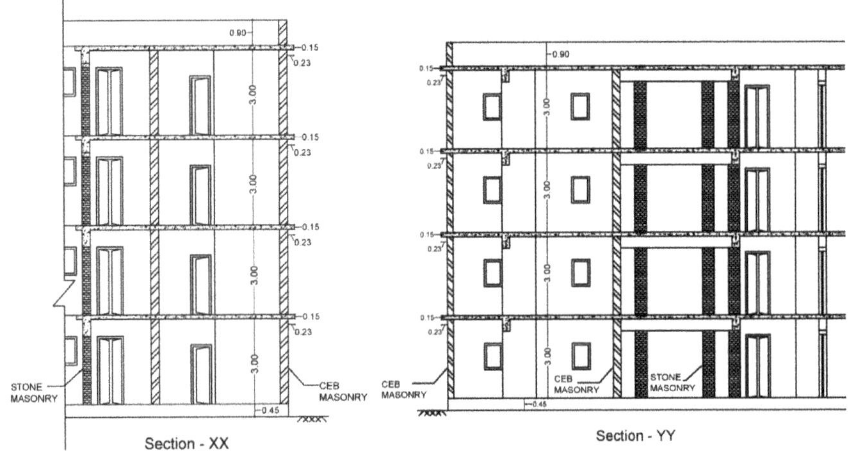

Fig. 8.2 Sections of the dormitory building

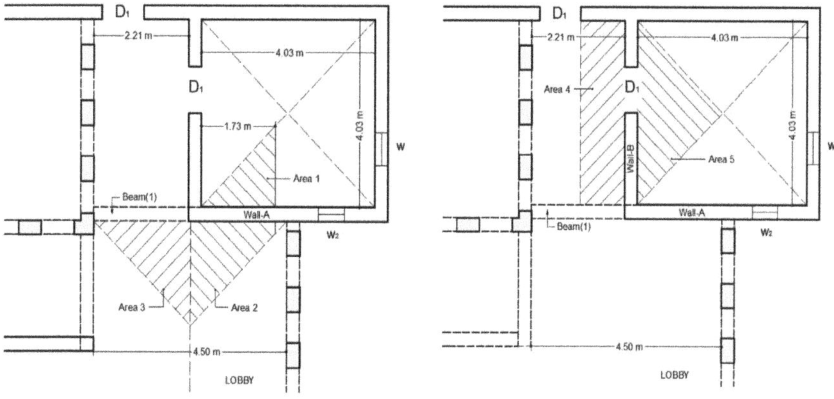

Fig. 8.3 Roof and floor slab areas transferring loads to Wall-A and Wall-B

Floor
The slab thickness: 150 mm including the floor and ceiling finishes

The bulk density of the reinforced concrete slab and finishes: 25 kN/m^3

Dead load = 0.15 m × 25 kN/m^3 = 3.75 kN/m^2

Live load or imposed load = 2.0 kN/m^2 (Clause 3.3.1, NBC 2016, Group 2, Part 2, Sect. 1)

Walls
The CEB walls of 200 mm thick, assuming for design purposes a 12 mm thick plaster layer on either face.

Self-weight of the wall = (0.20 + 2 × 0.012) m × 1.0 m × 20 kN/m^3 = 4.48 kN/m^2

Loads realised on the walls
Figure 8.3 shows the areas of the roof/floor slab transferring the loads to Wall-A and Wall-B.

Wall-A
The wall receives uniformly distributed loads from Areas (1) and (2) and a concentrated reaction from Beam (1). Due to the concentrated load reaction from the Beam (1), the masonry wall length to be considered as per the code (angle of dispersion 30°) is illustrated in Fig. 8.4.

The masonry wall length to be considered = 1.73 m

Slab area A1 (triangle) transferring the load on to the wall (Fig. 8.3) = 0.5(1.73)^2 = 1.50 m^2

Slab area A2 (trapezium) transferring the load on to the wall (Fig. 8.3) = 0.5(2.25 + 0.315) × (1.73 + 0.16) = 2.42 m^2

Total slab area transferring the load = 1.50 + 2.42 = 3.92 m^2

Reaction from Beam (1)
The beam is receiving a triangular slab load as shown in Fig. 8.5.

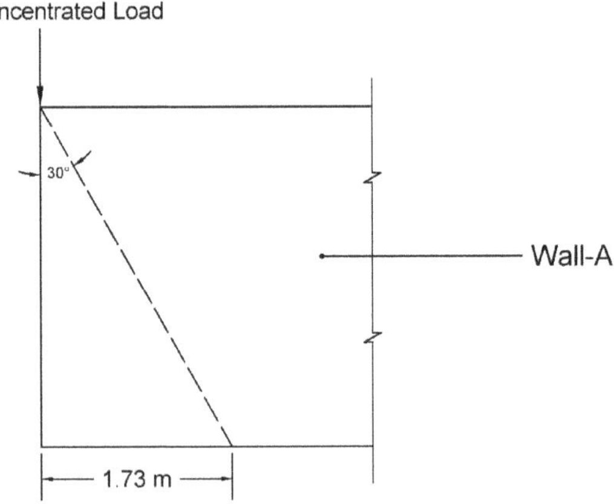

Fig. 8.4 Length of the masonry Wall-A due to reaction from Beam (1)

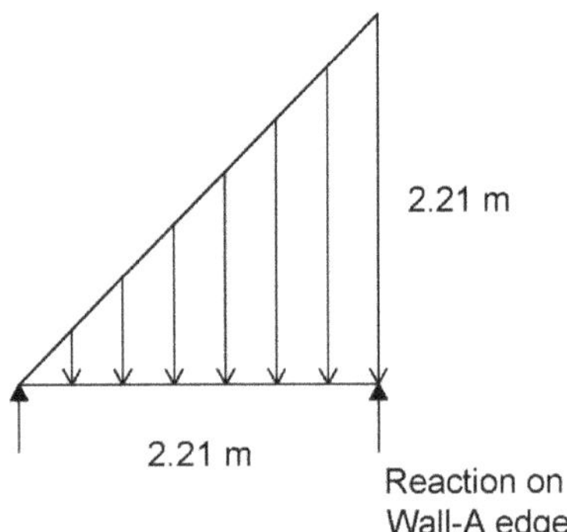

Fig. 8.5 Triangular load distribution area on beam (1)

The reaction (point load) on the edge of Wall-A due to the reaction from Beam (1) (refer Fig. 8.5) = (1.63 m^2) × load intensity

The calculations of vertical loading on Wall-A are illustrated in Table 8.3.

Wall-B (with an opening shown in Fig. 8.6)

The wall receives uniformly distributed loads from roof and floor slab portions (4) and (5). Area (4) is one way slab. It is assumed that the reaction from Beam (1) is not realised on the Wall-B. Area (5) is part of the two-way slab. The masonry wall has a door opening, and hence, the entire wall length can be considered for the design calculations.

The masonry wall length considered = 4.03 m

Slab area A4 (one way slab) transferring the load to the wall = 0.5(2.21) × (4.03) = 4.45 m^2

Slab area A5 (two-way slab) transferring the load to the wall = 0.5(4.03 × 2.015) = 4.06 m^2

Total slab area transferring the load = 4.45 + 4.06 = 8.51 m^2

The calculations of vertical loading on Wall-B are illustrated in the Table 8.4

Table 8.3 Calculation of vertical loading on Wall-A

Floor level considered		Load per 1.73 m run (kN)		Cumulative design load to floor (kN)
		Dead	Imposed	
3rd Floor				
Dead weight of roof: 3.92 m^2 × 3.75 kN/m^2	=14.70 kN	48.02	8.33	56.35
Weight of wall: 3.0 m × (1.73 + 0.2) m × 4.48 kN/m^2	=25.94 kN			
	=40.64 kN			
Reaction from Beam (1): (1.63 m^2) × 3.75 kN/m^2	=6.11 kN			
Self-weight of beam: 0.5(0.20 m × 0.23 m × 2.21 m)25 kN/m^3	=1.27 kN			
	48.02 kN			
Imposed load: (3.92 + 1.63) m^2 × 1.5 kN/m^2	8.33 kN			
2nd Floor				
Dead weight of floor: 3.92 m^2 × 3.75 kN/m^2	=14.70 kN	96.04	17.48	113.52
Weight of wall: 3.0 m × (1.73 + 0.2) m × 4.48 kN/m^2	=25.94 kN			
	=40.64 kN			
Reaction from Beam (1) + self-weight: (6.11 + 1.27)	=7.38 kN			
Dead weight from above:	=48.02 kN			
	96.04 kN			
90% Imposed load: 0.90(3.92 + 1.63) m^2 × (2 + 1.5) kN/m^2	=17.48 kN			
1st Floor				
Dead weight (as in 2nd floor): (40.64 + 7.38) kN	=48.02 kN	144.06	24.42	168.48
Dead weight from above:	=96.04 kN			
	144.06 kN			
80% Imposed load: 0.80(3.92 + 1.63) m^2 × (2 + 2 + 1.5) kN/m^2	=24.42 kN			
Ground Floor				
Dead weight (as in 2nd floor): (40.64 + 7.38) kN	=48.02 kN	192.08	29.14	221.22
Dead weight from above:	=144.06 kN			
	192.08 kN			

(continued)

Table 8.3 (continued)

Floor level considered		Load per 1.73 m run (kN)		Cumulative design load to floor (kN)
		Dead	Imposed	
70% Imposed load: 0.70(3.92 + 1.63) m^2 × (2 + 2 + 2 + 1.5) kN/m^2	=29.14 kN			

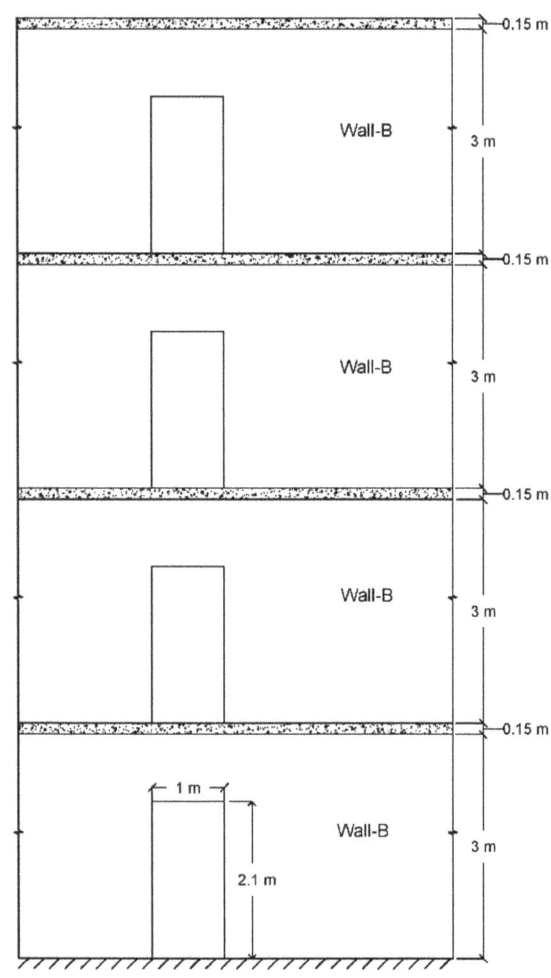

Fig. 8.6 Wall-B with an opening

Table 8.4 Calculation of vertical loading on Wall-B

Floor level considered		Load per 4.03 m run (kN)		Cumulative design load to floor (kN)
		Dead	Imposed	
3rd Floor				
Dead weight of roof: 8.51 $m^2 \times 3.75$ kN/m^2	=31.91 kN	86.07	12.77	98.84
Weight of wall: 3.0 m × 4.03 m × 4.48 kN/m^2	=54.16 kN			
	86.07 kN			
Imposed load: 8.51 m^2 × 1.5 kN/m^2	=12.77 kN			
2nd Floor				
Dead weight of floor: 8.51 $m^2 \times 3.75$ kN/m^2	=31.91 kN	172.14	26.81	198.95
Weight of wall: 3.0 m × 4.03 m × 4.48 kN/m^2	=54.16 kN			
	86.07 kN			
Dead weight from above:	=86.07 kN			
	172.14 kN			
90% imposed load: 0.90 × 8.51 m^2 × (2 + 1.5) kN/m^2	=26.81 kN			
1st Floor				
Dead weight (as in 2nd floor):	=86.07 kN	258.21	37.44	295.65
Dead weight from above:	=172.14 kN			
	=258.21 kN			
80% Imposed load: 0.80 × 8.51 m^2 × (2 + 2 + 1.5) kN/m^2	=37.44 kN			
Ground Floor				
Dead weight (as in 2nd floor):	=86.07 kN	344.28	44.68	388.96
Dead weight from above:	=258.21 kN			
	344.28 kN			
70% Imposed load: 0.7 × 8.51 m^2 × (2 + 2 + 2 + 1.5) kN/m^2	=44.68 kN			

Wind loading

General stability is met as per sec. 4.2.2.2 of NBC (2016) (Part 6, Group 2, Sect. 4). Height to width ratio of the building is about 0.60 < 2.0, and cross walls spacings are within the limits specified in the code. The solid reinforced concrete slab rests on the cross walls. Hence, the general stability need not be checked.

Selection of the strength of cement stabilised CEB using M2 grade mortar for Wall-A

CEB size: 230 × 200 × 100 mm (Length × width × height)

The mortar, 1:1:6 (cement:lime:sand, by volume) belongs to M2 grade (Tables 1 and 2 NBC 2016, Part 6-Sect. 4). The minimum compressive strength of the mortar, at 28 days, is 3 MPa

$$\text{Basic compressive stress} = \left[(\text{Design load}) \div (\text{Area of wall})\right] \times (1 \div \beta)$$

where $\beta = (k_a k_s k_p)$

k_a = Area reduction factor = 1, since area of the wall > 0.20 m^2 (Clause 5.4.1.2, NBC 2016, Part 6, Group 2, Sect. 4)

k_p = Shape modification factor = 1, since height-to-width ratio of the block = 100 ÷ 200 = 0.5 (Table 11 of NBC 2016, Part 6, Group 2, Sect. 4)

k_s = Stress reduction factor for slenderness ratio and load eccentricity

Effective height of the wall = 0.75 × (3.15) m = 2.36 m.

Effective thickness = 0.20 m.

Slenderness ratio = (2.36) ÷ (0.20) = 11.80

The load eccentricity = 0, since the floor slab is continuous, and the span of the slab on one side does not exceed by more than 15% of the slab span on the other side (Clause 4.7 and Annex A of NBC 2016, Group 2, Part 6 Sect. 4).

For slenderness ratio of 11.8 and load eccentricity = 0, k_S = 0.845 (Table 10 of NBC 2016, Part 6, Group 2, Sect. 4)

The wet compressive strength of the cement stabilised CEB (size: 230 × 200 × 100 mm), considering M2 grade mortar and 200 mm wall thickness, is given in Table 8.5, for each floor. The CEB strength required in the top floor wall is 3.5 MPa, which is the minimum strength required as per NBC (2016). Whereas the strength required for the ground floor wall is about 8.5 MPa.

The cement stabilised CEB compressive strength ascertained from the design is the wet compressive strength.

Selection of the strength of cement stabilised CEB using M2 grade mortar for Wall-B

CEB size: 230 × 200 × 100 mm (Length × width × height)

k_a = Area reduction factor = 1, since area of the wall > 0.20 m^2 (Clause 5.4.1.2 NBC 2016, Part 6, Group 2, Sect. 4)

Table 8.5 Basic compressive stress and brickwork strength (Wall-A)

Floor	Design load (kN)	Axial design stress (MPa)	Basic compressive stress (MPa)	*CEB strength (MPa) with M2 mortar
3rd Floor	56.35	0.15	0.18	3.5
2nd Floor	113.52	0.29	0.35	3.5
1st Floor	168.48	0.44	0.51	6.2
Ground floor	221.22	0.57	0.68	8.5

*Selection of CEB compressive strength using M2 grade mortar from Table 9 of NBC (2016) (Part 6, Sect. 4) code
Area of Wall-A = (1.73 + 0.20) m × 1000 × 200 mm = 386,000 mm^2

k_p = Shape modification factor = 1, since height-to-width ratio of the block = 100 ÷ 200 = 0.5 (Table 11 of NBC 2016, Part 6, Group 2, Sect. 4)

k_s = Stress reduction factor for slenderness ratio and load eccentricity

Effective height of the wall = 0.75 × (3.15) m = 2.36 m

Effective thickness = 0.20 m

Slenderness ratio = (2.36) ÷ (0.20) = 11.80

The Wall-B receives load from the slabs, whose spans differ by >15% as shown in Fig. 8.7.

The reactions from the slab loads are displaced by (1/6) of the bearing width from the centre of the wall. The reactions from the slabs A4 and A5 along with the concentric load from self-weight of the wall are indicated in Fig. 8.8.

Where P_S: Wall self-weight, P_4: Reaction from slab A4, P_5: Reaction from slab A5, t = wall thickness

$$W = P_S + P_4 + P_5$$

$$\text{Eccentricity of load} = e = [(P_4 - P_5)(t \div 12)] \div (W)$$

The loads realised on Wall-B are given in Table 8.4.

Consider the ground floor: Cumulative design load to the floor = 388.96 kN = Total load = W

P_S = 216.64 kN, P_4 = 90.11 kN, P_5 = 82.21 kN

Therefore, e = [(90.11 – 82.21)(200 ÷ 12)] ÷ (388.96) = 0.34 mm

The eccentricity is a negligible value, and therefore, consider e = 0. Also, clause 4.7 of NBC (2016) (Part 6, Group 2, Sect. 4), Annex A4 states that the interior walls carrying continuous floors are assumed to be axially loaded.

Stress reduction factor = k_s = 0.845, for slenderness ratio of 11.80 and $e = 0$ (Clause 5.4.1.1 Table 10, NBC 2016, Part 6, Group 2, Sect. 4).

The wet compressive strength of the cement stabilised CEB (size: 230 × 200 × 100 mm), considering M2 grade mortar and 200 mm wall thickness, is given in Table 8.6, for each floor. The CEB compressive strength required for wall in the top

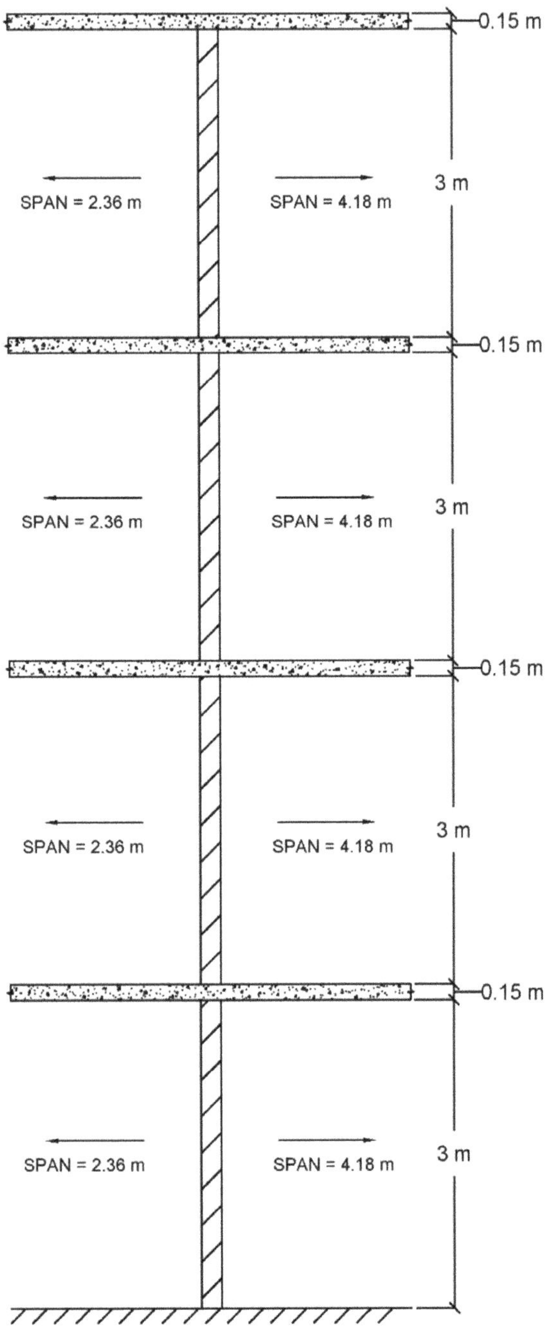

Fig. 8.7 Wall-B with continuous sold slab of unequal spans on either side

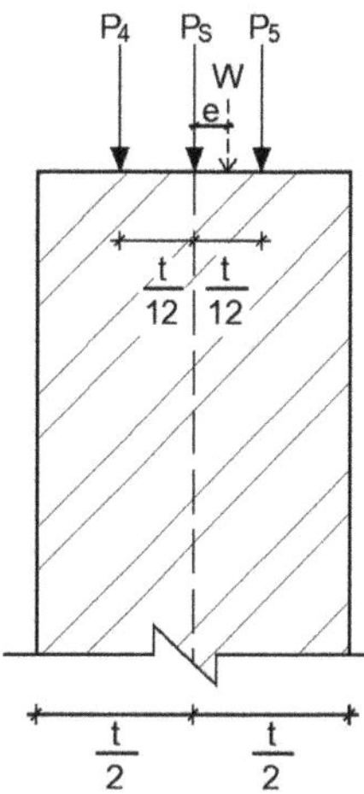

Fig. 8.8 Unequal reactions and the eccentricity of the total load reaction on Wall-B (bearing width = $t/2$)

Table 8.6 Basic compressive stress and brickwork strength (Wall-B)

Floor	Design load (kN)	Axial design stress (MPa)	Basic compressive stress (MPa)	*CEB compressive strength (MPa) with M2 mortar
3rd Floor	98.84	0.16	0.19	3.5
2nd Floor	198.95	0.33	0.39	4.2
1st Floor	295.65	0.49	0.58	7.3
Ground floor	388.96	0.64	0.76	9.4

* Selection of CEB compressive strength using M2 grade mortar from the Table 9 of NBC (2016) (Part 6, Group 2, Sect. 4)
Area of Wall-B = (4.03 − 1.0) m × 1000 × 200 mm = 606,000 mm^2

floor wall is 3.5 MPa, which is the minimum strength required as per NBC (2016). Whereas the CEB compressive strength required for the ground floor wall is about 9.4 MPa.

The cement stabilised CEB compressive strength ascertained from the design is the wet compressive strength.

8.4.2 Example 2—Six Storey Load Bearing Structure

Cement stabilised CEB masonry design using working stress method

The masonry design calculations are based on the working stress design principles of NBC (2016) code of India. The plan and sections of the six storey load bearing cement stabilised CEB masonry building are shown in Figs. 8.9 and 8.10, respectively. It is part of a hostel complex, located close to Bangalore city, India. The roof and

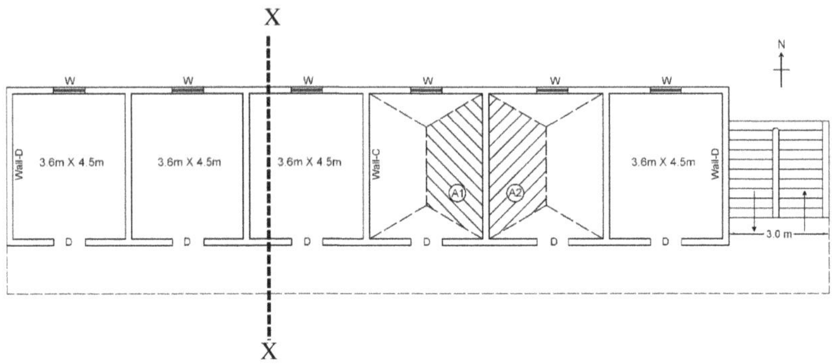

Fig. 8.9 Typical plan of six storey load bearing masonry building

floor slabs are continuous reinforced concrete in-situ construction. The design of the critical wall in the building indicted as Wall-C has been considered.

Loadings

Roof

The slab thickness: 150 mm including the weatherproof course and ceiling finishes

The bulk density of the reinforced concrete slab and finishes: 25 kN/m^3

Dead load = 0.15 m × 25 kN/m^3 = 3.75 kN/m^2

Live load or imposed load = 1.5 kN/m^2 (Clause 3.4 of NBC 2016, Part 6, Group 2, Sect. 4)

Floor

The slab thickness: 150 mm including the floor and ceiling finishes

The bulk density of the reinforced concrete slab and finishes: 25 kN/m^3

Dead load = 0.15 m × 25 kN/m^3 = 3.75 kN/m^2

Live load or imposed load = 2.0 kN/m^2 (Clause 3.3 of NBC 2016, Part 6, Group 2, Sect. 4)

Walls

CEB walls of 200 mm thick, assuming for design purposes a 12 mm plaster layer on either face of the walls.

$$\text{Self - weight of wall} = (0.20 + 2 \times 0.012)\ \text{m} \times 1.0\ m \times 20\ \text{kN/m}^3 = 4.48\ \text{kN/m}^2$$

Loads on the Wall-C

The shaded portion in the Fig. 8.9 (plan) shows the areas of the roof/floor slab transferring the loads to Wall-C. The wall receives uniformly distributed loads from the slab with Areas (1) and (2).

Consider masonry wall of length 1.0 m in the middle of the room.

Slab area (A1 and A2) transferring the load on to the wall = 2(3.6 m ÷ 2) × 1.0 m = 3.6 m^2

Fig. 8.10 Cross section at XX

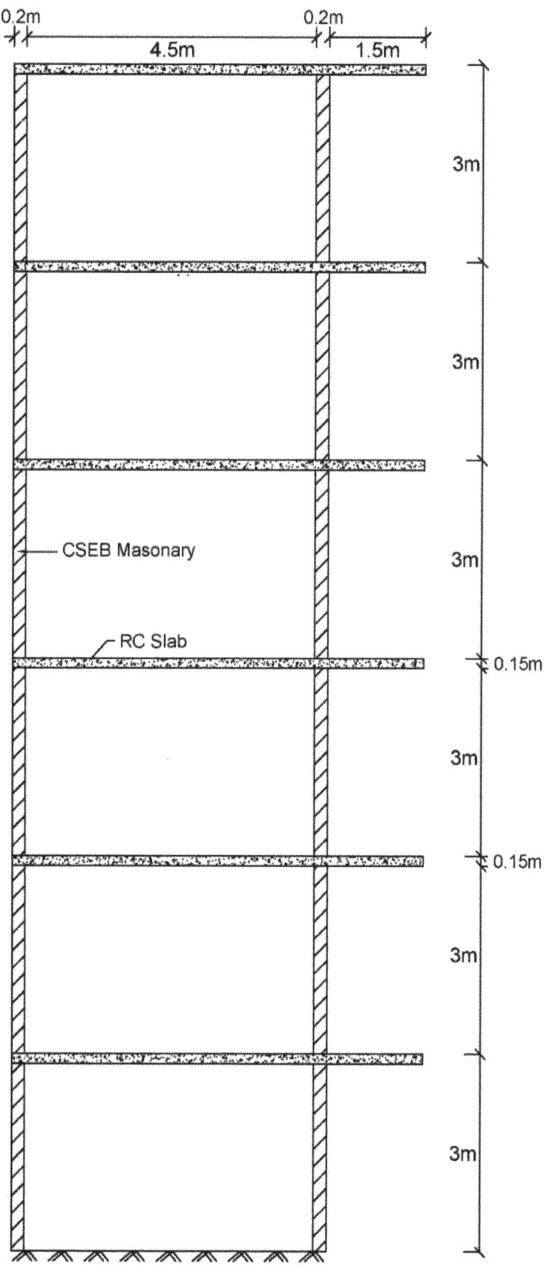

Table 8.7 Calculation of vertical loading on Wall-C

Floor level considered		Load per m run (kN)		Cumulative design load to floor (kN)
		Dead	Imposed	
5th Floor				
Dead weight of roof: 3.6 m^2 × 3.75 kN/m^2	=13.50 kN	26.27	5.40	31.67
Weight of wall: 2.85 m × 1.0 m × 4.48 kN/m^2	=12.77 kN			
	26.27 kN			
Imposed load: 3.6 m^2 × 1.5 kN/m^2	=5.40 kN			
4th Floor				
Dead weight of floor: 3.6 m^2 × 3.75 kN/m^2	=13.50 kN	52.54	11.34	63.88
Weight of wall: 2.85 m × 1.0 m × 4.48 kN/m^2	=12.77 kN			
Dead weight from above	=26.27 kN			
	52.54 kN			
90% Imposed load: 0.90 × 3.6 m^2 × (2 + 1.5) kN/m^2	=11.34 kN			
3rd Floor				
Dead weight of floor: 3.6 m^2 × 3.75 kN/m^2	=13.50 kN	78.81	15.84	94.65
Weight of wall: 2.85 m × 1.0 m × 4.48 kN/m^2	=12.77 kN			
Dead weight from above	=52.54 kN			
	78.81 kN			
80% Imposed load: 0.80 × 3.6 m^2 × (2 + 2 + 1.5) kN/m^2	=15.84 kN			
2nd Floor				
Dead weight of floor: 3.6 m^2 × 3.75 kN/m^2	=13.50 kN	105.08	18.90	123.98
Weight of wall: 2.85 m × 1.0 m × 4.48 kN/m^2	=12.77 kN			
Dead weight from above	=78.81 kN			
	105.08 kN			
70% Imposed load: 0.70 × 3.6 m^2 × (2 + 2 + 2 + 1.5) kN/m^2	=18.90 kN			
1st Floor				

(continued)

Table 8.7 (continued)

Floor level considered		Load per m run (kN)		Cumulative design load to floor (kN)
		Dead	Imposed	
Dead weight of floor: 3.6 m^2 × 3.75 kN/m^2	=13.50 kN	131.35	20.52	151.87
Weight of wall: 2.85 m × 1.0 m × 4.48 kN/m^2	=12.77 kN			
Dead weight from above	=105.08 kN			
	131.35 kN			
60% Imposed load: 0.60 × 3.6 m^2 × (2 + 2 + 2 + 2 + 1.5) kN/m^2	=20.52 kN			
Ground Floor				
Dead weight of floor: 3.6 m^2 × 3.75 kN/m^2	=13.50 kN	157.62	20.70	178.32
Weight of wall: 2.85 m × 1.0 m × 4.48 kN/m^2	=12.77 kN			
Dead weight from above	=131.35 kN			
	157.62 kN			
50% Imposed load: 0.50 × 3.6 m^2 × (2 + 2 + 2 + 2 + 2 + 1.5) kN/m^2	=20.70 kN			

The calculations of vertical loading on Wall-C are illustrated in Table 8.7.

Wind loading

General stability is not met as per clause 4.2.2.2 of NBC (2016) (Part 6, Group 2, Sect. 4). Height-to-width ratio of the building is about 3.83 > 2.0, and cross walls spacings are within the limits specified in the code. The solid reinforced concrete slab rests on the cross walls. The design of the walls receiving lateral wind and axial loads has been illustrated in the following sections.

The building orientation is shown in Fig. 8.9. The wind blowing in the north–south direction is critical. The stresses caused due to wind blowing parallel to the planes of the cross walls (Wall-C) are considered. Also, it has been assumed that the walls act as independent cantilevers and hence the moments and forces are apportioned according to their stiffnesses.

Wind loads as per the NBC (2016) **(Group 2, Part 6, Sect. 1) for the building location chosen**

Design wind pressure (P_d):

597 N/m^2 for height $\leq$ 10 m
678 N/m^2 for height = 10–15 m

735 N/m^2 for height = 15–20 m

$$\text{Total wind force} = F = C_{pe} A P_d$$

C_{pe} = Pressure coefficient = 1.2 as per clause 4.5.3.1 of NBC (2016) (Group 2, Part 6, Sect. 1)

A = Surface area of the structural element in m^2

P_d = Design wind pressure in N/mm^2

$F = 1.2AP_d$

Total maximum bending moment = $F(h \div 2)$, where h is the height under consideration

Total bending moment just above the floor level (wind direction north–south) has been calculated and given in Table 8.8

Table 8.8 Total bending moment just above the floor level

Details of bending moment (kN-m)		Total bending moment just above floor level (kN m)
5th Floor		
(1.2 × 23.0 m × 3.0 m × 0.735 kN/m^2) × (3.0 m ÷ 2)	=91.29	91.29
4th Floor		
(1.2 × 23.0 m × 3.0 m × 0.735 kN/m^2) × (4.5 m)	=273.86	358.07
(1.2 × 23.0 m × 3.0 m × 0.678 kN/m^2) × (3.0 m ÷ 2)	=84.21	
3rd Floor		
(1.2 × 23.0 m × 3.0 m × 0.735 kN/m^2) × (7.5 m)	=456.44	793.27
(1.2 × 23.0 m × 6.0 m × 0.678 kN/m^2) × (6.0 m ÷ 2)	=336.83	
2nd Floor		
(1.2 × 23.0 m × 3.0 m × 0.735 kN/m^2) × (10.5 m)	=639.01	1386.82
(1.2 × 23.0 m × 6.0 m × 0.678 kN/m^2) × (6.0 m)	=673.66	
(1.2 × 23.0 m × 3.0 m × 0.597 kN/m^2) × (3.0 m ÷ 2)	=74.15	
1st Floor		
(1.2 × 23.0 m × 3.0 m × 0.735 kN/m^2) × (13.50 m)	=821.58	1980.37
(1.2 × 23.0 m × 6.0 m × 0.678 kN/m^2) × (9.0 m)	=1010.49	
(1.2 × 23.0 m × 6.0 m × 0.597 kN/m^2) × (6.0 m ÷ 2)	=148.30	
Ground Floor		
(1.2 × 23.0 m × 3.0 m × 0.735 kN/m^2) × (16.50 m)	=1004.16	3018.81
(1.2 × 23.0 m × 6.0 m × 0.678 kN/m^2) × (12.0 m)	=1347.32	
(1.2 × 23.0 m × 9.0 m × 0.597 kN/m^2) × (9.0 m ÷ 2)	=667.33	

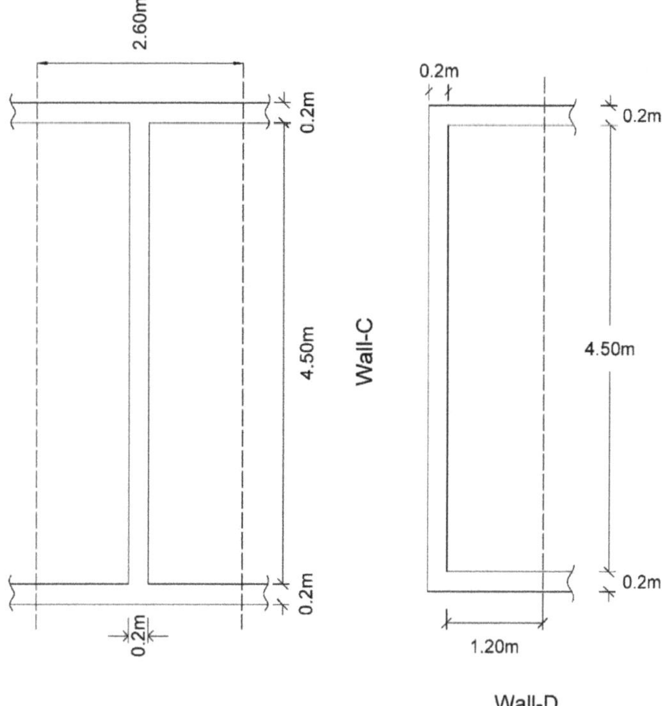

Fig. 8.11 Plan sections of Wall-C and Wall-D along with the flanges

Assumed section of the wall resisting the wind moment

The flanges which act together with the web of "I"-section and "["-section were considered as per clause 4.2.2.5 of NBC (2016) (Group 2, Part 6, Sect. 4). The Fig. 8.11 shows the dimensions of these wall sections.

Wall-C: Calculation of second moment of inertia

$$I_C = 2\left[\left\{\left(2.6 \times 0.20^3\right) \div 12\right\} + \left\{(2.6 \times 0.20) \times (4.7 \div 2)^2\right\}\right] + \left\{0.20 \times (4.5)^3 \div 12\right\} = 7.266\ \text{m}^4$$

Wall-D: Calculation of second moment of inertia

$$I_D = 2\left[\left\{\left(1.2 \times 0.20^3\right) \div 12\right\} + \left\{(1.2 \times 0.20) \times (4.7 \div 2)^2\right\}\right] + \left\{0.20 \times (4.5)^3 \div 12\right\} = 4.171\ \text{m}^4$$

Total second moment of inertia of the building = I

$$I = 5I_C + 2I_D = 5(7.266) + 2(4.171) = 44.672\ \text{m}^4$$

Moment carried by Wall-C, M_C = (Total moment) × ($I_C \div I$) = (M) × ($I_C \div I$) = 0.163 M

Moment carried by Wall-D, M_D = (Total moment) × ($I_D \div I$) = (M) × ($I_D \div I$) = 0.093 M.

Selection of the strength of cement stabilised CEB using M2 mortar for Wall-C

CEB size: 230 × 200 × 100 mm (Length × width × height)

The mortar, 1:1:6 (cement:lime:sand, by volume) belong to M2 grade (Tables 1 and 2 of NBC (2016) (Group 2, Part 6, Sect. 4). The minimum compressive strength of the mortar, at 28 days, is 3 MPa

$$\text{Basic compressive stress} = \left[(\text{Design load}) \div (\text{Area of wall})\right] \times (1 \div \beta)$$

where $\beta = (k_a k_s k_p)$

k_a = Area reduction factor = 1, since area of the wall > 0.20 m^2 (Clause 5.4.1.2 of NBC (2016) (Group 2, Part 6, Sect. 4)

k_p = Shape modification factor = 1, since height-to-width ratio of the block = 100 ÷ 200 = 0.5 (Table 10 of NBC 2016, Group 2, Part 6, Sect. 4)

k_s = Stress reduction factor for slenderness ratio and load eccentricity

Effective height of the wall = 0.75 × (3.0) m = 2.25 m (centre to centre of supports = 3.0 m)

Effective thickness = 0.20 m

Slenderness ratio = (2.25) ÷ (0.20) = 11.25

The eccentricity $e = 0$, NBC (2016) clause 4.7, Annex A4 states that the interior walls carrying continuous floors are assumed to be axially loaded.

For slenderness ratio of 11.25 and load eccentricity = 0, $k_S = 0.859$ (Table 10 of NBC 2016, Group 2, Part 6, Sect. 4).

The wet compressive strength of the cement stabilised CEB (size: 230 × 200 × 100 mm), considering M2 grade mortar and 200 mm wall thickness, is given in Table 8.9, for each floor. The CEB strength required in the 4th floor wall is 3.9 MPa, whereas the strength required for the ground floor wall is about 15.3 MPa.

Table 8.9 Basic compressive stress and brickwork strength (Wall-C)

Floor	Design load/m (kN)	Design stress (MPa)	Combined design stress (MPa)	Basic compressive stress (MPa)	*CEB strength using M2 mortar (MPa)
		Axial ± bending			
5th Floor	31.67	0.158 ± 0.005	0.163 or 0.153	0.184	3.5
4th Floor	63.88	0.319 ± 0.018	0.337 or 0.301	0.371	3.9
3rd Floor	94.65	0.473 ± 0.040	0.513 or 0.433	0.551	6.9
2nd Floor	123.98	0.620 ± 0.070	0.690 or 0.550	0.722	9.0
1st Floor	151.87	0.759 ± 0.100	0.859 or 0.659	0.884	11.4
Ground floor	178.32	0.892 ± 0.152	1.044 or 0.740	1.038	15.3

*Selection of CEB compressive strength using M2 grade mortar from the Table 9 of NBC (2016) (Group 2, Part 6, Sect. 4)

Note

(1) The CEB compressive strength ascertained is the wet compressive strength

(2) No tension develops anywhere (safe)

(3) Combined basic stress for ground floor = (0.892 + 0.152) ÷ (0.859) = 1.215 MPa < [(1.25) × (1.038) = 1.30 MPa]

(4) The combined axial and wind loading stresses will not create any problem as 25% increase in permissible design stress is allowed (clause 5.4.1.4 NBC 2016, Group 2, Part 6, Section 4)

8.5 Stabilised CEB Masonry Design Using Limit State Method

8.5.1 Example 3—Design of a Dormitory

This design example was dealt in the Sect. 8.2, using working stress design method. In this section, the design exercise has been illustrated using limit state design based on the design guidelines of Eurocode 6 (BS EN 1996-1-1:2005+A1:2012). The plan and sections of the four storey load bearing CEB building are shown in Figs. 8.1 and 8.2, respectively. It is a dormitory building, located close to Bangalore city, India. The roof and floor slabs are continuous reinforced concrete in-situ construction. The structural design has been shown for the two critical walls in the building indicted as Wall-A and Wall-B. The load calculations for these two walls have been illustrated in the Sect. 8.2, under Example 1, and the details are given in Tables 8.3 and 8.4.

In the limit state design, the design load (actions) shall be matched with design resistance as detailed in Eqs. (8.2) and (8.3). The design exercise involves assessing the values for the following.

(a) Partial safety factors for the ultimate limit state of strength
(b) Partial safety factors for the materials
(c) Capacity reduction factor (Φ) accounting for slenderness and load eccentricity.

The partial safety factors for the permanent load actions (γ_G) and variable load actions (γ_Q) can be taken conservatively as 1.35 and 1.5, respectively, as per the Eurocode (BS EN 1990-2002) code guidelines. Assessing the material partial safety factors (from the Tables given in the relevant EC codes) is difficult, because of the absence of stabilised earth block masonry units in those Tables. Based on the clause 2.4.3 of Eurocode-6, the material partial safety factor = γ_M = 2.5 for category II and class of control 3.

Capacity reduction factor (Φ) for Wall-A and Wall-B (Clause 6.1.2.2 Eurocode-6)

(i) At top and bottom of the wall (Φ_i)

Assumed CEB size: 230 × 200 × 100 mm, and the wall thickness = t = 200 mm.

$(\Phi_i) = 1 - 2(e_i \div t)$

e_i = eccentricity at top and bottom of the wall given by the following equation

$e_i = [(M_{id} \div N_{id}) + e_{he} + e_{init}] \geq 0.05t$

M_{id} = Design value of the bending moment at top or bottom of the wall resulting from the eccentricity of the floor load at the support

N_{id} = Design value of the vertical load at top or bottom of the wall

e_{he} = Eccentricity at the top or bottom of the wall, if any, resulting from horizontal loads

e_{init} = Initial eccentricity (construction defects), assumed to be [(slenderness ratio) ÷ 450]

$M_{id} \div N_{id} = 0$, since the walls are axially loaded

$e_{he} = 0$, since there is no horizontal load

$e_{init} = (h_{ef} \div 450) = 0.75\ (3000) \div 450 = 5$ mm, (Wall height = 3000 mm)

Slenderness ratio = $h_{ef}/t_{ef} = (0.75 \times 3000) \div (200) = 11.25$; $t_{ef} = 200$ mm

$e_i = 0 + 0 + 5 \geq 0.05t = 0.05 \times 200 = 10$ mm

$(\Phi_i) = 1 - 2(10 \div 200) = 0.9$

(i) In the middle of the wall (Φ_m)

Φ_m may be determined from the equations given in Sect. 6.1 and Annex G of Eurocode-6

$\Phi_m = A_1 e^{(x)}$, Where $x = (-u^2/2)$

$A_1 = 1 - 2(e_{mk}/t)$

$e_{mk} = (e_m + e_k) \geq 0.05t$

e_m = eccentricity due to loads = $(M_{md} \div N_{md}) + e_{hm} + e_{init}$

e_k = eccentricity due to creep = $0.002\ \Phi_\infty (h_{ef}/t_{ef})(t e_m)^{0.5}$

Φ_∞ = final creep coefficient

$u = [\lambda - 0.063] \div [0.73 - 1.17(e_{mk}/t)]$

$\lambda = (h_{ef}/t_{ef}) \times (f_k/E)^{0.5}$

h_{ef} = effective height, t_{ef} = effective thickness, f_k = characteristic compressive strength, E = modulus

M_{md}, N_{md} and e_{hm} correspond to mid height greatest value design moment, vertical load and eccentricity

$$(M_{md} + N_{md}) = 0$$

e_{hm} = eccentricity at mid height resulting from horizontal loads = 0 since there is no horizontal load
$e_{init} = (h_{ef} \div 450) = (2250) \div 450 = 5$ mm
$e_m = 0 + 0 + 5 \geq 0.05t = 0.05 \times 200 = 10$ mm
$e_k = 0$ for walls having slenderness ratio of 11.25 (clause 6.1.2)
Total eccentricity in the middle of the wall = $e_{mk} = e_m + e_k = 10 + 0 = 10$ mm
Assuming modulus of elasticity of masonry $E = 700f_k$ (Annex G of Eurocode-6)
$u = [h_{ef}/t_{ef} - 1.67] \div [19.3 - 31(e_{mk}/t)] = 0.54$
$A_1 = 1 - 2(e_{mk}/t) = 1 - 2(10/200) = 0.9$
$\Phi_m = A_1 e^{(x)} = 0.9e^{-0.1458} = 0.78$ (critical); Where $x = (-u^2/2)$
Consider capacity reduction factor (Φ) = 0.78 for Wall-A and Wall-B

Small plan area modification factor does not apply since horizontal cross-sectional area of the Walls A and B exceeds >0.10 m^2 (clause 6.1.2.1 of Eurocode-6)

$$N_{Rd} = \Phi t f_d = (0.78 \times 200 \times f_k) \div (\gamma_M = 2.5) = 62.4 f_k$$

$f_d = f_k \div \gamma_M$ (clause 2.4.1 of Eurocode-6)

$$N_{Rd} \geq N_{Ed}$$

For $N_{Rd} = N_{Ed}, f_k = (N_{Ed} \div 62.4)$ MPa, where N_{Ed} is in N/mm

$$f_k = K(f_b)^{0.7}(f_m)^{0.3}$$

$K = 0.55$ for Group 1 bricks and general purpose mortar (Table 3.3 Eurocode-6), $f_m = 4$ MPa

$(f_b)^{0.7} = (f_k) \div [(0.55) \times (4)^{0.3}] = 1.2(f_k)$

The design loads and the design details for the Wall-A and Wall-B are given in Table 8.10. Table 8.11 gives the comparison of the mean wet compressive strength of the cement stabilised CEB needed for the walls A and B using the working stress and limit state design approaches. In the WSD, minimum brick or block compressive strength needed is 3.5 MPa. The strengths shown in the Table clearly show that the CEB compressive strength ascertained in LSD is much lower when compared to the values obtained using WSD approach. The difficulty in the LSD approach for CEB walls is lack of information on the partial safety factors for the loads and the materials (to be used in CEB wall design calculations).

Table 8.10 Design details for Wall-A and Wall-B using Eurocode-6 code guidelines

Wall-A

Ultimate design load (N_{Ed}) for 1.73 m wall length	N_{Ed} (N/mm)	$f_k = (N_{Ed}) \div (62.4)$ (MPa)	f_b for M4 mortar (MPa)	Mean comp. strength of CEB (MPa)*
3rd Floor 1.35(48.02) + 1.5(8.33) = 77.32 kN	44.70	0.72	0.81	1.01
2nd Floor 1.35(96.04) + 1.5(17.48) = 155.87 kN	90.11	1.44	2.19	2.74
1st Floor 1.35(144.06) + 1.5(24.42) = 231.11 kN	133.59	2.14	3.85	4.81
Ground Floor 1.35(192.08) + 1.5(29.14) = 303.02 kN	175.16	2.81	5.67	7.09

Wall-B

Ultimate design load (N_{Ed}) for 4.03 m wall length	N_{Ed} (N/mm)	$f_k = (N_{Ed}) \div (62.4)$ (MPa)	f_b for M4 mortar (MPa)	Mean comp. strength of CEB (MPa)*
3rd Floor 1.35(86.07) + 1.5(12.77) = 135.35 kN	33.59	0.54	0.54	0.67
2nd Floor 1.35(172.14) + 1.5(26.81) = 272.60 kN	67.64	1.08	1.46	1.82
1st Floor 1.35(258.21) + 1.5(37.44) = 404.74 kN	100.43	1.61	2.56	3.20
Ground Floor 1.35(344.28) + 1.5(44.68) = 531.80 kN	131.96	2.11	3.78	4.73

* Wet compressive strength, as per BS EN 772-1:2011+A1:2015

Table 8.11 Comparison of compressive strengths of stabilised CEB required for Wall-A and Wall-B (CEB size: 230 × 200 × 100 mm)

Details of wall and floor	Mean wet compresive strength of CEB (MPa)	
	WSD	LSD
Mortar	M2, strength = 3 MPa	M4, strength = 4 MPa
Wall-A		
3rd Floor	3.5	1.0
2nd Floor	3.5	2.7
1st Floor	6.2	4.8
Ground Floor	8.5	7.1
Wall-B		
3rd Floor	3.5	0.7
2nd Floor	4.2	1.8
1st Floor	7.3	3.2
Ground Floor	9.4	4.7

References

BS EN 1052-1:1999 (1999) Methods of test for masonry—determination of compressive strength. The British Standards Institution, UK

BS EN 1990:2002+A1:2005 (2005) Eurocode. Basis of structural design. The British Standards Institution, UK

BS EN 1991-1-1:2002 (2002) (Eurocode 1), Actions on structures: general actions—densities, self-weight, imposed loads for buildings. The British Standards Institution, UK

BS EN 1996-1-1:2005+A1:2012 (2012) Eurocode 6: design of masonry structures. Part 1—1: general rules for reinforced and unreinforced masonry structures. The British Standards Institution, UK

BS EN 1996-2:2006 (2006) Design of masonry structures—part 2: design considerations, selection of materials and execution of masonry. The British Standards Institution, UK

BS EN 771-1:2011+A1:2015 (2015) Part 1: specification for masonry units part 1: clay masonry units. The British Standards Institution, UK

BS EN 771-2:2011+A1:2015 (2015) Part 2: specification for masonry units part 2: calcium silicate masonry units. The British Standards Institution, UK

BS EN 771-3:2011+A1:2015 (2015) Part 3: aggregate concrete masonry units (dense and lightweight aggregates). The British Standards Institution, UK

BS EN 771-4:2011+A1:2015 (2015) Part 4: autoclaved aerated concrete masonry units. The British Standards Institution, UK

BS EN 771-5:2011+A1:2015 (2015) Part 5: manufactured stone masonry units. The British Standards Institution, UK

BS EN 771-6:2011+A1:2015 (2015) Part 6: natural stone masonry units. The British Standards Institution, UK

BS EN 772-1:2011+A1:2015 (2015) Methods of test for masonry units part 1: determination of compressive strength. The British Standards Institution, UK

BS EN 998-2:2016 (2019) Specification for mortar for masonry—part 2: masonry mortar. The British Standards Institution, UK

NBC (2016) National Building Code of India 2016, Group 2, Sections 1 and 4. Bureau of Indian Standards, New Delhi, India

Chapter 9
Geopolymer or Alkali Activated Stabilised Earth Bricks

9.1 Geopolymers or Alkali Activated Binders

The alkali activated materials are alternative binders for the Portland cement, which are formed due to the reaction between an alkali and the silica/alumina precursor. Such binders are called inorganic polymers. The concept of alkali activated binders was first patented by German chemist and engineer Kühl in 1908 (Kühl 1908). Later on, Purdon (1935, 1940) developed the scientific basis for the alkali activated binders through many experiments using different blast furnace slags with alkalis. Purdon noted the enhanced tensile and flexural strength, low heat evolution and better acid resistance for slag-alkali cements compared to the Portland cements of similar compressive strength. After 1940s, the R&D on alkali activation technology moved to Soviet Union and China, specifically utilising metallurgical slags to produce alternatives to the Portland cement. Glukhovsky (1959) initiated research work on alkali activated binders in the former Soviet Union, during 1950s. Glukhovsky named the alkali activated products as "Gruntosilikat" or "Gruntocement-geocement". Then, the research on alkali activated materials was quite limited until the 1980s (Roy 1999). Davidovits patented many aluminosilicate-based formulations for niche applications from the early 1980s onwards and coined the name "geopolymer" to the alkali activated binders (Davidovits 1982, 2015). The patented alkali activated binder was sold under the brand name Pyrament, which was used for repairing concrete runways (Malone et al. 1985; Davidovits 2015).

B. V. V. Reddy, *Compressed Earth Block & Rammed Earth Structures*,
Springer Transactions in Civil and Environmental Engineering,
https://doi.org/10.1007/978-981-16-7877-6_9

9.2 Reaction Mechanisms in Geopolymerisation or Alkali Activation Processes

The reactions taking place between the solid aluminosilicate precursors and the alkali activating solution define the nature of binder products formed. The solid aluminosilicate precursors primarily contain reactive silica and alumina. The alkali activation or geoploymerisation can be visualised in three stages: (a) dissolution of aluminosilicate precursor into aluminate and silicate species through alkaline hydrolysis (b) gel formation containing silicate, aluminate and aluminosilicate mixture, which is in the aqueous phase (c) formation of the three-dimensional aluminosilicate network attributed to geopolymers (Duxson et al. 2007; Provis 2014; Davidovits 2015). The formation of geopolymer or alkali activated binder is mainly due to the poly-condensation of aluminates and silicates with alkali activating metals yielding polymeric Si–O–Al bonds. The alkali activating metals such as sodium hydroxide (NaOH) or potassium hydroxide (KOH) along with sodium silicate solution are generally used as an alkali activator in producing the geopolymer products (Davidovits 1989, 1991, 2015).

Geopolymerisation or alkali activation process involves the chemical reaction between aluminate and silicate oxides with the alkali metals resulting Si–O–Al bonds. The classification of geopolymer bonds is based on the mineral oxide units such as siloxo (–Si–O–Si–O–), sialate (Si–O–Al–O–), poly-sialate (Si–O–Al–O–)n, sialate-siloxo (–Si–O–Al–O–Si–O–), poly-sialate-siloxo (–Si–O–Al–O–Si–O–)n and sialate-disiloxo (–Si–O–Al–O–Si–O–Si–O–) (Davidovits 2015), which are called as the backbone of geopolymers. The geopolymerisation is an exothermic reaction, and the schematic representation of geopolymerisation is as follows:

$$\underset{\text{(Si-Al materials)}}{n(Si_2O_5,Al_2O_2)} + 2nSiO_2 + 4nH_2O + NaOH \longrightarrow (Na^+) + n(OH)_3\text{-}Si\text{-}O\text{-}Al\text{-}O\text{-}Si\text{-}(OH)_3 \;\; [\text{Al}\;|\;(OH)_2]$$

(Geopolymer precursor)

$$n(OH)_3\text{-}Si\text{-}O\text{-}Al\text{-}O\text{-}Si\text{-}(OH)_3 \;[\text{Al}\;|\;(OH)_2] + NaOH \longrightarrow (Na^+) - (\text{-}Si\text{-}O\text{-}Al\text{-}O\text{-}Si\text{-}O)n \;[\text{each}\;|\;O] + 4nH_2O$$

Poly (Sialate-Siloxo)
(Geopolymer backbone)

The geopolymer binders are temperature-resistant, acid-resistant and are unaffected in alkaline conditions. The concrete with geopolymer binders attains high early strength, lower shrinkage and good resistance to freeze–thaw cycles when compared with the properties of the concrete using Portland cement.

The sources for the alkali activated materials can be grouped under two categories based on the calcium content, as (a) low-calcium material and (b) high-calcium material. The low-calcium aluminosilicate precursors such as fly ash and metakaolin can be activated using alkali metal hydroxide or silicate solutions. In such systems, there is a possibility of leaching of alkalis related to the efflorescence in the form of alkali carbonates on the surfaces (Najafi et al. 2012; Skvara et al. 2009; Smith and Osborne 1977). The efflorescence can be controlled by the addition of Al-rich secondary binder components such as metakaolin (Najafi et al. 2012). The activation of calcium-rich materials such as ground granulated blast furnace slag (GGBS) mainly depends on the type of activator and the characteristics/composition of the calcium-rich materials (Provis and Deventer 2013). The range of activators for such materials includes alkali metal carbonate and sulphate solutions and hydroxides and silicates. The commonly used activators for the production of activated GGBS binders are sodium hydroxide (NaOH), sodium silicate ($Na_2O{\cdot}rSiO_2$), sodium carbonate (Na_2CO_3) and sodium sulphate (Na_2SO_4) (Provis and Deventer 2013). Higher alkalinity activators such as silicates provide strength development under ambient curing conditions (Provis 2014). The reaction of GGBS with water, over a long-extended period of time, will result in the formation of a hardened binder. Therefore, the critical role of the alkaline activator in an alkali activated product is to accelerate the reactions to take place within a reasonable time frame (Provis and Deventer 2013).

9.3 Alkali Activated/geopolymer Binders for Stabilised Compressed Earth Bricks

Soil, fine aggregates and stabiliser additives are the basic ingredients needed for the stabilised compressed earth brick production. Inorganic binders such as the cement and the lime are commonly used as stabilising binders. Such inorganic binders can be replaced with alkali activated/geopolymer binders. The clay minerals present in the soil can be exploited for the development of the geopolymer binder. Major parameters affecting the properties of the geopolymer stabilised compressed earth bricks are as follows.

(a) Type and quantity of clay mineral in the soil
(b) Concentration of the alkali (sodium or potassium hydroxide)
(c) Quantity of silica/alumina from an additional source
(d) Curing temperature and duration
(e) Dry density of the compressed earth brick.

Montmorillonite, illite and kaolinite are the most commonly occurring clay minerals in the soils. The direct alkali activation in transforming the natural clay minerals (precursors) present in the soils for the production of compressed earth blocks has not been adequately explored. The investigations of Heah et al. (2011), Davidovits (2015) and Marsh et al. (2018a, b, 2019a, b) show attempts to understand the alkali activation of natural clay minerals, to produce binders, which can be used for the production of geoploymer stabilised compressed earth blocks.

The investigations of Maskell et al. (2014), Muñoz et al. (2015) and Marsh et al. (2018a) show attempts to facilitate production of geopolymer stabilised compressed earth bricks. Preethi and Reddy (2019) attempted to examine the properties of the

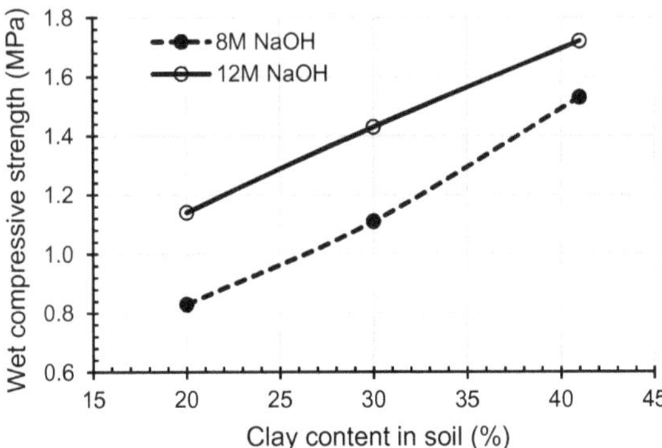

Fig. 9.1 Variation in strength with clay content and alkali concentration

compressed earth specimens stabilised with geopolymer binder using a natural soil having kaolin clay mineral. Figure 9.1 shows relationships between the strength of the geopolymer stabilised compressed earth specimen and the clay content of the soil. The natural soil had 42% clay. The compressed earth cylindrical specimen (38 mm diameter and 76 mm height, dry density: 1800 kg/m^3) prepared with alkali (NaOH) solution has been cured at 80 °C for 72 h in hot air oven. The clay content of the soil has been varied by reconstituting the soil with river sand. The wet compressive strength increased with the increase in the clay fraction of the soil mix as well as the molarity of the alkali activator solution. But the wet strength achieved was in the range of 1–1.5 MPa, which is low. At higher clay content (>30%) and at higher molar alkali solution (12 M), the mix forms into lumps (Fig. 9.2). It becomes difficult to break the lumps and to compact the mix into a high-density product. Hence, the clay content and the molarity of the alkali solution in the compressed earth specimen should be restricted to <30% and $\leq$12 M, respectively, while producing compressed earth bricks/blocks.

The compressive strength of the geopolymer stabilised compressed earth using the natural soil can be improved by the incorporation of silica/alumina-rich materials such as the fly ash and the GGBS. The addition of nano-calcite and synthetic nano-aluminosilicates significantly improve the compressive strength of the alkali activated compressed earth specimen (Muñoz et al. 2015). The strength improvement in the geopolymer stabilised compressed earth specimen, due to the incorporation of class-F fly ash and GGBS into the soil, is shown in Fig. 9.3. The strength is for compressed earth cylindrical specimen (38 mm diameter and 76 mm height, dry density: 1800 kg/m^3) prepared with 12M NaOH solution and cured at 80 °C for 72 h in an oven. Addition of 15% fly ash and 15% GGBS results in 2–3 times increase in the wet compressive strength. The compressed earth brick (Fig. 9.4) strength (of size 230 × 100 × 75 mm) with similar mixes will be three times higher than the cylinder strength (Preethi 2020).

Fig. 9.2 Formation of lumps with soil having 42% clay fraction and 12 M NaOH solution

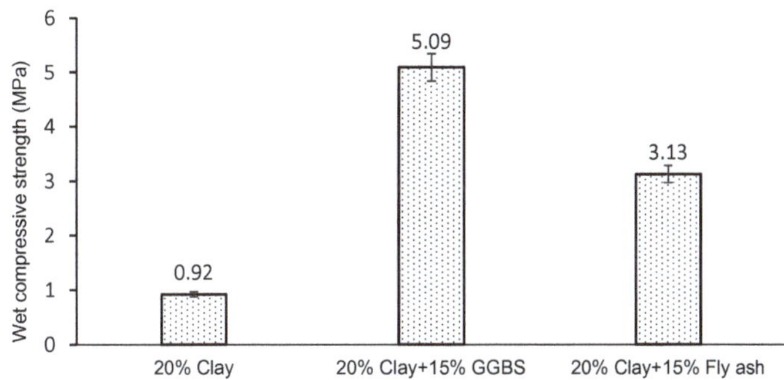

Fig. 9.3 Comparison of strengths of geopolymer stabilised compressed earth cylindrical specimen with and without fly ash and GGBS

Fig. 9.4 Compressed earth geopolymer stabilised cylinder and the brick specimen

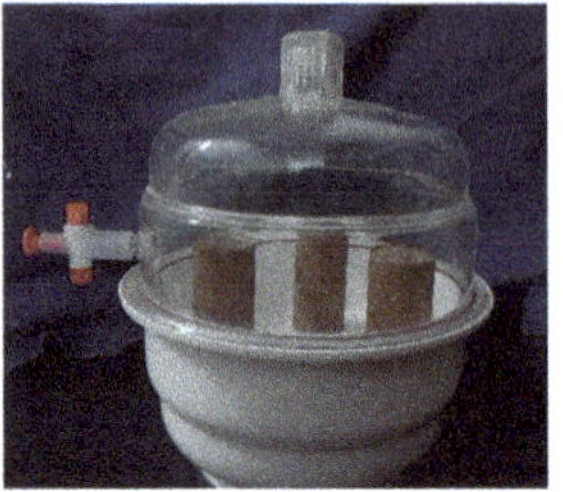

Fig. 9.5 **a** Efflorescence after 24 h of casting, **b** No efflorescence on specimens stored in desiccator, **c** No efflorescence on specimens cured immediately after casting (Preethi 2020)

9.4 Efflorescence in the Geopolymer Stabilised Compressed Earth Products

The manufacturing process of geopolymer mixed compressed earth cylindrical or brick specimens requires certain amount of moulding moisture to impart green strength for handling, and also, water is needed for dissolving the solid alkali (NaOH) pellets, though the water is not required for alkali activation process. The fresh specimens have poor handling (green) strength for initiating the curing immediately after casting in an oven. The green specimens attain enough handling strength after drying for 24 h in the open air inside a production unit. During this initial drying process, there will be evaporation of water from the green specimens, and this movement of water results in the deposition of alkali salts (efflorescence) on the surface (Fig. 9.5).

The sodium-rich solution evaporates from the exposed surface of the cylinders, leaving behind the salts as white deposit. The evaporation can be minimised by keeping the compressed cylindrical specimen in the desiccator with the relative humidity of 95%. No efflorescence was observed when the specimens were kept in a high humidity chamber for 3 days. Also, curing the specimens in the oven, immediately after casting, drastically minimises the efflorescence on the surface. Figure 9.5 shows the specimens without efflorescence due to change in the curing protocol.

9.5 Effect of Salt Leaching on the Strength of Geopolymer Stabilised Compressed Earth Specimens

The geopolymer stabilised compressed earth products when exposed to cyclic wetting and drying will shed the salts through leaching process. The concentration of salts in the leachate can be quantified through the measurement of total dissolved salts (TDS). Preethi (2020) examined the influence of salt leaching on the strength of geopolymer stabilised compressed earth specimens. Figure 9.6 shows a plot of TDS in the leachate

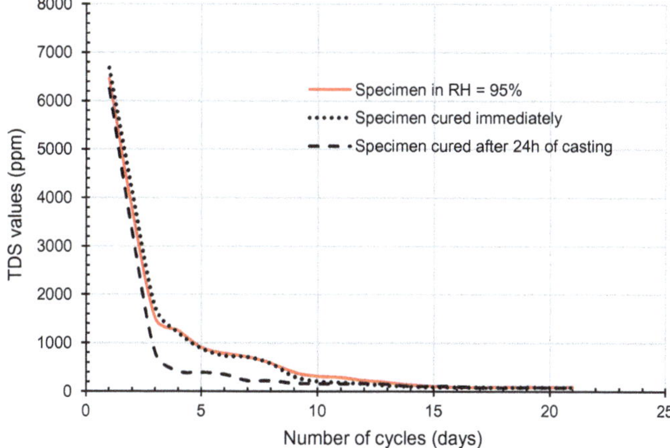

Fig. 9.6 Variation in TDS values of the leachate after cyclic leaching tests (Preethi and Reddy 2020, Copyright Elsevier)

versus number of cycles of leaching on geopolymer stabilised compressed earth cylindrical specimen using 12M NaOH solution for the three cases of (a) curing after 24 h of casting, (b) curing immediately after casting and (c) curing the specimen stored for a day in the desiccator having 95% RH. The plots show that there is rapid leaching within four cycles and nearly complete leaching of salts in about 10 cycles. The specimens from the desiccator (95% RH) and immediately cured after casting showed higher TDS value in the first two cycles. This can be attributed to the absence of efflorescence on the surface of the specimens, but the salts were there inside the specimen. The leachate contains carbonates of sodium called thermonatrites. Figure 9.7 shows the comparison of compressive strengths of geopolymer stabilised compressed earth specimen before and after 20 cycles of the cyclic leaching test. The strengths of the cylindrical specimen using the natural soil, soil + GGBS and soil + fly ash have been compared. The Figure shows only marginal difference in the compressive strengths of the leached and the reference specimens. Hence, the leaching of salts

Fig. 9.7 Comparison of strengths of geopolymer compressed earth cylinders before and after leaching (Preethi and Reddy 2020, Copyright Elsevier)

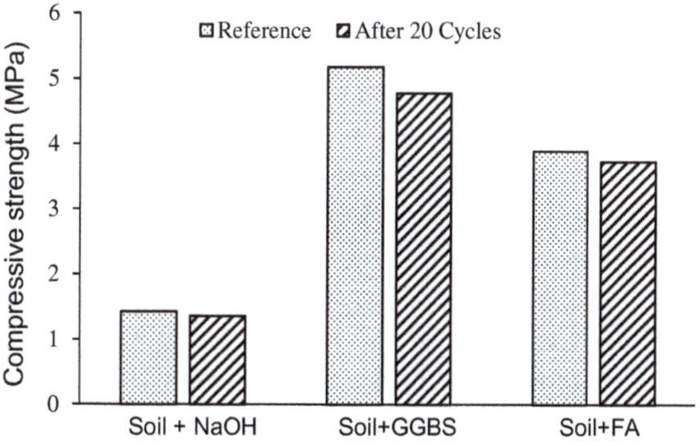

does not affect the strength of the alkali activated specimens. The efflorescence in the geopolymer stabilised materials can be minimised by using alumina-rich additional materials (such as metakaolin) and use of potassium hydroxide as alkali activator (Ebrahim et al. 2012; Longhi et al. 2019; Allahverdi et al. 2015).

9.6 Characteristics of Geopolymer Stabilised Compressed Earth Bricks

There are some attempts to characterise the properties of geopolymer stabilised compressed earth products (Maskell et al. 2012; 2014; Elert et al. 2015; Muñoz et al. 2015; Silva et al. 2015; Mirinda et al. 2017; Omar et al. 2018; Preethi and Reddy 2019). These investigations demonstrate the potential of the geopolymers or the alkali activation processes for the production of the stabilised compressed earth block (CEB). The following characteristics of the geopolymer stabilised CEB need to be evaluated.

(a) Strength
(b) Dimensional stability
(c) Durability
(d) Absorption characteristics
(e) Bond development with the mortar.

The detailed and comprehensive investigations on geopolymer CEB were conducted by Preethi (2020) and Preethi and Reddy (2020). The results of these studies are summarised in the following sections.

9.6.1 Production of Geopolymer Stabilised CEB

The basic ingredients required for the production of geopolymer stabilised CEB include (a) soil or earth (b) alkali solution (sodium or potassium hydroxide) and (c) fly ash/GGBS. The manufacturing process involves:

(a) preparation of the alkali solution
(b) reconstitution of the soil or the earth using sand or any such inert material such that the mix has about 20–30% clay. Higher clay content poses problems in achieving uniform mix due to the formation of lumps as described in the previous sections
(c) blending the soil and the fly ash or GGBS in a mixer
(d) mixing the soil/fly ash or GGBS mixture with the requisite quantity of alkali solution
(e) compressing the processed geopolymer-soil-fly ash/GGBS mix into a dense brick using a machine

(f) curing the geopolymer CEB at 80 °C.

Figure 9.8 shows the equipment and the processes involved in the geopolymer stabilised CEB (GCEB) production. Figure 9.9 shows the GCEB bricks. The bulk of the brick comprises stabilised soil–sand mixture. The soils having higher percentage of clay fractions (>20%) should be reconstituted by mixing with the sand or other inert materials such as C&D waste (CDW) and blast furnace slag.

Fig. 9.8 Geopolymer CEB production process: **a** addition of alkali solution into the pan mixer containing soil-fly ash/GGBS mixture **b** processed soil-fly ash/GGBS mixture **c** ejecting the pressed GCEB

Fig. 9.9 Geopolymer stabilised CEBs

9.6.2 *Characteristics of Geopolymer Stabilised Compressed Earth Bricks*

The characteristics of the geopolymer stabilised CEB have been examined by Narayanaswamy et al. (2020a, b), Preethi and Reddy (2020), and Reddy and Monto (2021). Two types of red soils (Soil-A and Soil-B) (with a clay size fractions of 42 and 25%) were reconstituted by mixing with the sand, CDW and processed granulated blast furnace slag (PGBS) such that the reconstituted mixtures contain clay content in the range of 10–20%. The grain size distribution curves of the two types of soils, sand, CDW and PGBS are shown in Fig. 9.10. Figure 9.11 shows the shape (the SEM images) of sand, CDW and PGBS particles. The loose bulk densities of sand, CDW and PGBS were 1450, 1500 and 1350 kg/m^3, respectively. The geopolymer binder used in these studies consists of 12M NaOH solution and fly ash or GGBS. The proportions of the reconstituted mixes and the details of the geopolymer binders are given in Table 9.1. Also, the Table gives brick designation. There are six types of geopolymer stabilised compressed earth bricks (GCEB's) as detailed in Table 9.1. Table 9.2 gives the details of the strength, absorption properties, durability character-

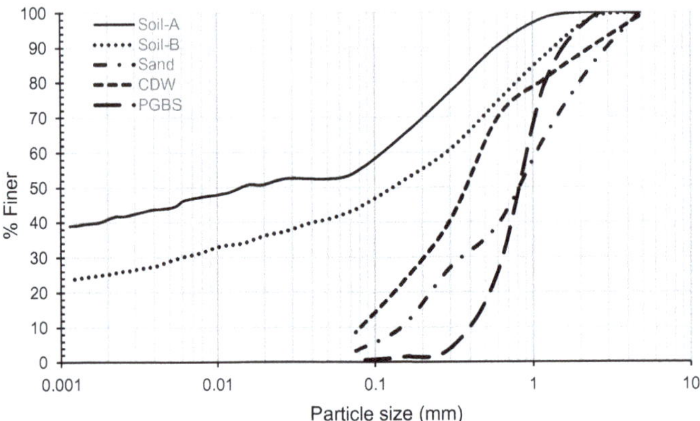

Fig. 9.10 Grain size distribution curves for soils, sand, CDW and PGBS

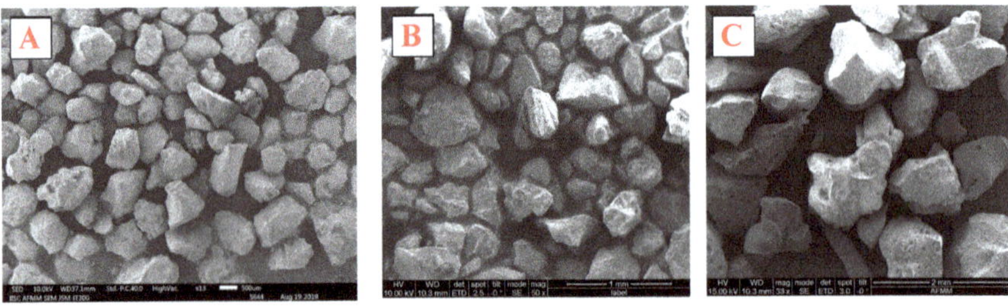

Fig. 9.11 Particle shape (SEM images) of (A) CDW, (B) Sand and (C) PGBS

Table 9.1 Mix details for geopolymer stabilised compressed earth bricks

Type of soil	Mix details				Clay (%) in mix	Geopolymer binder details			Brick designation	References
	Soil	Sand	CDW	PGBS		NaOH	Fly ash (%)	GGBS (%)		
Soil-A	1	1	0	0	21.0	12M	–	15	GCEB1	Preethi and Reddy (2020), Reddy and Monto (2021), Narayanaswamy et al. (2020a)
Soil-A	1	1	0	0	21.0	12M	30	–	GCEB2	
Soil-B	1	1	0	0	12.5	12M	15	–	GCEB3	
Soil-B	1	0	1	0	12.5	12M	15	–	GCEB4	
Soil-B	1	0	0	1	12.5	12M	15	–	GCEB5	
Soil-A	1	0	3	0	10.5	10M	15	–	GCEB6	

Table 9.2 Characteristics of the geopolymer stabilised CEBs

Characteristics	GCEB1	GCEB2	GCEB3	GCEB4	GCEB5	GCEB6
Dry density (kg/m^3)	1800	1800	1890	1860	1850	1850
Wet compressive strength (MPa)	20.1	18.7	17.0	15.0	21.9	17.6
Dry compressive strength (MPa)	36.0	38.9	34.9	31.0	31.4	31.5
Wet to dry strength ratio	0.56	0.48	0.49	0.48	0.70	0.56
Water absorption (%)	11.81	11.90	9.30	10.50	9.50	11.87
IRA (kg/m^2/min)	3.81	2.07				3.41
Linear expansion on saturation (%)	0.081	0.07	0.096	0.07	0.08	0.053
Mass loss after durability test (%)	0.67	0.89	1.39	1.35	0.73	1.84
Surface porosity (%)	10.43	6.65	13.52	11.58	17.04	–
Stress–strain characteristics						
Initial tangent modulus (MPa)	6.0	9.6	7.49	7.00	8.77	7.80
Poisson's ratio	0.10	0.11	0.16	0.20	0.16	0.12
Strain at peak stress	0.0027	0.0025	0.0022	0.0022	0.0017	0.0018
Source for stress–strain characteristics	Preethi and Reddy 2020, Narayanaswamy et al. 2020a, Reddy and Monto 2021					

istics and surface porosity of the GCEB's. The wet compressive strength represents the strength determined after 48 h of soaking in water prior to the test. The dry strength represents the strength determined after curing the specimen in an oven at 80 °C for 72 h. The dry density of the GCEBs was in the range of 1800–1890 kg/m^3.

Compressive strength: The wet compressive strength of GCEB's using different types of geopolymer and fly ash/GGBS dosages is in the range of 15–22 MPa. The dry compressive is in the range 31–39 MPa. The wet to dry strength ratio is in the range 0.50–0.70. There is considerable difference between the wet and the dry strengths. The trend is similar to the one noticed for the cement stabilised CEB. In the case of stabilised CEBs, there will be residual clay (Reddy and Latha 2014a), and hence, the strength reduces as the CEB absorbs moisture. The slag-based GCEBs show higher strength. The compressive strength of GCEBs is an order of magnitude higher when compared with the strength of the cement stabilised CEB. Generally, the cement (6–8%) stabilised CEBs show a wet compressive strength in the range of 4–7 MPa.

Absorption characteristics: The water absorption values for the different types of GCEB's are in a range of 9–12%. These values are slightly lower than the ones noticed for the cement stabilised CEBs. The IRA values are in the range of 2–4 kg/m^2/minute.

Durability characteristics: The linear expansion on saturation for GCEBs is well within the permissible limit of 0.10% suggested in the IS-1725 (2013) code for the stabilised CEB's. The mass loss after the cyclic wet–dry test for the GCEB's is in the range of 0.70–1.40%. This is again well within the permissible limit of 3% (IS-1725 2013) specified for the stabilised CEBs. These values and the experience on the performance of the cement stabilised CEBs suggest that the geopolymer stabilised CEBs will have good durability performance.

Surface porosity: The surface porosity of the GCEBs varies in the range of 7–13%, except the GCEB with PGBS shows higher porosity, mainly attributed to porous nature of the slag. The PGBS shows higher water absorption (5.5%) when compared with the water absorption values of sand and CDW aggregates. The studies of Reddy and Latha (2014b) showed that the surface porosity for the cement stabilised CEBs using 7% cement and 1800 kg/m^3 dry density, in the range of 15–20%.

Stress–strain characteristics of GCEB's

Figure 9.12 shows typical stress–strain curves for GCEB5 and GCEB6 in the wet state. Table 9.2 gives the stress–strain characteristics for the GCEBs. The initial tangent modulus for the GCEBs in the wet state is in the range of 6.0–9.6 GPa and the Poisson's ratio in the range of 0.10–0.20. The strain at peak stress for the GCEBs is in the range of 0.0018–0.0027.

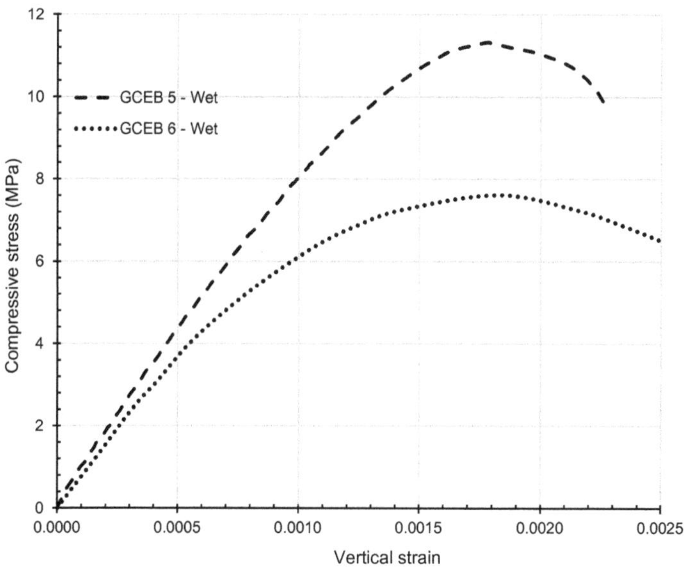

Fig. 9.12 Stress–strain curves for the GCEBs

References

Allahverdi A, Kani EN, Hossain KMA, Lachemi M (2015) Methods to control efflorescence in alkali-activated cement-based materials. In: Pacheco-Torgal F, Labrincha JA, Leonelli C, Palomo A, Chindaprasirt P (eds) Handbook of alkali-activated cements, mortars and concretes. Woodhead Publishing, Oxford, pp 463–483. https://doi.org/10.1533/9781782422884.3.463

Davidovits J (1982) Mineral polymers and methods of making them. US Patent 4,349,386

Davidovits J (1989) Geopolymers and geopolymeric materials. J Therm Anal 35:429–441. https://doi.org/10.1007/BF01904446

Davidovits J (1991) Geopolymers—inorganic polymeric new materials. J Therm Anal 37:1633–1656. https://doi.org/10.1007/BF01912193

Davidovits J (2015) Geopolymer chemistry and applications, 4th edn. Institut Géopolymère, Saint-Quentin, France

Duxson P, Fernandez-Jimenez A, Provis JL, Lukey GC, Palomo A, van Deventer JSJ (2007) Geopolymer technology: the current state of the art. J Mater Sci 42:2917–2933

Ebrahim NK, Allahverdi A, Provis JL (2012) Efflorescence control in geopolymers binders based on natural pozzolana. Cement Concr Compos 34:25–33

Elert K, Pardo ES, Rodriguez-Navarro C (2015) Alkaline activation as an alternative method for the consolidation of earthen architecture. J Cult Herit 16(4):461–469. https://doi.org/10.1016/j.culher.2014.09.012

Glukhovsky VD (1959) Gruntosilikaty (soil silicates). Gosstroyizdat, Kiev

Heah CY, Kamarudin H, Mustafa Al Bakri AM, Binhussain M, Luqman M, Khairul Nizar I, Ruzaidi CM, Liew YM (2011) Effect of curing profile on kaolin-based geopolymers. In: Physics Procedia. Elsevier B.V., Amsterdam, pp 305–311

IS-1725 (2013) Specifications for soil based blocks used in general building construction. Bureau of Indian Standards, New Delhi, India

Kühl H (1908) Slag cement and process of making the same. US Patent 900,939

Longhi MA, Zhang Z, Rodríguez ED, Kirchheim AP, Wang H (2019) Efflorescence of alkali-activated cements (geopolymers) and the impacts on material structures: a critical analysis. Front Mater 6:89. https://doi.org/10.3389/fmats.2019.00089

Malone PG, Randall CJ, Kirkpatrick T (1985) Potential applications of alkali-activated alumi-nosilicate binders in military operations. Geotechnical Laboratory, Department of the Army, GL-85-15

Marsh A, Heath A, Patureau P, Evernden M, Walker P (2018a) Alkali activation behaviour of un-calcined montmorillonite and illite clay minerals. Appl Clay Sci 166:250–261. https://doi.org/10.1016/j.clay.2018.09.011

Marsh A, Heath A, Patureau P, Evernden M, Walker P (2018b) A mild conditions synthesis route to produce hydrosodalite from kaolinite, compatible with extrusion processing. Microporous Mesoporous Mater 264(2018):125–132

Marsh A, Heath A, Patureau P, Evernden M, Walker P (2019a) Phase formation behaviour in alkali activation of clay mixtures. Appl Clay Sci 175:10–21

Marsh A, Heath A, Patureau P, Evernden M, Walker P (2019b) Influence of clay minerals and associated minerals in alkali activation of soils. Constr Build Mater 229:116816

Maskell D, Heath A, Walker P (2014) Geopolymer stabilisation of unfired earth masonry units. Key Eng Mater 600:175–185

Maskell D, Andrew H, Walker P (2012) The compressive strength of lignosulphonate stabilised extruded earth masonry units. In: 11th international conference on the study and conservation of earthen architecture heritage, Lima, Peru

Miranda T, Silva RA, Oliveira DV, Leitão D, Cristelo N, Oliveira J, Soares E (2017) ICEBs stabilised with alkali-activated fly ash as a renewed approach for green building: exploitation of the masonry mechanical performance. Constr Build Mater 155:65–78

Muñoz JF, Easton T, Dahmen J (2015) Using alkali-activated natural aluminosilicate minerals to produce compressed masonry construction materials. Constr Build Mater 95:86–95. https://doi.org/10.1016/j.conbuildmat.2015.07.144

Najafi KE, Allahverdi A, Provis JL (2012) Efflorescence control in geopolymer binders based on natural pozzolan. Cem Concr Compos 34(1):25–33

Narayanaswamy AH, Peter Walker, Reddy BVV, Heath A, Maskell D (2020b) Compressive strength of novel alkali activated stabilised earth materials incorporating solid wastes. J Mater Civ Eng 32(6):1–8. https://doi.org/10.1061/(ASCE)MT.1943-5533.0003188

Narayanaswamy AH, Walker P, Reddy BVV, Heath A, Maskell D (2020a) Mechanical and thermal properties, and comparative life-cycle impacts, of stabilised earth building products. Constr Build Mater 243. https://doi.org/10.1016/j.conbuildmat.2020.118096

Omar SS, Messan A, Prud'homme E, Escadeillas G, François T (2018) Stabilization of compressed earth blocks by geopolymer binder based on local materials from Burkina Faso. Constr Build Mater 165:333–345

Preethi RK (2020) Studies on alkali activated compressed earth bricks for structural masonry. PhD thesis, Department of Civil Engineering, Indian Institute of Science, Bangalore, India

Preethi RK, Reddy BVV (2019) Studies on geopolymer based earthen compacts. In: Venkatarama Reddy BV, Mani M, Walker P (eds) Earthen dwellings and structures: current status in their adoption. Springer Transactions in Civil & Environmental Engineering, Singapore, pp 3–14

Preethi RK, Reddy BVV (2020) Experimental investigations on geopolymer stabilised compressed earth products. Constr Build Mater 257:119563. https://doi.org/10.1016/j.conbuildmat.2020.119563

Provis JL, van Deventer JSJ (eds) (2013) Alkali-activated materials: state of the art report, RILEM TC 224-AAM. Springer/RILEM, Berlin

Provis JL (2014) Geopolymers and other alkali activated materials: why, how and what? Mater Struct (Materiaux et Constructions) 47:11–25. https://doi.org/10.1617/s11527-013-0211-5

Purdon AO (1935) Improvements in processes of manufacturing cement, mortars and concretes. British Patent GB427227

Purdon AO (1940) The action of alkalis on blast-furnace slag. J Soc Chem Ind Trans Commun 59:191–202

Reddy BVV, Latha MS (2014b) Influence of soil grading on the characteristics of cement stabilised soil compacts. Mater Struct 47(10):1633–1645. https://doi.org/10.1617/s11527-013-0142-1

Reddy BVV, Monto M (2021) A project report on "Developing alkali-activated low carbon bricks using construction and demolition wastes for energy efficient walling envelopes". DST sponsored project, Indian Institute of Science, Bangalore, India

Reddy BVV, Latha MS (2014a) Retrieving clay minerals from stabilised soil compacts. Appl Clay Sci 101:362–368

Roy D (1999) Alkali-activated cements—opportunities and challenges. Cem Concr Res 29(2):249–254

Silva RA, Soares E, Oliveira DV, Miranda T, Cristelo NM, Leitão D (2015) Mechanical characterisation of dry-stack masonry made of CEBs stabilised with alkaline activation. Constr Build Mater 75(2015):349–358

Skvara F, Kopecky L, Myskova L, Smilauer V, Alberovska L, Vinsova L (2009) Aluminosilicate polymers influence of elevated temperatures efflorescence. Ceram Silik 53(4):276–282

Smith MA, Osborne GJ (1977) Slag/fly ash cements. World Cem Technol 1(6):223–233

Chapter 10
Compressed Earth Blocks Using Non-organic Solid Wastes

10.1 Non-organic Solid Wastes

Industrial and mining activities generate large quantities of non-organic solid wastes (NOSWs). The NOSWs are basically by-products from industrial and mining activity. Such by-products can be recycled to produce stabilised compressed earth blocks. Some of the NOSWs useful for the stabilised compressed earth blocks (CEB's) are as follows:

- Mine tailings (from mines such as iron ore, copper, manganese and gold)
- Dust and debris from coal mines, kiln dust and stone dust
- Coal combustion residues (fly ash, bottom ash and pond ash)
- By-products from fertiliser and chemical industries (lime sludge, phosphogypsum)
- Slag, red-mud from metal industries
- Construction and demolition waste

Some of the NOSWs can be directly used in the production of stabilised CEBs, and others need processing while using them in the production of stabilised CEBs. Also, some of the NOSWs have pozzolanic properties, which can be exploited to produce CEB's using the non-OPC binders such as calcium hydroxide (lime). The benefits of using NOSWs for the production of stabilised CEBs are as follows:

(a) Conservation of soil or earth, i.e. minimising the use of precious earth
(b) Opportunity to exploit non-OPC binders
(c) Reducing the consumption of mined materials such as fine aggregates
(d) Scope for producing low embodied carbon walling materials
(e) Environmentally safe disposal of solid wastes.

B. V. V. Reddy, *Compressed Earth Block & Rammed Earth Structures*,
Springer Transactions in Civil and Environmental Engineering,
https://doi.org/10.1007/978-981-16-7877-6_10

10.2 Characteristics of Stabilised Compressed Earth Blocks Using NOSW's

The stabilised compressed earth blocks can be manufactured using NOSWs. The characteristics of the stabilised CEBs manufactured using NOSWs such as fly ash, iron ore tailings, slag, gold mine tailings and construction and demolition waste are discussed in the subsequent sections.

10.2.1 Stabilised Compressed Earth Blocks Using Iron Ore Tailings

Generally, the raw iron ore extracted from the mines contains impurities apart from the iron. The iron ore is crushed and refined before palletising it for further processing in steel mills. The iron ore tailings are the residues left after refining and synthesising the iron ore. The iron ore tailings in certain cases are stored and contained in specially built structures. Figures 10.1 and 10.2 show the iron ore tailings (IOT) stored in a specially built dam (Lakya dam) in a place called Kudremukh, Karnataka state, India (Latitude 13° 14′ 56″ N and Longitude 75° 13′ 30″ E). The depth of the IOT storage here is 100 m, spreading across an area of about 2 square kilometres. The estimated quantity of IOT in this dam is over 200 million tonnes. The typical chemical composition of the IOT sample collected from this dam is given in Table 10.1. The major elements in the IOT are silica (66%) and iron (26%). The IOT contains basically the inert materials. The pH is about 8.0, and the total dissolved salts (TDS) is low.

Fig. 10.1 Top view of Lakya dam showing storage of iron ore tailings

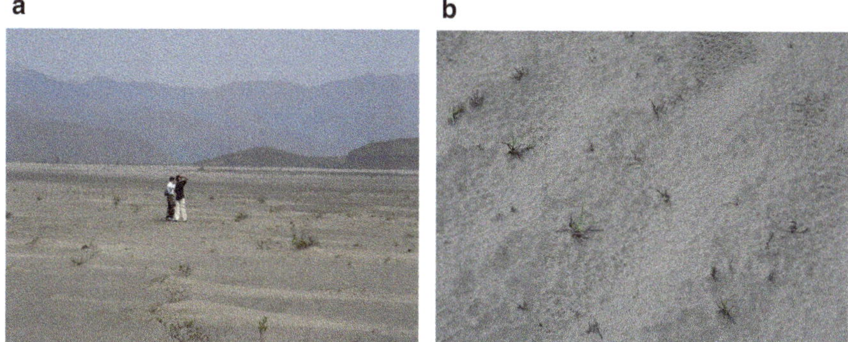

Fig. 10.2 Spread of iron ore tailings in the dam

Table 10.1 Chemical properties of IOT samples (Ullas 2014)

Compound	Composition (%)
SiO_2	65.89
Fe_2O_3	26.83
Al_2O_3	1.08
CaO	1.59
MgO	1.42
Na_2O	0.10
K_2O	0.07
TiO_2	0.19
MnO	0.13
SO_3	Traces
TDS	134 ppm
LOI	1.31
pH	8.16

The typical grain size distribution curve for the IOT is shown in Fig. 10.3. The IOT contains fine particles, and the size ranges between 0.10 and 1.0 mm. Ullas (2014) investigated the characteristics of the cement stabilised compressed earth blocks using the IOT. The use of IOT in the production of the cement stabilised CEBs and their characteristics are discussed here.

Figure 10.3 shows the grain size curve for the soil, sand and IOT used for the manufacture of CEBs. The soil had sand, silt and clay size fractions of 48.6%, 22.5% and 28.9%, respectively. The specific gravity of the soil was 2.62, and the predominant clay mineral was kaolinite. The optimum clay content of the soil should be in the range of 10–15% (Reddy et al. 2007; Reddy and Latha 2014). Hence, the clay content of the soil was brought down by mixing with either the river sand or the IOT in the ratio of 1 : 1 (soil: sand or IOT, by mass). The reconstituted soil contains 14.5% clay size fraction. The details of the mix proportions are given in Table 10.2.

The CEBs were prepared using a manual press, and the density was controlled by controlling the mass of the mix fed into the mould. The stabilised CEBs were cured for 28 days under wet burlap and then dried inside the laboratory for couple of weeks. The wet compressive strength of the stabilised CEBs was determined by

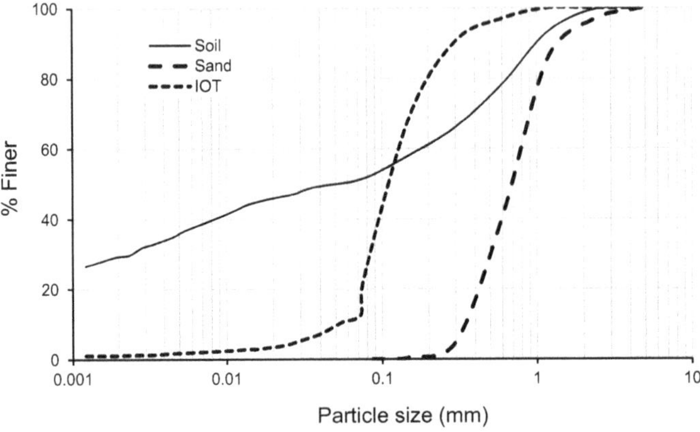

Fig. 10.3 Grain size distribution curves for IOT, natural sand and soil

Table 10.2 Mix proportions for cement stabilised compressed earth blocks (Ullas 2014)

Mix type	Mix proportion by mass			Cement content (%)	Clay content (%)	Std. Proctor OMC (%)	Moulding moisture content (%)
	Soil (%)	Sand (%)	IOT (%)				
A	50	50	0	7	14.5	11.05	12.15
B	50	0	50	7	14.5	13.63	15.00

soaking them in water for 48 h prior to testing. Table 10.3 gives the characteristics of the CEBs. Both the reference block with the river sand and the one with IOT have similar dry density. The wet compressive strength of the blocks was 6.9 and 6.6 MPa for the cement stabilised CEB using the sand and the IOT, respectively. The water absorption of CEB with IOT is about 15% as against 12% for the one with natural sand. Both the types of the blocks have similar IRA and linear expansion on saturation values. The linear expansion on saturation was found to be well within the upper limit of 0.10% suggested in IS 1725 (2013) code. These results clearly demonstrate that the IOT can be used instead of the river sand for the production of the cement stabilised CEBs. Figure 10.4 shows a load bearing masonry building using the cement stabilised CEB made with IOT.

Table 10.3 Characteristics of CEB using sand and IOT (Ullas 2014) (Block size: 230 × 190 × 100 mm)

Mix type	Wet compressive strength (MPa)	Water absorption (%)	IRA ($kg/m^2/min$)	Linear expansion (%)	Dry density (kg/m^3)
A	6.89	12.13	0.59	0.046	1830
B	6.63	15.07	0.58	0.035	1820

Fig. 10.4 Load bearing exposed masonry building with cement stabilised CEB having IOT

10.2.2 Stabilised Compressed Earth Blocks Using C&D Waste

Large quantities of construction and demolition waste (CDW) is generated. There is a wide variation in the estimates of annual CDW generation. The annual global generation of CDW is over 3 billion tonnes (Ali and Ajit 2018) and over 10 billion tonnes (Wu et al. 2019). The composition of CDW varies from region to region. Bulk of the CDW consists of coarse aggregate, concrete, mortar and masonry units. Majority of the efforts in utilising the CDW are focused on recycled aggregates for the concrete. The CDW (including the coarse aggregate) can be crushed into fine aggregate form matching the grain size distribution of the natural river sand (Reddy et al. 2016). Such a fine aggregate can be used in concrete, mortars and the manufacture of the stabilised CEBs (Reddy et al. 2016).

The compressed earth blocks are manufactured using a mixture of soil and sand, such that the reconstituted mix contains the optimum clay content (12–15%). The sand fraction of the mix can be replaced by CDW. The particle size distribution (PSD) curves for the crushed CDW, soil and a mixture of soil and CDW (1 soil: 1 CDW) are shown in Fig. 10.5. The soil had 29% clay content and hence has been reconstituted to bring down the clay content of the mix to about 15%. The properties of the CDW are given in Table 10.4. The crushed CDW contains mainly sand size fraction with a bulk density and the fineness modulus of 1530 kg/m^3 and 2.16, respectively. It is alkaline (pH = 9.43) and had poor lime reactivity. It is mainly an inert material.

The characteristics of the stabilised compressed earth bricks manufactured using the soil-CDW and the soil-sand mixtures are given in Table 10.5. The bricks have been cured under moist burlap for 28 days, dried in air for a week and then oven dried at 60 $^{\circ}$C till constant mass was achieved, before testing. The CEBs with a stabiliser content of 7% cement + 2% lime show similar strengths for the CDW and the sand mixes. The wet compressive strength is about 7 MPa. The stabiliser combination of 15% GGBS and 9% lime gives higher wet compressive strength of 10 MPa. The dry compressive strength of the CEB is much higher than that of the wet compressive strength. The durability characteristics of mass loss and linear expansion on saturation are well within the limiting values specified in IS 1725 (2013) code. The results clearly demonstrate the possibility of using crushed CDW as a fine aggregate in producing the stabilised compressed earth bricks.

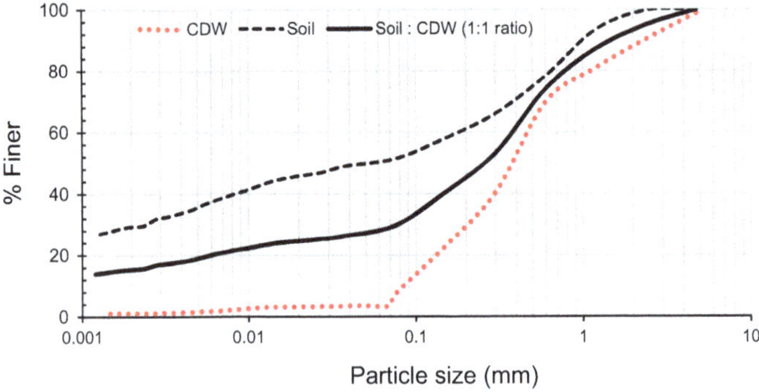

Fig. 10.5 Particle size distribution curve for the crushed CDW

Table 10.4 Characteristics of CDW

Properties	CDW
Textural composition	
Sand (%) (4.75–0.075 mm)	91.4
Silt (%) (0.075–0.002 mm)	8.6
Physical properties	
Fineness modulus	2.16
Specific gravity	2.49
Bulk density (kg/m^3)	1530
pH	9.43
Lime reactivity (MPa)	0.17

Table 10.5 Properties of compressed earth bricks using CDW and sand (*GGBS: Ground granulated blast furnace slag)

Property	Details and Results		
Clay content in the mix (%)	12.5 (with natural sand)	12.5 (with CDW)	12.5 (with CDW)
Brick size (mm)	230 × 110 × 70	230 × 110 × 70	230 × 110 × 70
Stabiliser (by mass)	7% cement + 2% $Ca(OH)_2$	7% cement + 2% $Ca(OH)_2$	15% GGBS* + 9% $Ca(OH)_2$
Dry density (kg/m^3)	1850	1860	1820
Wet compressive strength (MPa)	6.7	7.8	10.4
Dry compressive strength (MPa)	17.3	18.3	19.2
Water absorption (%)	11.2	10.8	12.0
Mass loss after durability test (%)	1.50	2.33	0.24
Linear expansion on saturation (%)	0.03	0.03	0.06

10.2.3 *Stabilised Compressed Blocks Using Gold Mine Tailings*

The gold mine tailings is another unutilised resource, which can be explored for the compressed stabilised earth block production. Generally, millions of tonnes of tailings are piled up near the gold mine sites. The Bharat Gold Mines Limited at Kolar gold fields (KGF) in India has 32 million tonnes of tailings (Fig. 10.6). Sudhakar Rao and Reddy (2006) investigated the environmental impact and the bulk utilisation of KGF mine tailings for the construction purposes.

The typical particle size distribution curves for the samples collected from the KGF mine tailings are shown in Fig. 10.7. The characteristics of the KGF gold mine tailings are given in Table 10.6. The particle size analysis indicates that the mine tailings are very fine and non-plastic crushed rock samples. Though there are particles in the clay size range (<2 μm), the commonly occurring clay minerals found in the soils were absent in the mine tailings. The tailing samples have pH in the range of 6.9–8.6.

The KGF mine tailings contain very high percentage (>60%) of fines (<75 μm). The mine tailings are cohesion-less particles. Compacting such fine materials to produce compressed earth blocks of higher density will be difficult. Addition of soil/earth, coarse sand and industrial by-products such as fly ash help in achieving higher density for the CEB using mine tailings. The variation in the 28-day wet compressive strength with the fly ash content of the cement stabilised compressed specimen (size: 38 × 76 mm) is displayed in Fig. 10.8. The results show that a mix containing 20% fly ash and 80% gold mine tailings yield maximum compressive strength. There is significant difference in the strength values for the specimens with

Fig. 10.6 Gold mine tailings dump at Kolar Gold Fields, India

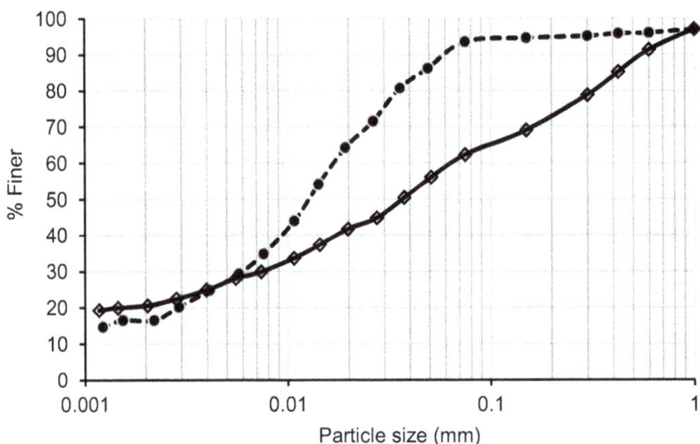

Fig. 10.7 Particle size distribution curves for gold mine tailing from KGF, India

Table 10.6 Properties of KGF gold mine tailings

Properties	Values
Textural composition	
Sand size fraction (4.75–0.075 mm) (%)	6–69
Silt size fraction (0.075–0.002 mm) (%)	16–77
Clay size fraction (<0.002 mm) (%)	6–24
Physical properties	
Specific gravity	2.49
Bulk density (kg/m^3)	1530
pH	6.9–8.6

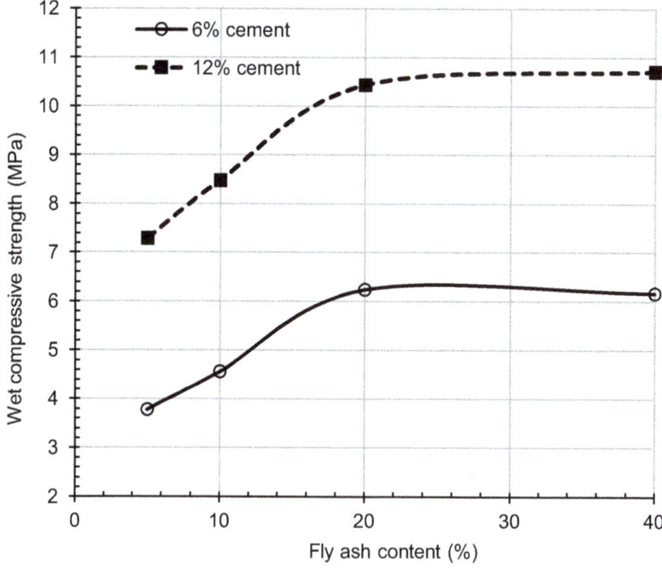

Fig. 10.8 Strength versus fly ash content for compressed specimen using gold mine tailings

and without fly ash. The strength at optimum fly ash content (20%) is double that of the strength at zero fly ash content. The Class-F fly ash has been used in these experiments. The compressed blocks of size 230 × 190 × 100 mm (Fig. 10.9) made using KGF mine tailings and coarse sand (1:1 ratio by mass) with 8% Portland cement showed a wet compressive strength of 5.3 MPa after 28 days moist curing. This illustrates the potential of the gold mine tailings for the production of the stabilised CEB's.

Fig. 10.9 Stabilised compressed blocks using gold mine tailings

10.2.4 Stabilised Compressed Earth Blocks Using Fly Ash

Generally, sandy soils containing predominantly non-expansive clays are ideally suited for the cement stabilised compressed earth blocks. Cement stabilisation is not well suited for the production of the stabilised earth blocks using soils containing expansive clay minerals (example montmorillonite) and soils with higher clay content. The lime (calcium hydroxide) stabilisation is best suited for the production of stabilised CEBs using such soils. In the case of lime stabilised CEBs, the strength is derived mainly from the cementitious material produced due to the lime-clay reactions. The calcium hydroxide reacts with reactive alumino-silicates present in the clay minerals, and such reactions are nothing but lime-pozzolana (LP) reactions. The process of lime stabilisation of the soils was discussed in Sect. 3.5. The lime-clay reactions are slow at ambient moist curing temperatures, and also, the lime reactivity of the clay minerals is not as good as other pozzolanic materials (such as fly ash, ground slag and rice husk ash). It is difficult to get satisfactory compressive strength for the lime stabilised compressed earth bricks. In such situations, supplementing pozzolanic materials such as fly ash in addition to the clay minerals will facilitate achieving satisfactory strength for the lime stabilised compressed earth bricks. The investigations of Reddy and Jagadish (1984), Reddy and Lokras (1998) and Reddy and Hubli (2002) deal with lime-fly ash stabilised compressed earth blocks. These studies and the literature on lime stabilisation of soils (discussed in Sects. 3.5 and 5.5.2) shows that (1) the optimum lime content is dependent upon the type and percentage of clay mineral in the soil and (2) lime-clay and lime-fly ash reactions can be accelerated through steam curing for short durations. These studies indicate the stabiliser clay ratio of >1.0 and >0.75 for the soils with expansive (montmorillonite) and non-expansive (kaolinite) clays, respectively, for the stable long-term strength performance of the lime stabilised compressed earth products.

The characteristics of the lime-fly ash stabilised compressed earth specimens and the blocks from the investigations of Reddy and Lokras (1998) and Reddy and Hubli (2002) are compiled in Table 10.7. Two types of soils have been used in these investigations, designated as BCS and TB. The black cotton soil (BCS) had 36% clay size fraction (montmorillonite) with a liquid limit and plasticity index of 53.1% and 27.4%, respectively. The tank bed (TB) soil had 41.8% clay size fraction (kaolinite) with a liquid limit and plasticity index of 45% and 21%, respectively. These soils have been reconstituted by mixing with sand to bring down the clay contents to optimum levels. The Class-F fly ash has been used in these studies. The results given in the Table clearly indicate the following.

(1) There is a drastic difference in the compressive strength of the lime stabilised compressed earth specimens cured under moist conditions for 28 days and at 80 ^{o}C steam curing for 18–24 h.
(2) Incorporation of fly ash (10–15%) into the lime-soil mixtures results in doubling of the compressive strength of steam cured specimens.

Good quality stabilised compressed earth bricks can be manufactured utilising the fly ash and the lime (calcium hydroxide). The optimum mix proportions can be

Table 10.7 Characteristics of lime-fly ash stabilised compressed earth blocks

Sl. No.	Soil designation	Clay content of soil-sand mix	Type of clay	Stabiliser content (%)			Type of curing	Block or specimen		Wet comp. strength (MPa)	Source
				Lime	cement	Fly ash		size (mm)	Dry density (kg/m^3)		
1	BCS	12.0	Montmorillonite	10.0	0.0	0.0	28 d moist	76 × 76 × 76	1840	1.05	Reddy and Hubli (2002)
2	BCS	12.0	Montmorillonite	10.0	0.0	0.0	24 h steam	76 × 76 × 76	1840	2.65	Reddy and Hubli (2002)
3	BCS	10.3	Montmorillonite	10.0	0.0	10.0	24 h steam	76 × 76 × 76	1840	4.23	Reddy and Hubli (2002)
4	BCS	10.3	Montmorillonite	11.0	0.0	0.0	18 h steam	230 × 190 × 100	1780	2.81	Reddy and Lokras (1998
5	BCS	10.3	Montmorillonite	11.0	0.0	15.0	18 h steam	230 × 190 × 100	1780	7.07	Reddy and Lokras (1998)
6	BCS	10.3	Montmorillonite	8.0	3.0	15.0	18 h steam	230 × 190 × 100	1780	11.00	Reddy and Lokras (1998)

(continued)

Table 10.7 (continued)

Sl. No.	Soil designation	Clay content of soil-sand mix	Type of clay	Stabiliser content (%)			Type of curing	Block or specimen		Wet comp. strength (MPa)	Source
				Lime	cement	Fly ash		size (mm)	Dry density (kg/m^3)		
7	TB	13.9	Kaolinite	10.0	0.0	0.0	28 d moist	76 × 76 × 76	1840	0.49	Reddy and Hubli (2002)
8	TB	13.9	Kaolinite	10.0	0.0	0.0	24 h steam	76 × 76 × 76	1840	2.74	Reddy and Hubli (2002)
9	TB	13.9	Kaolinite	8.0	2.0	10.0	20 h steam	230 × 190 × 100	1800	6.89	Reddy and Hubli (2002)

Moisture absorption (%): 15–17%; moist: moist curing under wet burlap; steam: steam curing at 80 °C

Fig. 10.10 Exposed surface of the masonry wall using steam cured lime-fly ash stabilised compressed earth block after 22 years

designed using lime-fly ash-clay systems to promote the production of compressed earth blocks. Figure 5.23 shows a small-scale production unit to produce lime-fly ash compressed earth blocks. Figure 5.25 shows a three-storey load bearing exposed masonry building, and the ground floor was built (in 1999) using steam cured lime-cement-fly ash (8% lime + 2% Portland cement: + 10% fly ash) stabilised compressed earth blocks manufactured with TB soil and Class-F fly ash. The average wet compressive strength: of the block, steam cured at 80 °C for 20 h (block size: 230 × 190 × 100 mm), was 6.89 MPa. A close-up picture of the masonry wall surface after 22 years of weathering is shown in Fig. 10.10. The performance of the building against weathering actions is excellent over the last two-decade period.

10.2.5 Stabilised Compressed Earth Blocks Using Slag

The slag is a by-product of the metallurgical and the steel industry. The slag produced in the blast furnace, during the manufacturing of the pig iron or the crude iron, is called granulated blast furnace slag (GBS). The molten GBS from the blast furnace is cooled/quenched in different ways. The method used and the rate of cooling adopted will result in different forms of hardened blast furnace slag (BFS). Typically, for iron ore feed containing 60–65% of iron, the BFS output varies from 300 to 540 kg per tonne of hot metal produced (Indian Bureau of Mines 2018). About 40 million tonnes of BFS is produced in India, out of which about 3 million tonnes is consumed by the cement industry. The remaining BFS is either stored at the steel plants in open fields or in the landfills.

The pozzolanic reactivity of GBS is governed by the chemical composition, degree of vitrification or glass content and the fineness. The granulated slag with high glass content, when ground to fine powder form (<2 μm size), shows pozzolanic properties. The ground granulated blast furnace slag (GGBS) in the presence of a suitable activator has the ability to form hydration products similar to those formed during hydration of ordinary Portland cement (Mehta and Monteiro 2014). IS 383 (2016) code identifies GBS as a fine aggregate that can be used in concrete. The GBS in its native state has sharp edges (Fig. 10.11) and lower loose bulk density (~1200 kg/m^3). Such particles pose problems in handling as well as workability of the concrete. The sharp edges can be smudged by processing the GBS through a vertical shaft impactor. Figure 10.12 shows the rounded edges of the processed GBS (PGBS). The processed GBS shows higher loose bulk density of 1350 kg/m^3. The PGBS is more suitable as a fine aggregate. The grain size distribution curves for the GBS, the PGBS and the natural sand are shown in Fig. 10.13. Table 10.8 gives the chemical composition of PGBS. The physical properties of PGBS are given in Table 10.9.

The particle size distribution curve of the PGBS matches with that of the natural river sand, with a fineness modulus of 2.40. The chemical composition shows that the PGBS comprises silica/alumina (56%) and about 26% of calcium oxide. It has poor lime reactivity, mainly attributed to coarser particles and a higher value of water absorption (6%) compared to other natural fine aggregates. These results are based

Fig. 10.11 SEM images of GBS particles

Fig. 10.12 SEM image of processed GBS (PGBS) particles

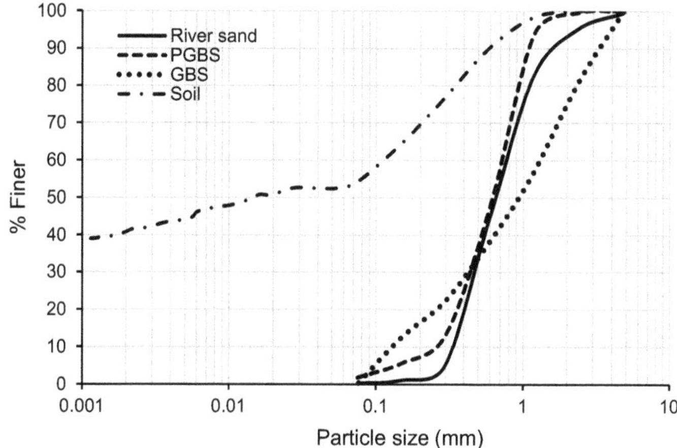

Fig. 10.13 Particle size distribution curves for soil, GBS, PGBS and natural sand

Table 10.8 Chemical composition of PGBS

Compound	SiO_2	Al_2O_3	CaO	MgO	NaO	Fe_2O_3	MnO_2	TiO_2	LoI and others
(% by mass)	33.6	22.3	26.3	5.4	0.9	1.1	0.8	0.6	9

Table 10.9 Properties of PGBS

Property	Values
Fineness modulus	2.40
Specific gravity	2.71
Bulk density (kg/m^3)	1350
pH	10.67
Lime reactivity (MPa)	0.22
Water absorption (%)	5.96
Loss on ignition (%)	1.79

on a slag from a steel industry in Karnataka state of India, and the properties of the PGBS might slightly differ for other sources of slag.

The stabilised CEBs are manufactured using local soils. Generally, the soils have higher clay content than the optimum clay content of 10–15% (Reddy and Latha 2014). Hence, the soils are reconstituted by mixing with inert materials such as sand. The soil-sand ratio depends upon the clay content of the soil. Instead of natural or manufactured sand, PGBS can be used for reconstituting the soil to attain optimum clay in the mix used for CEB. Figure 10.14 shows a comparison of wet compressive strengths of the compressed earth cylindrical specimen (38 mm diameter and 76 mm height) made from the mixes from the natural sand and the PGBS using cement and cement-lime as stabiliser. The specimens had a dry density in the range of 1800–1820

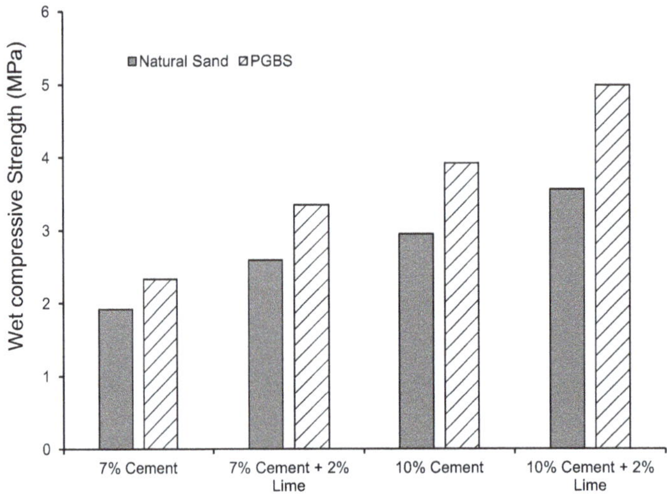

Fig. 10.14 Comparison of strengths of compressed earth cylindrical specimen using sand and PGBS

kg/m^3. The soil containing 41% clay (Fig. 10.13) was used in these castings. The soil was reconstituted by mixing either with the sand or the PGBS in the ratio of 1: 1.6 (soil: sand/PGBS, by mass) such that the reconstituted mix had about 15% optimum clay fraction. Figure 10.14 clearly shows that the wet compressive strength does not get altered while the sand was replaced by the PGBS. In fact, there is 20–30% increase in the wet compressive strength when the sand was replaced by the PGBS.

Figure 10.15 shows a stabilised CEB (size: 230 × 110 × 70 mm) manufactured using soil-PGBS mix and using 7% cement + 2% lime (calcium hydroxide). The dry density of the CEB was 1890 kg/m^3. The wet and dry compressive strengths of the CEB were 11.60 and 19.10 MPa, respectively. The water absorption of the CEBs was 11.2%. The linear expansion on saturation and the mass loss after 12 cycles of

Fig. 10.15 Stabilised CEB using soil-PGBS mix

weathering test (as per IS 1725 2013) was 0.01% and 1.50%, respectively. These values are well within the limits specified in the IS 1725 (2013) code.

10.3 Summary

Annually billions of tonnes of NOSWs are generated across the world, which has a great potential to be utilised in the construction industry. The utilisation of the NOSWs such as fly ash, CDW, slag and mine tailings (iron and gold mines tailings) for the production of stabilised CEB was illustrated. The very high quality CEBs can be manufactured using the NOSWs. There is immense scope to explore many other types of NOSWs for the production of stabilised CEBs. The consumption of NOSW in the construction industry has great potential to mitigate the adverse environmental impacts caused by the disposal of NOSW's.

References

Ali A, Ajit KS (2018) Construction and demolition waste generation and properties of recycled aggregate concrete: a global perspective. J Clean Prod 186:262–281. https://doi.org/10.1016/j.jclepro.2018.03.085

Indian Bureau of Mines (2018) Indian minerals yearbook part 2—metals & alloys, New Delhi. https://ibm.nic.in/index.php?c=pages&m=index&id=1007

IS 1725 (2013) Specification for soil-based blocks used in general building construction. Bureau of Indian Standards, New Delhi, India

IS 383 (2016) Coarse and fine aggregate for concrete—specification, Third revision. Standards. Bureau of Indian Standards, New Delhi, India

Mehta PK, Monteiro PJM (2014) Concrete—microstructure, properties, and materials. Mc Graw Hill, New Delhi

Reddy BVV, Hubli R (2002) Properties of lime stabilised steam-cured blocks for masonry. Mater Struct 35:293–300

Reddy BVV, Jagadish KS (1984) Pressed soil-lime blocks for building construction. Mason Int 3:80–84

Reddy BVV, Latha MS (2014) Influence of soil grading on the characteristics of cement stabilised soil compacts. Mater Struct 47(10):1633–1645

Reddy BVV, Lokras SS (1998) Steam-cured stabilised soil blocks for masonry construction. Energy Build 29:29–33

Reddy BVV, Hemanth Kumar H, Ullas SN, Gourav K (2016) Non-organic solid wastes—potential resource for construction materials. Curr Sci 111(12):1968–1976

Reddy BVV, Lal R, Nanjunda Rao KS (2007) Optimum soil grading for the soil-cement blocks. J Mater Civil Eng 19(2):139–148

Sudhakar Rao M, Reddy BVV (2006) Characterisation of Kolar Gold Fields mine tailings for cyanide and acid drainage. Geotech Geol Eng 24(6):1545–1559

Ullas SN (2014) Studies on utilisation of iron ore tailings as fine aggregate in mortar and concrete. PhD thesis, Centre for Sustainable Technologies, Indian Institute of Science, Bangalore, India

Wu H, Zuo J, Zillante G, Wang J, Yuan H (2019) Status quo and future directions of construction and demolition waste research: a critical review. J Clean Prod 240:118163. https://doi.org/10.1016/j.jclepro.2019.118163

Part III
Stabilised Rammed Earth

Chapter 11
Introduction to Rammed Earth

11.1 Rammed Earth

The rammed earth is a monolithic construction and finds applications in the construction of walls, roofs/floors, foundations, built-in furniture, embankments, earthen bunds, etc. The rammed earth construction process mainly involves the compaction of the processed partially saturated soil in progressive layers in a rigid formwork (Figs. 11.1 and 11.2). The rammed earth construction is primarily an in-situ operation, though there are attempts to popularise the precast rammed earth elements for use in the construction of the walls (Lindsay 2012; Otto Kapfinger and Marko Sauer 2015).

The rammed earth constructions can be classified into two broad categories: (1) stabilised rammed earth and (2) unstabilised rammed earth. The unstabilised rammed earth elements are constructed, mainly using natural materials (soil and aggregates). Generally, the stabilised rammed earth construction uses inorganic stabilisers (such as cement or lime) in addition to the natural materials such as soil, gravel and aggregates. The rammed earth structures have several distinct advantages when compared to the conventional types of constructions:

(a) The rammed earth is an environment-friendly and low embodied carbon material
(b) Bulk of the raw materials used are local and available within a short distance from the construction site
(c) The rammed earth structures possess aesthetically pleasing appearance, resembling closed to a sedimentary rock with stratified layers (Fig. 11.2).
(d) Varieties of texture and colour finishes can be achieved for the rammed earth using specifically selected raw materials and special construction processes (Figs. 11.3).
(e) The plan form as well as the vertical cross section can be varied easily through proper design of the formwork (Fig. 11.4). There is scope for creating built-in artwork into the rammed earth surfaces (Figs. 11.5 and 11.6).

B. V. V. Reddy, *Compressed Earth Block & Rammed Earth Structures*,
Springer Transactions in Civil and Environmental Engineering,
https://doi.org/10.1007/978-981-16-7877-6_11

(f) The rammed earth buildings can have better living environmental conditions with better indoor air quality. The walls can be designed to be dense and bulky having considerable thermal mass offering scope for passive environmental performance design.
(g) The compressive strength of the rammed earth wall is higher when compared to the strength of a masonry wall using masonry units made from a similar material composition and density.
(h) Thinner (150 mm or more) rammed earth walls can be designed for load bearing purposes. It is easy to implement different wall thicknesses (of desired odd sizes) across different floor heights of a building.

A comparison between the stabilised and unstabilised rammed earth construction is provided in Table 11.1.

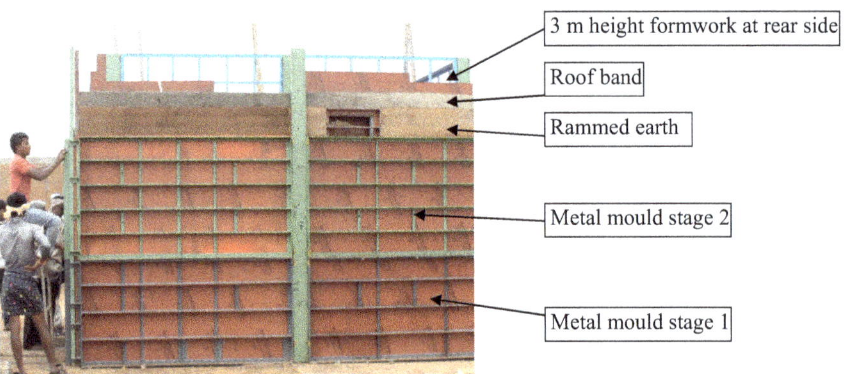

Fig. 11.1 Metal formwork for rammed earth wall construction

Fig. 11.2 Compacted rammed earth layers

Fig. 11.3 Different coloured textures in rammed earth using natural soils

Fig. 11.4 Curve shaped (in plan) rammed earth wall

Fig. 11.5 Artwork in cement stabilised rammed earth walls at Hunnarashala Bhuj, India (Pictures: Tejas Kotak)

Fig. 11.6 Carved artwork on a load bearing cement stabilised rammed earth pillar at Hunnarashala Bhuj, India (Picture: Tejas Kotak)

Table 11.1 Comparison between stabilised and unstabilised rammed earth

Unstabilised rammed earth	Stabilised rammed earth
• Use of natural materials and negligible environmental costs at the end of life • Good moisture buffering properties • Lower cost and scope for self-help construction • Low embodied carbon • Lower strength and hence thicker walls • Loss of strength and stiffness on moisture absorption • Prone for erosion due to rain impact • Prone for termite/insect infestation and damage especially in tropics and sub-tropics	• Higher strength, even in saturated condition • Easy to achieve desired strength and scope for taller structures • Possible to build thinner walls • Use of inorganic stabilisers such as cement and lime • Higher embodied carbon when compared with carbon in unstabilised rammed earth • Better durability and good erosion resistance against rain impact • Free from termite/insect damage

11.2 History and Developments in Rammed Earth Construction

The history of rammed earth construction was dealt in detail in Chap. 1. Different types of structures have been built using the rammed earth construction technique. The developments and applications of rammed earth construction, since BC period are detailed in the Table 11.2. Few examples of rammed earth constructions are illustrated in the Figs. 11.7, 11.8 and 11.9. The earlier rammed earth constructions utilised mainly the local soil/materials and resulted in environment-friendly structures. Buildings of 2–6 storey height have been successfully built with load bearing unstabilised rammed earth walls.

Because of the lower strength, the unstabilised rammed earth walls are made thicker (>400 mm), mainly to reduce compressive stresses developed due to gravity loads and for lateral stability. The inorganic additives such as cement have been explored for the rammed earth construction since 1940s. The cement stabilisation facilitates in building thinner structural walls (150–300 mm). Also, in the case of cement stabilised rammed earth, the wall strength can be easily varied by adjusting the cement content. Figures 11.10, 11.11, 11.12 1.34, 1.37–1.43 show some of the recent cement stabilised rammed earth buildings.

The details of the stabilised rammed earth constructions in twentieth and twenty-first century were discussed in Sect. 1.2 of Chap. 1. Section 1.2.2.1 gives a summary of stabilised earth including rammed earth constructions in India. There are several successful examples of cement stabilised rammed earth buildings in Australia, USA, Europe, Asia and many other countries (Verma and Mehra 1950; Easton 1982, 1982, 2008; Hall 2002; Houben and Guillaud 2003; Walker et al. 2005; Tejas 2007; Windstorm and Schmidt 2013; Reddy et al. 2014, 2019). The load bearing cement stabilised rammed earth buildings of 1–3 storeys have been built in India since 2000. Large numbers of cement stabilised rammed earth houses were built in the rehabilitation projects in Gujrat, India (Tejas 2007; Kiran and Tejas 2019). A three-storeyed cement

Table 11.2 History and developments in rammed earth construction

Circa	Place	References
475 BC	China	Jiyao and Weitung (1990)
300 BC	Great wall of China	Anger and Fontaine (2009)
246–209 BC	• Great Wall of China • Monuments by Assyrians, Babylonians, Persians & Sumerians	Easton (1982)
200 BC	Africa	Houben and Guillaud (2003)
300 AD	Watch towers—Europe	Easton (1982), Cody (1990)
650–815 AD	Monumental structures	Chayet et al. 1990
1368–1644 AD	Tulou rammed earth buildings, Fujian Province, China	https://eartharchitecture.org Jaquin (2012)
1600 AD	Buildings in Ladakh and Bhutan	Chayet et al. 1990
1660–1865 AD	Swiss pisé structures	Kleespies (2000)
1796–1850 AD	Six-storey apartment and residential buildings, Weilburg, Germany	Schick (1987)
1857 AD	Church of Holy Cross in Staatsburg	Cody (1990), Cointeraux (1971)
1900 AD	Rammed earth chalk buildings in UK	Walker et al. (2005)
1914–1942 AD	Exploration/research on rammed earth material	Kleespies (2000)
1948 AD	4000 rammed earth houses in Punjab province, India	Verma and Mehra (1950)
After 1970 AD	• Modern rammed earth constructions across the world • R&D in rammed earth	Hall (2002), Easton (1982, 2008), Houben and Guillaud (2003), Walker et al. (2005), Tejas (2007), Reddy et al. (2014, 2019)

stabilised rammed earth wall dormitory building is shown in Fig. 11.10. This building is located in the region where the outside temperatures in summer cross 43 °C. The load bearing walls are 300 mm thick, designed to have high thermal mass for passive cooling effect. Figure 11.11 shows a three-storey school building using load bearing cement stabilised rammed earth in Bangalore, India (Reddy et al. 2014). The building has 400 mm thick walls in the ground floor and reducing to 300 mm in the top floor. The floor slabs have a span of 7.8 m. Figure 11.12 shows a three-storey residential cement stabilised rammed earth building.

Fig. 11.7 Unstabilised rammed earth house, Weilburg, Germany, (Constructed in 1850, picture on 12 October 2008)

Fig. 11.8 Six storey load bearing unstabilised rammed earth building, Weilburg, Germany, (Constructed in 1826, picture on 12 October 2008)

Fig. 11.9 Old Pise-Haus (rammed earth house) in Weilburg (Built in 1826, picture taken on 12 October 2008)

Fig. 11.10 Three-storeyed load bearing cement stabilised rammed earth dormitory building, India, built in 2019

Fig. 11.11 Three-storeyed load bearing cement stabilised rammed earth school building, Bangalore, India, built in 2009, corridor has few RC columns for housing rainwater pipes

Fig. 11.12 Three-storeyed load bearing cement stabilised rammed earth residential building, Bangalore, India, built in 2017, designed by architect Mr. Harsha S

11.3 Method of Casting Rammed Earth

The rammed earth construction needs a dismountable rigid formwork. Some of the typical formworks are shown in Figs. 11.11, 11.13, 11.14 and 11.15. The moulds consist of two leaves linked by lengthy bolts. The rammed earth casting process involves the following steps.

(a) Setting the mould
(b) Processing the soil
(c) Mixing the soil, gravel/aggregates and stabiliser in dry state
(d) Mixing the materials with water
(e) Pouring the partially saturated soil-aggregate-stabiliser mixture into the mould
(f) Compacting the processed material into a desired density
(g) Dismantling the formwork
(h) Curing.

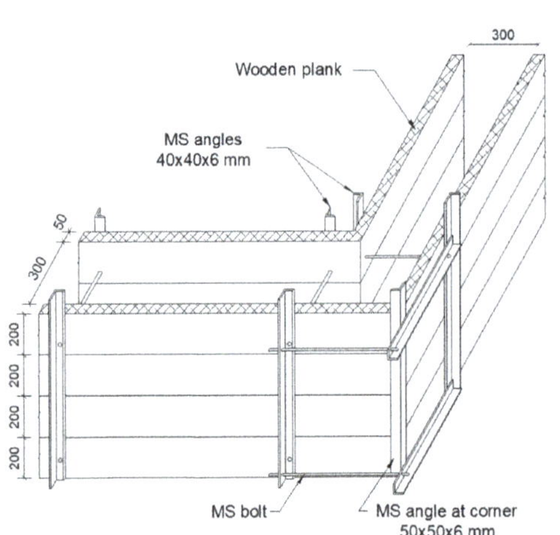

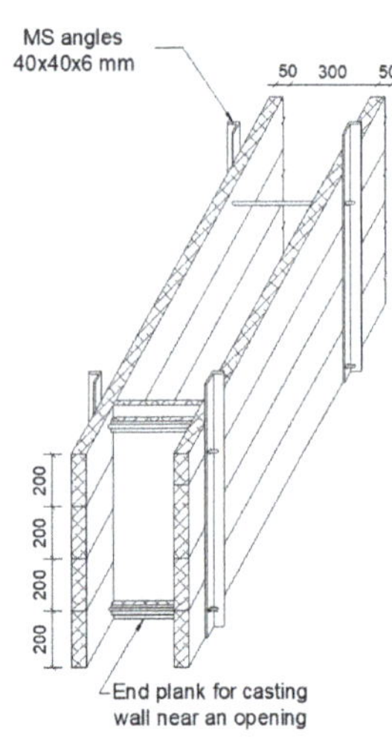

Fig. 11.13 Wooden formwork (as per IS 2110 code)

Fig. 11.14 Long continuous wooden formwork

Fig. 11.15 Formwork for the entire height and length of rammed earth wall (Courtesy: https://www.rammedearthworks.com)

11.3.1 Setting the Mould

Essentially any mould assembly for the rammed earth construction consists of plates or sheets, fastened mainly using bolt-nut system, resulting in a box like assemblage having stiffeners. Figure 11.16 shows one such assemblage. The mould assembly shown in Fig. 11.16 consists of a sheet/plate of full wall height (3 m) at one end and one third the height on the opposite side. As the ramming progresses, more sheets are added in stages to complete the full height of the wall. The mould should be set to have correct plumb line, before commencing the compaction of processed earth in layers.

11.3.2 Processing the Soil and Preparation of the Mix

The processing of soil and the mix preparation for the stabilised rammed earth involves the following activities:

(a) Pulverisation
(b) Sieving
(c) Modifying the soil composition
(d) Mixing the soil and the binders
(e) Mixing the soil-binder mixture and the water.

Fig. 11.16 Rammed earth mould assembly

Section 4.3 (Chap. 4) gives details of soil processing and mix preparation for the stabilised earth block production. A similar procedure is needed for preparing the mix for the rammed earth construction.

11.3.3 Compacting the Processed Material

Rammed earth compaction is carried out in layers. Generally, the compacted layer thickness varies between 60 and 150 mm. The investigations of Lepakshi (2017) and Lepakshi and Reddy (2018) have shown that the optimum layer thickness yielding maximum strength and stiffness for the CSRE is in the range 80–100 mm. The strength of CSRE greatly depends upon its dry density. The strength and density are linearly related (Reddy and Kumar 2011). Hence, the dry density of the compacted layers of CSRE should be controlled. Therefore, a definite quantity of the wetted processed mix should be poured into the mould and then compacted to the desired thickness in order to achieve a specified dry density. A sample calculation for arriving at the mass of the material in each layer is provided below.

Mould length: 3.0 m.

Wall thickness: 0.23 m.

Compacted layer thickness: 0.10 m.

Dry density: 1850 kg/m^3.

Moulding moisture content: 11% (by mass).

Quantity of processed soil mixture needed in one layer = $(3.0 \times 0.23 \times 0.10)(1850 \times 1.11)$.

= 141.69 kg.

Known quantity (141.69 kg) of the mix is poured into the mould, uniformly spreading the loose processed earth mixture, and then, compaction is carried out using the rammer such that a uniform layer thickness (100 mm) is achieved. After the compaction, dents are created on the freshly laid compacted layer as illustrated in Fig. 11.17. Such dents help in interlocking of the compacted layers in the CSRE. The compaction process continues till the desired height for the CSRE is achieved. Figure 11.18 shows a part of compacted CSRE wall where on one side 3.0 m height formwork sheet is present and on the opposite side the formwork is stripped.

Fig. 11.17 Dents made on the fresh compacted layer of CSRE wall

Fig. 11.18 Part of the CSRE wall exposed during formwork stripping

Fig. 11.19 Cement stabilised rammed earth wall covered with burlap for curing

11.3.4 Dismantling the Formwork and Curing

The CSRE wall compaction process in a particular segment of the building may take 3–5 days; hence, the formwork can be dismantled a day after completing the last layer of the wall. After dismantling the formwork, the CSRE wall should be covered with a wet gunny cloth or burlap and continue curing for four weeks. The water should be sprayed onto the burlap 3–4 times a day, until the curing period is completed. Figure 11.19 shows a CSRE wall covered with burlap, undergoing curing.

11.4 Rammed Earth Structures—Potential and Prospects

The rammed earth can be used for different components of the building or the structure. The historical rammed earth structures, though bulky, showed the potential of rammed earth for the construction of the buildings. With the advent of knowledge on the soil stabilisation and the construction techniques, the stabilised rammed earth has been explored for the building construction even in the seismically active areas. The R&D work on rammed earth pursued across the world touches upon the following aspects.

1. Optimisation of mix proportions and mix design procedures
2. Mechanical characteristics and durability of rammed earth
3. Behaviour of rammed earth under compression, shear and flexure
4. Fibre reinforced rammed earth
5. Reinforced rammed earth

6. Earthquake-resistant rammed earth structures
7. Prefabricated rammed earth elements.

Modern rammed earth construction has caught the attention of the architects, the engineers and the other building professionals, because of the inherent advantages of low embodied carbon and greenness, offering scope for aesthetically pleasing finishes and potential/scope to shape into different forms. Also, there is large scope for maximising the utilisation of local materials and resources, as well as the industrial by-products and the mining industry wastes.

References

Anger R, Fontaine L (2009) Bâtir en Terre. Publisher, Belin, Paris France. 2701152046

Cointeraux F (1971) Traite des constructions rurales et e leur disposition. Franca, Paris

Cody JW (1990) Earthen walls from France and England for North American farmers, 1806–1870. 6th international conference on the conservation of earthen architecture: Adobe 90 preprints: Las Cruces. New Mexico, USA, pp 35–43

Chayet A, Jest C, Sanday J (1990) Earth used for building in the Himalayas, the Karakoram, and central Asia-recent research and future trends. 6th international conference on the conservation of earthen architecture: Adobe 90 preprints: Las Cruces. New Mexico, USA, pp 29–34

Easton D (1982) The rammed Earth experience. Blue Mountain Press, Wilseyville, California, USA

Easton D (2008) The industrialisation of monolithic Earth walling for first world applications. In: Proceedings of the 5th international conference on building with Earth (LEHM 2008), Koblanz, Dachverband Lehm eV, Germany, pp 90–97

https://eartharchitecture.org. Accessed Jan 2019

https://www.rammedearthworks.com

Hall M (2002) Rammed Earth: traditional methods, modern techniques, sustainable future. Build Eng 77(11), pp 22–24

Houben H, Guillaud H (2003) Earth construction: a comprehensive guide. Intermediate Technology Publications/CRATerra-EAG, London

IS 2110 (1980) Code of practice for in situ construction of walls in buildings with soil-cement (first revision). Bureau of Indian Standards, New Delhi, India (reaffirmed 1998)

Jiyao H, Weitung J (1990) Earth-culture-architecture: the protection and development of rammed Earth and adobe architecture in China. In: 6th international conference on the conservation of earthen architecture: Adobe 90 preprints: Las Cruces. New Mexico, USA, pp 72–76

Jaquin P (2012) History of Earth building techniques. In: Hall MR, Lindsay R, Krayenhoff M (eds) Modern Earth buildings: materials, engineering, construction and applications. Woodhead Publishing Ltd. Cambridge, UK, pp 307–323

Kiran V, Tejas K (2019) Innovations in construction of cement-stabilized rammed Earth dwellings post Bhuj-2001 earthquake. In: Venkatarama Reddy BV, Mani M, Walker P (eds) Earthen dwellings and structures: current status in their adoption. Springer Transactions in Civil & Environmental Engineering, Singapore, pp 225–234

Kleespies T (2000) The history of rammed Earth buildings in Switzerland. In: Terra 2000: 8th international conference on the study and conservation of earthen architecture, Torquay, Devon, UK, pp 137–138

Lepakshi R (2017) Studies on characteristics of cement stabilised rammed Earth and flexural behaviour of plain and reinforced rammed Earth. PhD thesis, Department of Civil Engineering, Indian Institute of Science, Bangalore, India

Lepakshi R, Reddy BVV (2018) Influence of layer thickness and plasticisers on the characteristics of cement stabilised rammed Earth. J Mater Civil Eng 30(12):04018314,1–10. https://doi.org/10.1061/(ASCE)MT.1943-5533.0002539
Lindsay R (2012) Australian modern Earth construction. In: Hal MR, Lindsay R, Krayenhoff M (eds) Modern Earth buildings: materials, engineering, construction and applications. Woodhead Publishing Ltd. Cambridge, UK, pp 609–649
Kapfinger O, Sauer M (2015) Martin Rauch refined Earth construction and design with rammed Earth, Eberl Print GmbH. Immenstadt, Germany
Reddy BVV, Georg L, Sreeram VS (2014) Low embodied energy cement stabilised rammed Earth building—a case study. Energy Build 68, pp 541–546
Reddy BVV, Mani M, Walker P (eds) (2019) Earthen dwellings and structures—current status in their adoption. In: Springer transactions in civil and environmental engineering. Springer, Singapore, ISBN 978-981-13-5882-1
Schick W (1987) Der Pise–Bau Zu Weilburg an der Lahn. Burgerintiative, Alt Weilburg, Germany
Tejas K (2007) Constructing cement stabilised rammed Earth houses in Gujarat after 2001 Bhuj earthquake. In: Proceedings of international symposium on earthen structures, Interline Publishers, Bangalore, pp 62–71
Verma PL, Mehra SR (1950) Use of soil-cement in house construction in the Punjab. Indian Concr J 24(4), pp 91–96
Walker P, Keable R, Martin J, Maniatidis V (2005) Rammed Earth: design and construction guidelines. BRE Bookshop, Watford
Windstorm B, Schmidt A (2013) A report of contemporary rammed Earth construction and research in North America. Sustainability 5(2), pp 400–416

Chapter 12
Compressive Strength of Rammed Earth

12.1 Compressive Strength

The design of a structural element using a material such as rammed earth needs the information on strength under different types of loading conditions. Primarily, understanding the behaviour of rammed earth under compression, shear and tension is essential for the structural design of rammed earth structures. The strength of the rammed earth under compression, particularly the compressive strength, has been discussed in the following sections. The factors having significant influence on the strength of the rammed earth are:

(a) Composition of soil and gravel/aggregate mixture (grading, clay type, clay content)
(b) Moulding moisture content
(c) Dry density
(d) Layer thickness
(e) Type of stabiliser and the stabiliser content
(f) Moisture content

12.2 Optimum Soil Grading

There are varieties of soils and clay minerals as discussed in Chap. 2. Considering both the coarse and the fine-grained soils, the optimum soil grading for the production of compressed earth blocks was discussed in Chap. 5. The soil consists of four main particle types, classified based on the particle sizes. The grading, i.e. relative proportions of gravel, sand, silt and clay, governs the properties and the suitability of a soil for the rammed earth construction. All four types of particle sizes play an important role on the performance of the rammed earth. Large-sized particles such as gravel and sand provide the inert skeleton or the matrix. The clay particles

B. V. V. Reddy, *Compressed Earth Block & Rammed Earth Structures*,
Springer Transactions in Civil and Environmental Engineering,
https://doi.org/10.1007/978-981-16-7877-6_12

provide the main source of binder (in unstabilised rammed earth) and impart certain characteristics for the soil such as plasticity for moulding or shaping and the green strength. The clays swell upon wetting and shrink on drying. The soils containing less expansive (also called non-expansive) clay minerals such as kaolinite and illite are suitable for the rammed earth construction. The clay minerals present in the soil play a significant role in controlling the strength and the durability of the stabilised rammed earth.

Table 12.1 summarises the soil grading limits specified for the rammed earth construction from several scientific investigations and codes books/monographs. Walker et al. (2005) and Houben and Guillaud (2003) specify particle size ranges for soil–sand–aggregate mixtures suitable for the rammed earth construction. Figure 12.1 shows the Walker et al. (2005) suggested range for soil grading suitable for the rammed earth construction. The range for soil particle grading/sizes shown in this figure indicates proportions for sand and gravel: 45–80%, silt: 10–30% and clay: 5–20%.

The clay minerals control the soil behaviour and the soil characteristics. Hence, Kumar (2009), and Reddy and Kumar (2011b) attempted to arrive at optimum clay fraction in the soil–sand mixture yielding maximum strength for the cement stabilised rammed earth. Considering a coarse-grained natural soil having 31.6% clay fraction

Table 12.1 Summary of soil grading limits for rammed earth construction

Source	Soil composition (%, by mass)			
	Gravel (>4.75 mm)	Sand (4.75–0.075 mm)	Silt (0.075–0.002 mm)	Clay (<0.002 mm)
Patty and Minium (1945)	45	40–75	–	–
Verma and Mehra (1950)	–	>35	–	–
Easton (1982)	–	70	–	30
Middleton (1992)	Well graded soil and low clay content			
Lilley and Robinson (1995)	–	–	–	11–21
IS-2110 (2002)	–	>35	–	–
Houben and Guillaud (2003)	–	–	–	–
Clark and Walker (2003)	39	–	12	11
Walker et al. (2005)	45–80		10–30	5–20
Maniatidis and Walker (2008)	30	45	13	12
Reddy and Kumar (2011b)	–	73	12	15

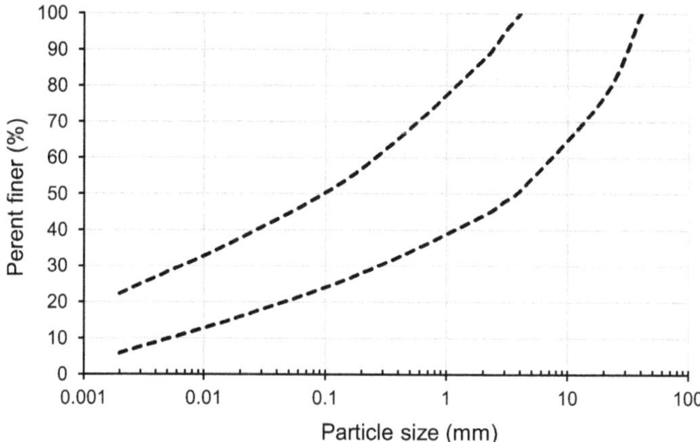

Fig. 12.1 Grain size distribution range for soils suitable for rammed earth proposed by Walker et al. (2005)

(kaolinite), the rammed earth prisms (150 mm diameter and 300 mm height) were cast by reconstituting the soil to have different clay fractions in the mix ranging between 9 and 31.6%. The dry density of the prisms was maintained at 1800 kg/m^3 across the clay fractions. A cement content of 8% (by mass) was used in these studies. The results of the investigations are plotted in Fig. 12.2. The figure shows relationships between the compressive strength (in wet and dry state) and the clay size fraction of the mix. Figure 12.2 reveals that the compressive strength of cement stabilised rammed earth increases with the increase in the clay content, reaches a peak value and then drops as the clay fraction increases further. The maximum strength can be noticed at the clay content of 15.8%. These results suggest that the optimum clay percentage yielding maximum compressive strength for the cement stabilised rammed earth is in

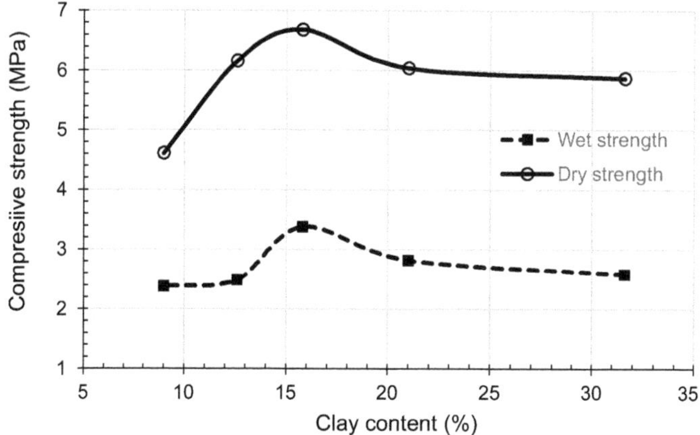

Fig. 12.2 Strength versus clay content for cement stabilised rammed earth

the vicinity of 16%. Studies of Reddy and Latha (2014) showed optimum clay content for the cement stabilised compressed earth blocks (CEB's) using coarse-grained soil at 15%, attributing it to lower void ratio at optimum clay content. This study shows 10% clay fraction as optimum for fine-grained soils to produce the cement stabilised CEB's. Like for the cement stabilised CEB's, the optimum clay content yielding maximum strength, for the cement stabilised rammed earth, can be in the range of 10–15% for soils containing non-expansive clay minerals. Even soils or soil–sand mixtures having less than 10% clay are suitable, but the soils with more than 15% clay might pose problems of durability and cracking due to shrinkage.

12.3 Optimum Moisture Content

The dynamic compaction process is adopted in the production/construction of rammed earth. The partly saturated soil–sand–aggregate mixtures are used for the rammed earth construction. There is a need for determining the quantity of water to be added to achieve a desired consistency for the mix, which facilitates compaction process as well as to achieve desired density for the rammed earth element. The standard Proctor test is the most widely used test to assess the optimum moisture content (OMC) for a soil to achieve maximum possible dry density. The standard Proctor test yields OMC and maximum dry density (MDD) for the soil. The influence of (a) soil grading and soil composition (particularly clay content) and (b) cement content, on the compaction characteristics (OMC and MDD) of the cement stabilised rammed earth (CSRE), is discussed here.

12.3.1 Influence of Soil Grading and Cement Content on the Compaction Characteristics

The results of the compaction tests performed by Kumar (2009), and Reddy and Kumar (2011a) are presented here. A natural soil containing 31.6% clay was reconstituted by diluting with sand such that four different soil grading were generated with clay contents ranging between 9 and 31.6%. The standard Proctor compaction tests were carried out on four soil compositions mixed with three different cement contents. The variation in OMC and MDD as the cement content was varied is shown Figs. 12.3 and 12.4, respectively. The results show that there is a marginal variation (2–3%) in OMC and MDD values as the cement content was varied between 5 and 12%. Thus, OMC and MDD are not sensitive to the variation in the cement content of the mix irrespective of the clay fraction of the mix.

The OMC and MDD are sensitive to the quantity of clay present in the cement stabilised soil mixture as illustrated in Figs. 12.5 and 12.6. The OMC increases steeply as the clay content of the mix increases. The MDD–clay content relationships

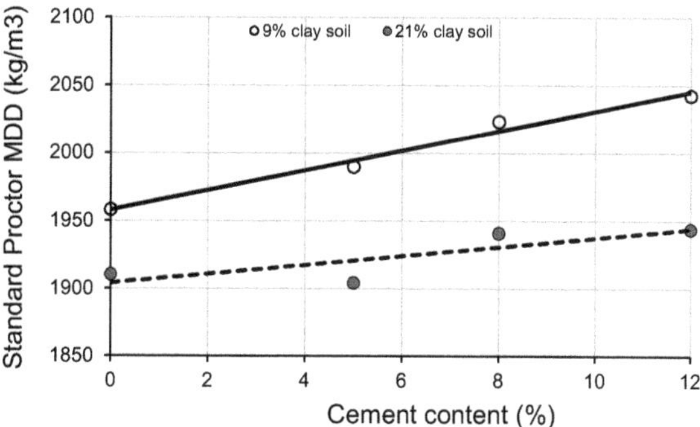

Fig. 12.3 Cement content versus standard Proctor OMC

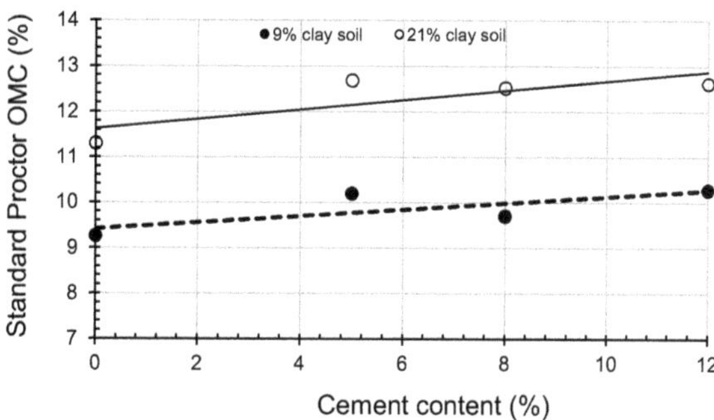

Fig. 12.4 Cement content versus standard Proctor MDD

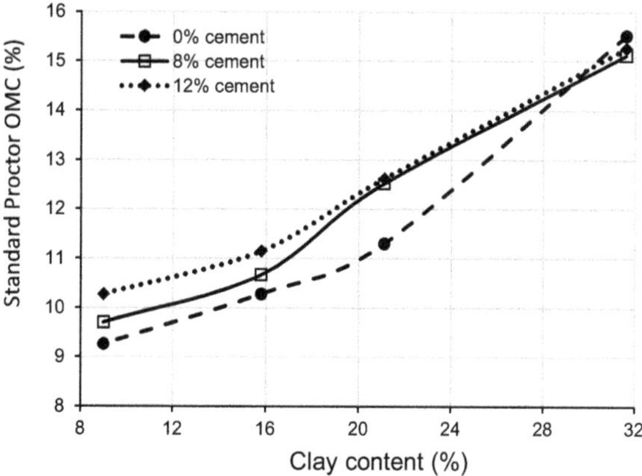

Fig. 12.5 Variation in standard Proctor OMC with the clay content of the soil (Reddy and Kumar 2011a)

(Fig. 12.6) show hardly any variation in MDD for the clay content in the range of 9–15.8%. For 15.8–31.6% clay content range, there is 8–10% decrease in MDD irrespective of the cement content of the mix. These results shed some light on the variations in the OMC and MDD values as the clay fraction in the mix and the cement content is varied.

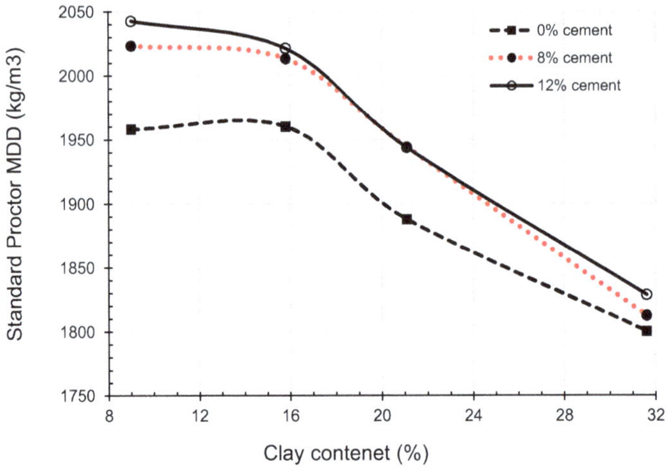

Fig. 12.6 Standard Proctor MDD versus clay content of the soil (Reddy and Kumar 2011a)

12.4 Moulding Moisture Content, Density and Strength

The compaction characteristics (OMC and MDD) of the cement–soil mixtures can be established using the standard Proctor test. The standard Proctor OMC can be used for the rammed earth construction to achieve corresponding MDD. It is possible to achieve higher or lower dry density than MDD, during rammed earth construction by varying the compaction energy as compared to the standard Proctor test energy. In such situations, the question arises regarding OMC to be used in order to achieve best possible strength for the cement stabilised rammed earth. The density plays a crucial role in controlling the strength of the compacted rammed earth. The strength, density and moulding moisture content (MMC) relationships for cement stabilised rammed earth (CSRE) were generated by Kumar (2009).

A soil with 15.8% clay was chosen to prepare CSRE prism specimen (150 × 150 × 300 mm). The standard Proctor OMC and MDD values for the natural soil and the soil with cement are given in Table 12.2. The OMC varies between 10.28 and 11.15% as the cement content varies between 0 and 12%. Similarly, the MDD values vary between 1992 and 2023 kg/m^3 as the cement content varies between 0 and 12%. Based

Table 12.2 Standard Proctor OMC and standard Proctor MDD values for the soil

Cement content (% by weight)							
0%		5%		8%		12%	
MDD (kg/m^3)	OMC (%)	MDD (kg/m^3)	OMC (%)	MDD (kg/m^3)	OMC (%)	MDD (kg/m^3)	OMC (%)
1992	10.28	2010	10.75	2023	10.67	2021	11.15

Clay fraction: 15.8%, silt: 11.6%, sand: 72.6%

on these results, three moulding moisture contents (MMC) have been selected for examining the influence of MMC on the strength of the compacted cement stabilised rammed earth specimens. The MMC of 8.5, 12 and 14.5% corresponding to dry of OMC, close to OMC and wet of OMC respectively, were used in the preparation of the samples. For each of these combinations, dry density of the specimens was varied over wide limits.

Figure 12.7 shows strength and density relationships as the MMC was varied. The dry density was varied between 1550 and 1950 kg/m^3. The compressive strength increases with the increase in the moulding water content in a linear fashion, irrespective of the MMC. There is a considerable increase in the strength for small changes in the dry density. For example, 10% increase in density from 1650 kg/m^3 results in about 100% increase in the compressive strength. The increase in the strength due to the increase in the density can be attributed to reduction in the porosity of the compacted specimen, resulting in better bonding due to cement hydration products. The void ratio and strength relationships for the stabilised CEB was discussed in Chap. 5.

Figure 12.8 shows strength and MMC relationships for the CSRE using three different cement contents, keeping the dry density at 1900 kg/m^3. The compressive strength increases with the increase in the moulding water content. For 5 and 8% cement, the strength increase is in the range of 20–30% as the moulding water content changes from 8 to 14.5%. Whereas for 12% cement, the strength increase is about 40%.

The dry soil–cement mixture contains cement and clay particles both having affinity for the water. When the water is added to the dry soil–cement mixture, the water will be shared by both the cement and the clay particles. When the samples are compacted using lower quantity of moulding water (say 8%), there could be insufficient supply of water for proper cement hydration. As the moulding water

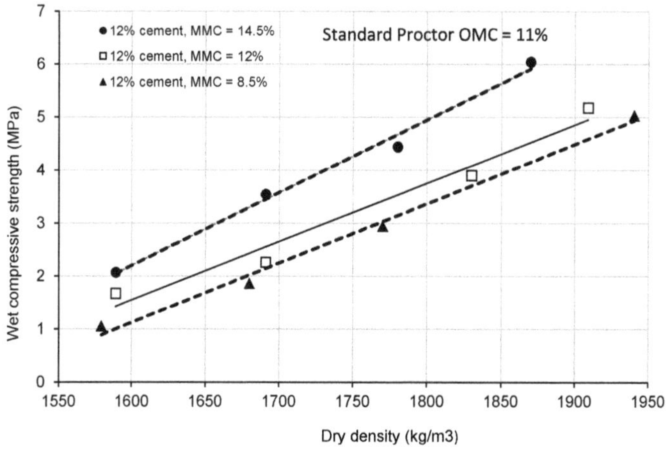

Fig. 12.7 Strength and density relationships for CSRE

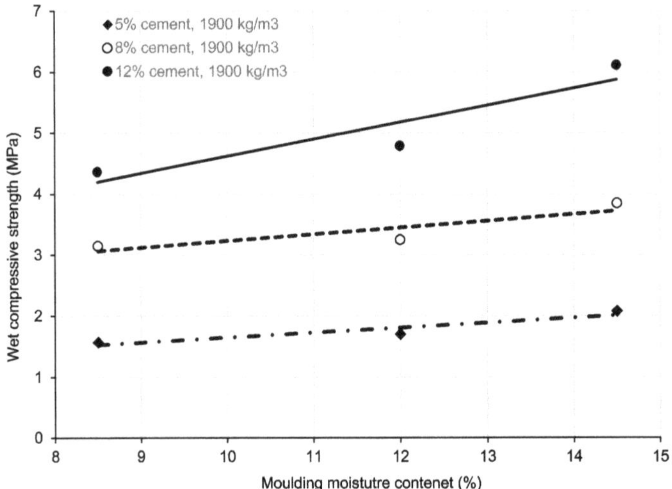

Fig. 12.8 Strength versus moulding moisture content for CSRE

content is increased (say 14.5%), more water is available for the cement hydration. This could be the reason for increased strength when higher percentage of moulding moisture was used.

The standard Proctor OMC is only an indication for choosing MMC for the CSRE construction. It is possible to achieve higher dry density for the CSRE than the MDD, using extra compaction energy either at lower or higher than the standard Proctor OMC. The strength and the density are linearly related. Higher density means higher strength. It is possible to achieve higher dry density for the CSRE by using higher MMC than indicated by the Proctor OMC.

12.5 Optimum Layer Thickness

In the construction of rammed earth elements, the processed soil is poured into the rigid formwork and then compacted using a rammer for achieving the desired layer thickness. Generally, the compacted layer thickness is in the range of 100–150 mm. Figure 12.9 shows typical compacted layers in a sample of the rammed earth wall. There is a visible variation in the density across the height of the compacted layer. Figure 12.10 shows typical variation in the dry density across the height of the compacted layer in a rammed earth cylindrical specimen (of size 150 mm diameter and 300 mm height), for the compacted layers of 150 and 100 mm thickness. The figure shows the variation in the density across the height of the compacted layer. The density is not uniform across the height of the compacted layer. The density is higher at the top than at the bottom of the layer. The difference in the densities is more pronounced for the thicker layer (150 mm) than the thinner layer (100 mm). The

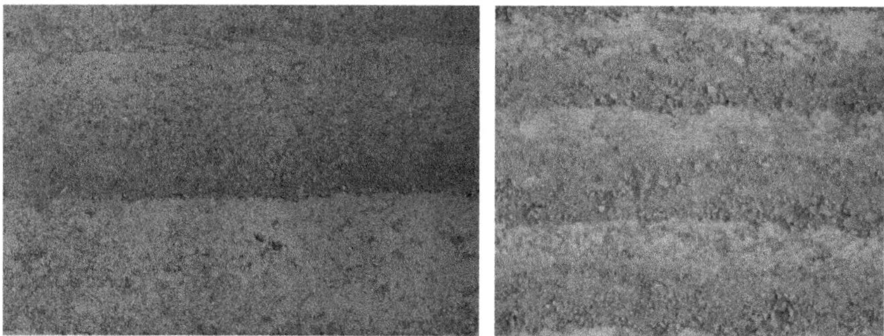

Fig. 12.9 Compacted layers in rammed earth

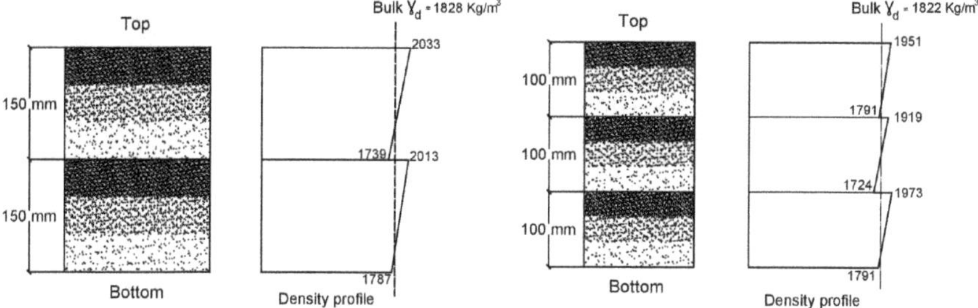

Fig. 12.10 Density variation across the height of compacted layers in CSRE

difference in density across the height of the 150 and 100 mm-thick compacted layer is about 15% and 10%, respectively. The bulk dry density of these CSRE cylindrical specimen is nearly constant (~1825 kg/m^3) as shown in Fig. 12.10. The difference in the density across the height of the compacted layer further reduces to about 6% as we go towards 70–80 mm layer thickness (Lepakshi and Reddy 2018). The difference in the density across the height of the compacted layer decreases with the decrease in the compacted layer thickness. Similar observations were made by Bui and Morel (2009) for unstabilised rammed earth specimens.

The influence of layer thickness on the strength, stiffness and other characteristics of rammed earth was examined by Lepakshi (2017), and Lepakshi and Reddy (2018). A soil with 72.6% sand, 11.6% silt and 15.8% clay size fractions and 7% cement (by mass) was used in casting CSRE cylindrical specimen (150 mm diameter and 300 mm height). Figure 12.11 shows a plot of compressive strength and the compacted layer thickness. Here, the compacted layer thickness has been varied between 75 and 300 mm. Beyond 100 mm layer thickness, the compressive strength reduces. The strength reduction is significant as the layer thickness goes beyond 150 mm. Between 300 and 100 mmlayer thickness, there is 140% difference in the compressive strength of the CSRE rammed earth cylinder. Similarly, there is 55% difference in the strength between 100 and 150 mmlayer thickness. The optimum layer thickness yielding

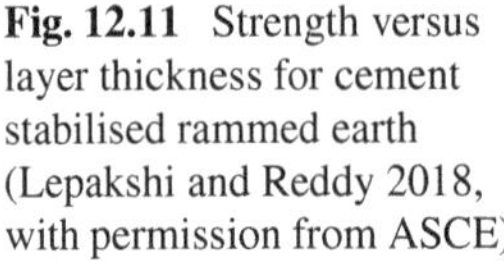

Fig. 12.11 Strength versus layer thickness for cement stabilised rammed earth (Lepakshi and Reddy 2018, with permission from ASCE)

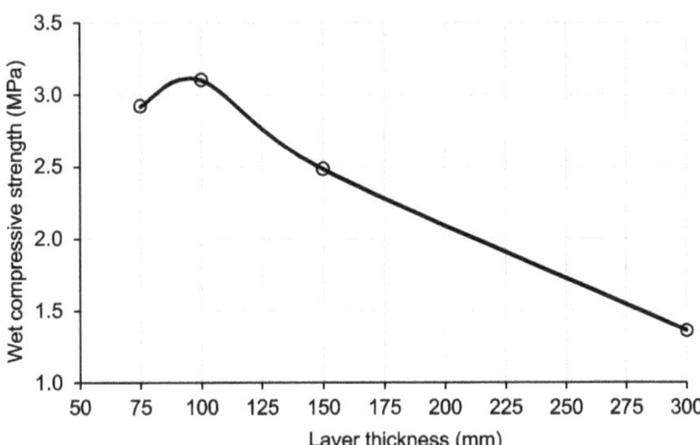

maximum strength is in the range of 90–100 mm. The strength reduction with the larger layer thickness is mainly attributed to the larger differences in the density across the height of the compacted layer. The results clearly show that the compacted layer thickness in the range of 75–100 mm is better for achieving maximum compressive strength for the CSRE.

12.6 Effect of Delayed Compaction on the Compaction Characteristics and the Strength of Cement Stabilised Rammed Earth

12.6.1 Compaction Characteristics

The dry cement–soil mix is doused with optimum quantity of water to prepare the partially saturated mix, used in the CSRE construction process. The elapsed time since the mixing of water and the starting of the compaction process is termed as "time lag". If the time lag is beyond the setting time of the cement, the resulting CSRE element can suffer from certain drawbacks. Kumar (2009) examined the influence of time lag on the strength, stiffness and compaction characteristics of CSRE. The salient conclusions of the Kumar's (2009) studies are as follows.

The standard Proctor OMC steadily increased with the increase in the time lag irrespective of the soil type and the cement content. The percentage increase in the OMC varied between 25 and 40% for the time lag between 0 and 10 h. There will be aggregation of cement-mixed soil particles as the elapsed time progresses, due to the hydration and the setting of the cement. Figure 12.12 illustrates the aggregated cement–soil particles with the time lag. The size of the aggregated lumps increased as the time lag increased. The energy supplied in the standard Proctor compaction test is fixed, and hence, some of this energy will be utilised to break the already established bonds due to the aggregation of the particles; the remaining energy may not be sufficient to properly compact the aggregated particles leading to creation of

Fig. 12.12 Aggregation of wetted cement–soil mixture particles with time lag

more porous structure. This could be the reason for higher OMC values with the increase in the time lag.

In contrast to the increase in OMC with the time lag, the standard Proctor MDD steadily decreases with the increase in the time lag. The decrease in the MDD was in the range from 10 to 16% for the time lag between 0 and 10 h. The decrease in the MDD could be attributed to increase in the porosity due to the aggregation of the cement mixed soil particles with the time lag, even before compaction. Graham West (1959) examined the influence of elapsed time on the density of medium clay stabilised with 10% cement. The study revealed ~15% reduction in the dry density for a time lag of 7 h.

The density of the compacted CSRE element reduces with the increase in the time lag. The density significantly influences the strength, and hence, the decrease in the density affects the strength and the absorption characteristics of the cement stabilised rammed earth. Hence, it is preferable to complete the compaction process of cement stabilised soil mix within 60 min (within the initial setting time of the cement) after mixing with the water.

12.6.2 Compressive Strength and Time Lag

The influence of the time lag on the compressive strength of cement stabilised rammed earth was examined by Kumar (2009) through the testing of rammed earth

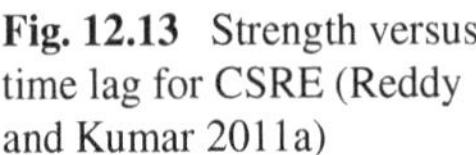

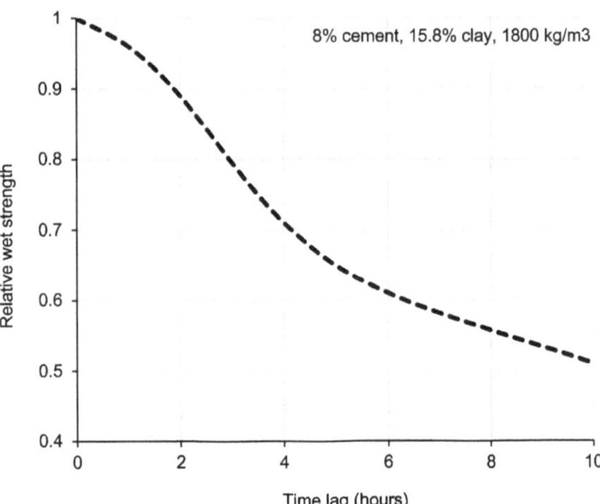

Fig. 12.13 Strength versus time lag for CSRE (Reddy and Kumar 2011a)

prisms (size 150 × 150 × 300 mm). The soil mix with 15.8% clay and 8% Portland cement was used for casting the prisms. The dry density of the prisms was controlled and maintained at 1800 kg/m^3, thus avoiding the interference of the density on the strength. Figure 12.13 shows a plot of strength versus time lag for the 28 day cured CSRE prisms.

For the time lag beyond 10 h, it becomes difficult to cast the prisms at the designated dry density of 1800 kg/m^3 due to the larger-sized aggregated hard cement–soil lumps. The time lag and strength relationship clearly shows that the compressive strength steadily decreases with the increase in the time lag. There is 50% strength reduction for a time lag of 10 h. This decrease in the strength can be attributed to the compaction of cement stabilised soil after the partially set cement, wherein already established cementitious bonds are broken during compaction. Graham West (1959) reports 37% reduction in the compressive strength for a time lag of 7 h, for a medium clay stabilised with 10% cement. These results indicate that the wetted cement–soil mixture should be rammed into the wall within 60 min (within initial setting time of cement) after mixing with the water.

12.7 Influence of Stabiliser Content on the Compressive Strength of Rammed Earth

The Portland cement and the lime are the commonly used inorganic stabilisers. The published work on the strength of lime stabilised rammed earth (Bui et al. 2014) is limited, whereas the strength of cement stabilised rammed earth has been pursued in greater detail (Lilley and Robinson 1995; Hall and Djerbib 2004; Walker et al. 2005; Jayasinghe and Kamaladasa 2007; Maniatidis and Walker 2008; Kumar 2009; Reddy and Kumar 2009, 2011a, b, c; Hall et al. 2012; Tripura and Singh 2014; Daniela and

Fig. 12.14 CSRE cylinder compression test set-up

Joshua 2012; Beckett and Ciancio 2014; Reddy et al. 2017; Lepakshi 2017; Lepakshi and Reddy 2018). The compressive strength of the CRSE can be determined either by using a cylinder/prism (height to width ratio 2) or using a wallette. Figure 12.14 shows a typical set-up for testing the CSRE cylinder under compression. Some of the typical failure patterns of the CSRE prism, the cylinder and the wallette are shown in Fig. 12.15. Since the rammed earth is monolithic, the rammed earth specimens show shear failure when tested under compression.

Kumar (2009), Lepakshi (2017) and Reddy et al. (2017) carried out systematic and controlled studies in understanding the strength and stabiliser content relationships. Some information from the studies of Kumar (2009) and Lepakshi (2017) was compiled and shown in Fig. 12.16. In both these studies, a soil with 15.8% clay (kaolinite) and ordinary Portland cement was used. The specimens (cylinder/prism) were of size 150 × 300 mm (aspect ratio 2) and cured for 28 days under the wet burlap. The wet strength was determined after soaking the cured and dried specimen in water for 48 h prior to the testing.

Figure 12.16 shows compressive strength and cement content relationships for three different cement contents and three dry densities as the cement content was varied between 5 and 12%. The strength increases with the increase in the cement content. The compressive strength and the cement content relationships are linear irrespective of the density and the moisture content during testing. The bi-linear or nonlinear relationships are also possible, especially for the range of cement contents stretching beyond 12%. It is difficult to arrive at a unique analytical equation for predicting the compressive strength of the CSRE with the cement content as a variable. Many factors (as highlighted in the earlier sections) other than the cement content control the compressive strength of the CSRE. Similar difficulties are faced

Fig. 12.15 Failure patterns for CSRE prism, cylinder and wallette specimen

in deriving an analytical expression for predicting the concrete compressive strength based on the cement content.

12.8 Moisture Absorption in Stabilised Rammed Earth

The stabilised rammed earth absorbs water due to two reasons (a) porosity and (b) residual clay content. The CSRE will have residual clay minerals in the matrix (Reddy and Latha 2014). The CSRE specimen absorbs water to fill the pores as well as to

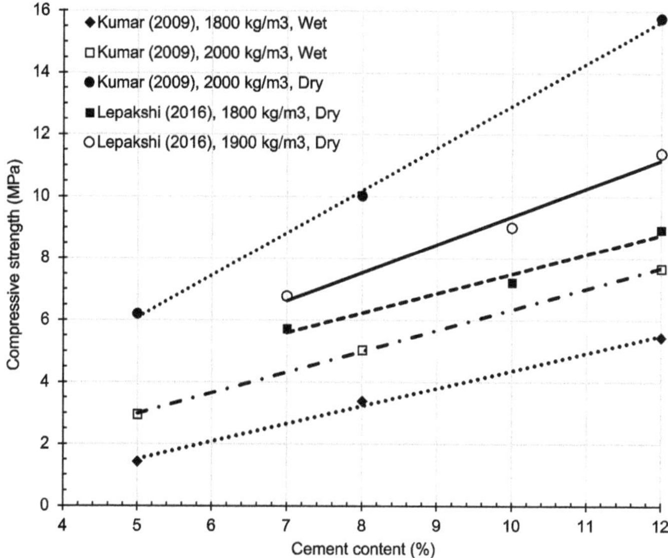

Fig. 12.16 Strength and cement content relationships

satiate the clay minerals affinity for the water. The quantity of water absorbed by the CSRE element depends upon the dry density (porosity) and the clay content of the mix used while producing the CSRE element. Figure 12.17 shows relationships between water absorption and the clay content of the mix used for the rammed earth.

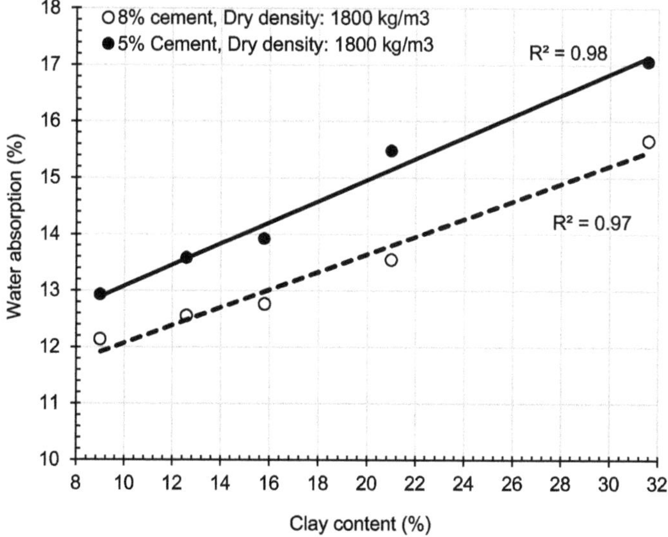

Fig. 12.17 Water absorption versus clay content of the mix for CSRE

Here, the dry density of rammed earth specimens has been maintained constant at 1800 kg/m^3. The water absorption means, the quantity of water absorbed by the rammed earth after 48 h of soaking in cold water at ambient temperature (20–30 °C). The specimen does not attain 100% saturation even after soaking for 72 h in cold water. The relationships in Fig. 12.17 show that the water absorption value increases with the increase in the clay fraction of the mix. The relationships are linear with good coefficient of determination (0.98). The optimum clay fraction for the CSRE will be in the range of 10–15%. At this clay content, the water absorption value will be in the range of 11–14%.

12.8.1 Influence of Density and Cement Content on Water Absorption

The water absorption (after 48 h of soaking in water) of cement stabilised rammed earth depends upon the density to a large extent. Figure 12.18 shows the relationship between the density and the water absorption for the CSRE for the two cases of 8 and 12% cement contents. The water absorption decreases as the dry density of the CSRE increases. About 35–40% reduction in water absorption can be noticed as the dry density increases from 1650 to 2000 kg/m^3. At 1650 kg/m^3 dry density, the void ratio will be about 0.60. The void ratio reduces to 0.34 as the dry density reaches 2000 kg/m^3. The CSRE is more porous at lower densities and hence can accommodate more water in the pores. At lower densities, the water absorption of the CSRE with 12% cement is marginally lower than that of the water absorbed by 8% cement CSRE. At higher dry density (>1800 kg/m^3), the CSRE with 12% cement absorbs about 10% less water when compared to the water absorption of the CSRE

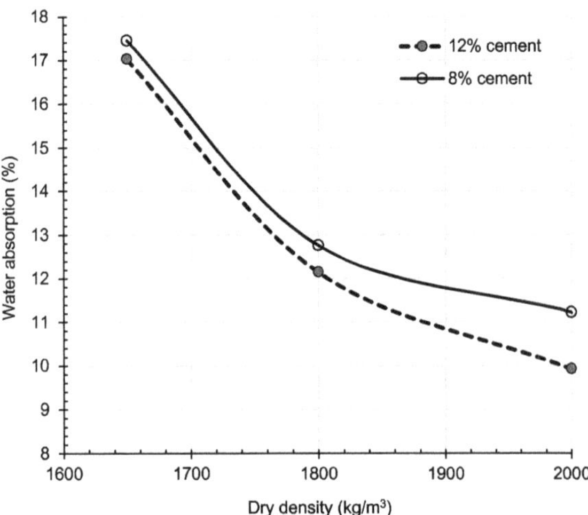

Fig. 12.18 Density versus water absorption for the CSRE

with 8% cement. It should be noted that the CSRE specimens do not get saturated even after 48 h of soaking in cold water at ambient room temperature. This can be mainly attributed to some entrapped air in the voids, which will not easily escape and make room for water to penetrate the entire void spaces in the dense CSRE.

12.9 Moisture Content and Strength of Stabilised Rammed Earth

The role of moulding moisture content on the compressive strength of the CSRE was discussed in the earlier sections. The moisture content of the CSRE at the time of the testing has significant influence on the compressive strength. The earth-based building materials exhibit different strength values in dry and wet conditions. This is mainly attributed to the presence of residual clay minerals in the stabilised compressed earth product. The moisture content and the strength relationships for the CSRE using soil with two different clay contents and cement contents (with a dry density of the CSRE at 1800 kg/m^3) are shown in Fig. 12.19. The compressive strength of the CSRE has been determined at the oven dry state, partially saturated state and nearly saturated state. There is a linear relationship between the strength and the moisture content of the CSRE at the time of testing, irrespective of the cement and the clay content of the mix.

The compressive strength of CSRE at the oven dry state and the nearly saturated state (48 h soaking in water) has been examined by Kumar (2009), Reddy and Kumar (2011b), Reddy et al. (2017) and Lepakshi (2017). The compressive strength has been determined by testing the cured and oven-dried (at 50 °C to constant weight) specimens. After oven drying and cooling to the rooming temperature at ambient conditions, the CSRE specimens will have a residual moisture of about 1.5%. While the specimens after oven drying soaked in water for 48 h prior to the testing will nearly saturate. The relationships shown in Fig. 12.20 are for these two extreme moisture states (called wet and dry states) during testing for the strength. The compressive

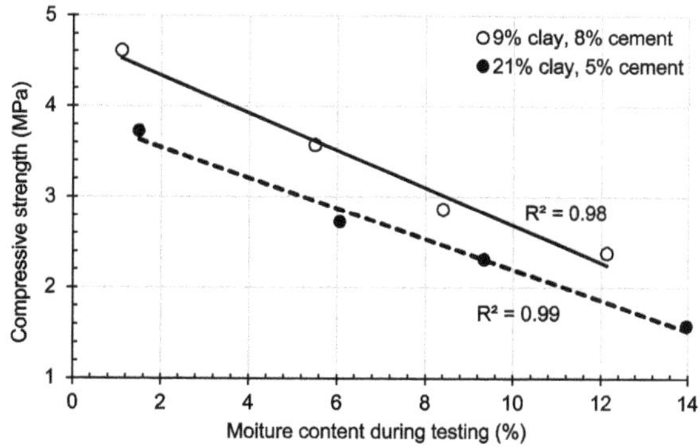

Fig. 12.19 Strength versus moisture content during testing

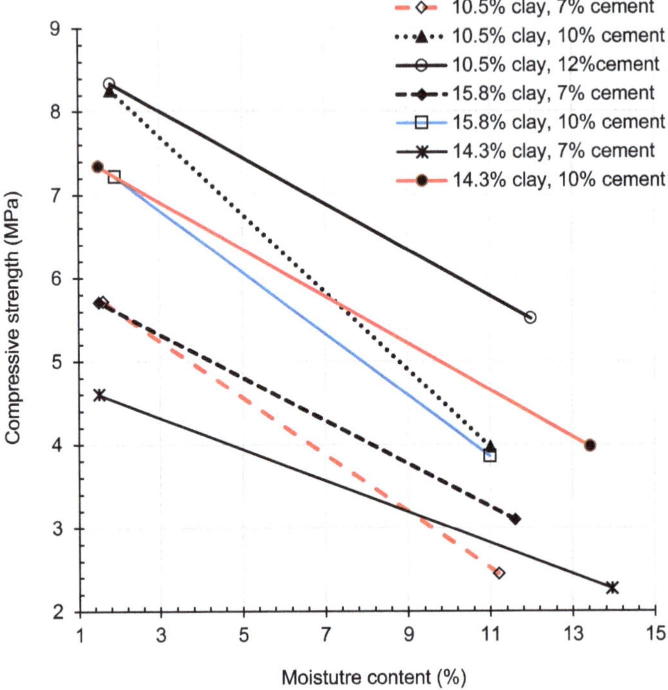

Fig. 12.20 Compressive strength and moisture content relationships

strength drastically reduces in the wet (nearly saturated) condition. The slopes of the linear lines vary for different clay, cement and density conditions. The wet-to-dry strength ratio for the CSRE will be in the range of 0.45–0.60, for 10–15% clay content, 7–12% cement and the dry density of about 1800 kg/m^3.

12.10 Comparison of Strengths of Rammed Earth and Masonry

The rammed earth is a monolithic material, whereas the masonry is a composite consisting of the masonry unit and the mortar joints. The strength and behaviour of masonry greatly depend upon the characteristics of the masonry unit and the mortar. It will be interesting to compare the compressive strength and the failure patterns of CSRE wallette with that of rammed earth brick masonry wallette. Kumar (2009), and Reddy and Kumar (2009) examined the behaviour of rammed earth and the rammed earth brick masonry through experimental studies. Brief procedure on the preparation and testing of the rammed earth wallette and the rammed earth brick masonry wallette is as follows.

CSRE wallette: The CSRE wallettes (size: 600 × 155 × 720) were cast using a soil with 15.8% clay and 8% cement, achieving a dry density of 1800 kg/m^3. After 28 days curing, the wallettes were dried in the air for four weeks and then tested for compressive strength.

CSRE brick masonry wallette: Rammed earth bricks of size 258 × 128 × 72 mm were prepared, using the same mix which was used for the CSRE wallette to achieve a dry density of 1800 kg/m^3 (Fig. 12.21). Single wythe rammed earth brick masonry wallettes (size: 625 × 128 × 700 mm) were prepared using cement–lime mortar (1:1:6 by volume) with a water–cement ratio of 1.75. The mortar bed joint thickness was 10 mm. The masonry wallettes were cured for 28 days and then dried in the air for four weeks before testing.

Table 12.2 gives the details of the sizes, moisture content during testing and the results of the strengths for the rammed earth wallette, the rammed earth brick, and the rammed earth brick masonry wallette. At a similar moisture content (6.5%) during testing, the rammed earth wallette shows one-third more strength than the rammed earth brick masonry wallette. The material composition and density used for the rammed earth and the rammed earth brick were similar. When the wallette specimen is under compression, the state of stresses developed in the jointed material like the masonry wallette is much different from the monolithic rammed earth. In the case of masonry, the brick will be under bi-axial tension and compression (Fig. 7.26), resulting in vertical splitting cracks at failure (Fig. 12.22). Monolithic material like rammed earth under compression ultimately fails in shear as shown in Fig. 12.22. The failure patterns of the monolithic rammed earth and the rammed earth brick masonry wallette are distinctly different (Table 12.3).

Fig. 12.21 Cement stabilised rammed earth bricks

Fig. 12.22 Failure patterns of rammed and rammed earth brick masonry wallettes

Table 12.3 Characteristics of rammed earth and masonry specimens (Kumar 2009)

Details	Rammed earth wallette	Rammed earth brick	Rammed earth brick masonry wallette
	Air dry	Air dry	Air dry
Size (mm)	600 × 155 × 720	258 × 128 × 72	625 × 128 × 700
Moisture content at the time of test (%)	6.61	6.60	6.43
Compressive strength (MPa)	5.47	9.19	4.05

References

Beckett C, Ciancio D (2014) Effect of compaction water content on the strength of cement-stabilised rammed earth materials. Can Geotech J 51(5):583–590

Bui QB, Morel JC (2009) Assessing the anisotropy of rammed earth. Constr Build Mater 23(9):3005–3011

Bui QB, Morel JC, Hans S, Walker P (2014) Effect of moisture content on the mechanical characteristics of rammed earth. Constr Build Mater 54:163–169

Clark D, Walker P (2003) The influence of soil properties on the behaviour of rammed earth. In: Proceeding 9th international conference on study and conservation of earthen architecture Terra 2003, Yazd, Iran, pp 93–101

Daniela C, Joshua G (2012) Experimental investigation on the compressive strength of cored and molded cement-stabilized rammed earth samples. Constr Build Mater 28(1):294–304

Easton D (1982) The rammed earth experience. Blue Mountain Press, Wilseyville, California, USA

Hall M, Djerbib Y (2004) Rammed earth sample production: context, recommendations and consistency. Constr Build Mater 18(4):281–286

Hall MR, Lindsay R, Krayenhoff M (2012) Modern earth buildings: materials, engineering, constructions and applications. Woodhead Publishing, Cambridge, UK

Houben H, Guillaud H (2003) Earth construction—a comprehensive guide. Intermediate Technology Publications, London, UK

IS 2110-1980 (reaffirmed 2002) Code of practice for in situ construction of walls in buildings with soil-cement. Bureau of Indian Standards, New Delhi, India

Jayasinghe C, Kamaladasa N (2007) Compressive strength characteristics of cement stabilised rammed earth walls. Constr Build Mater 21:1971–1976

Kumar PP (2009) Stabilised rammed earth for walls: materials, compressive strength and elastic properties. PhD thesis, Department of Civil Engineering, Indian Institute of Science, Bangalore, India

Lepakshi R, Reddy BVV (2018) Influence of layer thickness and plasticisers on the characteristics of cement stabilised rammed earth. J Mater Civ Eng 30(12):04018314, 1–10

Lepakshi R (2017) Studies on characteristics of cement stabilised rammed earth and flexural behaviour of plain and reinforced rammed earth. PhD thesis, Department of Civil Engineering, Indian Institute of Science, Bangalore, India

Lilley DM, Robinson J (1995) Ultimate strength of rammed earth walls with openings. Proc Inst Civ Eng, Struct Build 110(3):278–287

Maniatidis V, Walker P (2008) Structural capacity of rammed earth in compression. J Mater Civ Eng 20(3):230–238

Middleton GF (revised by Schneider LM) (1992) Earth-wall construction, Bulletin 5, 4th edn. Commonwealth Scientific and Industrial Research Organization, Sydney, Australia

Patty RL, Minium LW (1945) Rammed earth walls for farm buildings, Bulletin 277. Agricultural Engineering Department, Agricultural Experiment Station, South Dakota State College, Brookings, South Dakota, USA

Reddy BVV, Suresh V, Nanjunda Rao KS (2017) Characteristic compressive strength of cement stabilised rammed earth. J Mater Civ Eng 29(2):04016203-1–04016203-7

Reddy BVV, Kumar PP (2009) Compressive strength and elastic properties of stabilised rammed earth and masonry, Masonry International. J Int Mason Soc 22(2):39–46

Reddy BVV, Kumar PP (2011a) Cement stabilised rammed earth part A: compaction characteristics and physical properties of compacted cement stabilised soils. Mater Struct 44:681–693

Reddy BVV, Kumar PP (2011b) Cement stabilised rammed earth part B: compressive strength and stress–strain characteristics. Mater Struct 44:695–707

Reddy BVV, Kumar PP (2011c) Structural behaviour of story high cement stabilised rammed earth walls under compression. J Mater Civ Eng 23(3):240–247

Reddy BVV, Latha MS (2014) Influence of soil grading on the characteristics of cement stabilised soil compacts. Mater Struct 47(10):1633–1645

Tripura DD, Singh KD (2014) Characteristic properties of cement-stabilised rammed earth blocks. J Mater Civ Eng 27(7):04014214

Verma PL, Mehra SR (1950) Use of soil-cement in house construction in the Punjab. Indian Concr J 24(4):91–96

Walker P, Keable R, Martin J, Maniatidis V (2005) Rammed earth: design and construction guidelines. BRE Bookshop, Watford, UK

West G (1959) A laboratory investigation into the effect of elapsed time after mixing on the compaction and strength of soil-cement. J Geotech Eng (ASCE) 9(1):9–13

Chapter 13
Stress–Strain Characteristics of Cement Stabilised Rammed Earth

13.1 Introduction

The elastic properties and stress–strain characteristics of the materials such as cement stabilised rammed earth (CSRE) are essential in the context of analysis and design of rammed earth structures. The stress–strain characteristics include establishing the stress–strain relationships and determining the elastic properties such as modulus of elasticity, Poisson's ratio, strain at peak stress, ultimate/failure strain, etc. Some of the specific cases where stress–strain characteristics of rammed earth are useful can be listed as follows.

- Understanding the strength and stability of rammed earth structural elements such as walls, columns, etc.
- Predicting the buckling/failure loads of rammed earth structural elements
- Understanding the behaviour of rammed earth elements under flexure and shear
- The design of reinforced rammed earth structures, for establishing the stress blocks and stress block parameters
- Use of analytical and numerical tools for predicting the behaviour of rammed earth structures static and dynamic forces

The stress–strain characteristics of unstabilised rammed earth have been explored by Maniatidis and Walker (2008), Bui et al. (2009, 2014), Bui and Morel (2014), Lombillo et al. (2014) and Miccoli et al. (2014, 2015). These investigations reveal the initial tangent modulus and Poisson's ratio in the range of 100–1000 MPa and 0.22–0.37, respectively. The wide range for the modulus and Poisson's ratio for the unstabilised rammed earth can be attributed to the wide varieties of soils (having different characteristics) used in the investigations and the wide range for dry density (1850–2190 kg/m^3) for the rammed earth. Also, the tests have been performed on air dried and partially dried rammed earth samples, and hence, the moisture content of

B. V. V. Reddy, *Compressed Earth Block & Rammed Earth Structures*,
Springer Transactions in Civil and Environmental Engineering,
https://doi.org/10.1007/978-981-16-7877-6_13

the sample influences the stress–strain characteristics of unstabilised rammed earth. The stress–strain characteristics of cement stabilised rammed earth are discussed in the following sections.

13.2 Stress–Strain Characteristics of Cement Stabilised Rammed Earth

The stress–strain relationships for CSRE can be established by testing either the cylindrical specimens or the prisms. Generally, cylinder of size 150 mm diameter and 300 mm height is used. Also, smaller cylindrical specimen with height-to-width ratio of 2 can be used. A typical experimental set-up for the stress–strain measurements is illustrated in Fig. 12.14. The vertical or the longitudinal strains and the lateral strains can be measured using the extensometers or the electrical resistance strain gauges. The displacement-controlled servo-hydraulic machines are preferred to capture the post-peak response.

The stress–strain characteristics of the CSRE are influenced by (a) cement content, (b) soil composition/grading, (c) density and (d) strength. There are few investigations on the stress–strain characteristics of the cement stabilised rammed earth (Kumar 2009; Reddy and Kumar 2009, 2011; Lepakshi 2017). The data culled from the investigations of Kumar (2009) and Lepakshi (2017) was analysed and discussed in the following sections. Figure 13.1 shows typical stress–strain relationships for the CSRE in the wet and the dry state. The stress–strain relationships show initially a linear portion, followed by nonlinear portion till the peak stress. The post peak curves show drooping portion. The modulus and the strains at peak stress and at failure are higher in the dry state when compared with the values in the wet state.

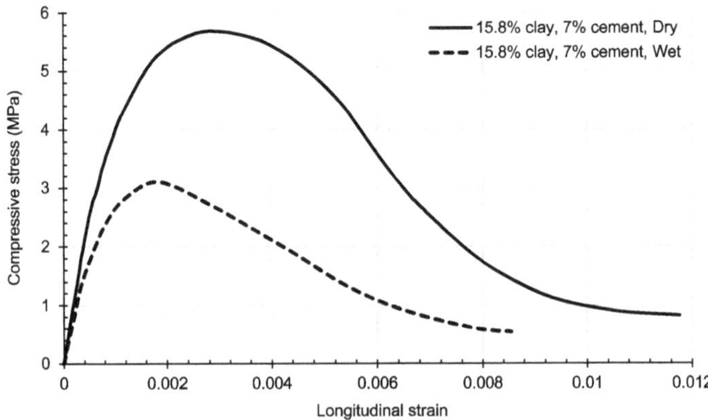

Fig. 13.1 Typical stress–strain relationships for CSRE

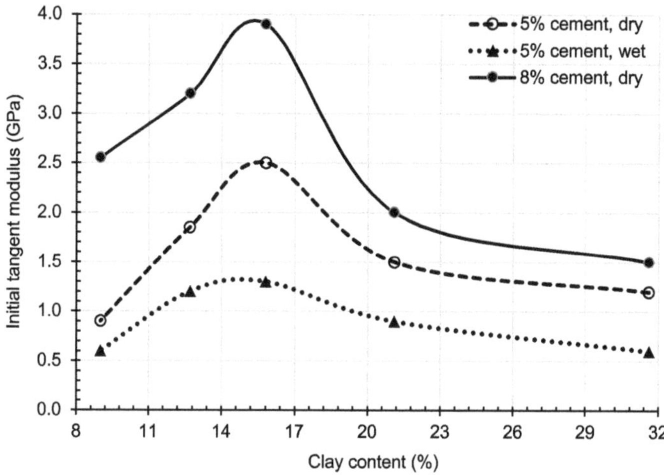

Fig. 13.2 Modulus versus clay content of CSRE

13.3 Influence of Soil Composition on the Stress–Strain Characteristics

The clay content and the type of clay mineral control the physical characteristics of a soil. The influence of the clay content on the strength of the CSRE was discussed in Chap. 12. The optimum clay content yielding maximum strength was in the range of 10–15%. Figure 13.2 shows variation in the initial tangent modulus with the clay content of the mix used for CSRE for the two different cement contents (5 and 8%). The modulus peaks at about 15% clay content. There is considerable difference in the modulus between the peak and the extreme values of the clay content. It should be noted here that the moulding moisture content (MMC) and the dry density have been maintained constant along the curves showing the relationships between the modulus and the clay content. The results are for CSRE using coarse-grained soil. The studies of Reddy and Latha (2014) showed optimum clay content for the cement stabilised compressed earth blocks in the range of 10–15% using both the coarse-grained and the fine-grained soils, attributing it to lower void ratio at the optimum clay content. Similar arguments can be put forward for the modulus and the clay content relationships in the case of CSRE also. The strength and the stiffness are directly related for in the materials such as the CSRE.

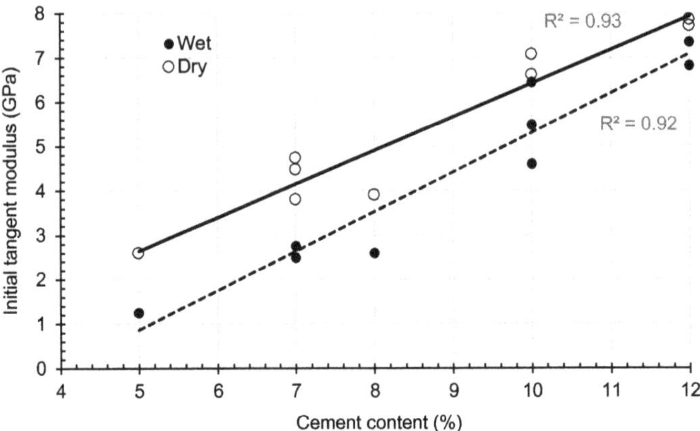

Fig. 13.3 Modulus versus cement content of CSRE (clay: 11–16%)

13.4 Cement Content and Modulus Relationships

Figure 13.3 shows plots for initial tangent modulus and cement content of CSRE. The results shown in these plots are for CSRE specimens using clay content in the range of 11–16% and a dry density of the specimen at 1800 kg/m^3. Thus, the interference of the density and the soil composition on the modulus of CSRE are eliminated. There is a linear relationship between the modulus and the cement content of the CSRE. The modulus increases with the increase in the cement content. The slope of the line representing the wet case is higher than the one for the dry case. The modulus increases by two and seven times as the cement content goes up from 5 to 12% for the dry and the wet cases, respectively. The difference in the modulus values of the CSRE in the wet and the dry state is maximum at lower cement contents (5%), and it narrows down at higher cement (12%). This can be attributed to the better stabilisation at higher cement contents. The modulus values of the CSRE are in the range of 1 to 7 GPa for the wet state as the cement content increases from 5 to 12%. For the dry condition, the modulus is in the range of 2.5–8.0 GPa. These discussions and conclusions can change for lower and higher dry densities of the CSRE when compared to the CSRE with the dry density of 1800 kg/m^3.

13.5 Modulus and Density Relationships

The density has significant influence on the characteristics of the CSRE. The density and strength relationships for CSRE were discussed in Chap. 12. The density and the modulus relationships are shown in Fig. 13.4 for the two cement contents in the wet and the dry state. The soil's clay content was maintained at 15.8% (optimum) for all the relationships shown in the figure. The density and the modulus are

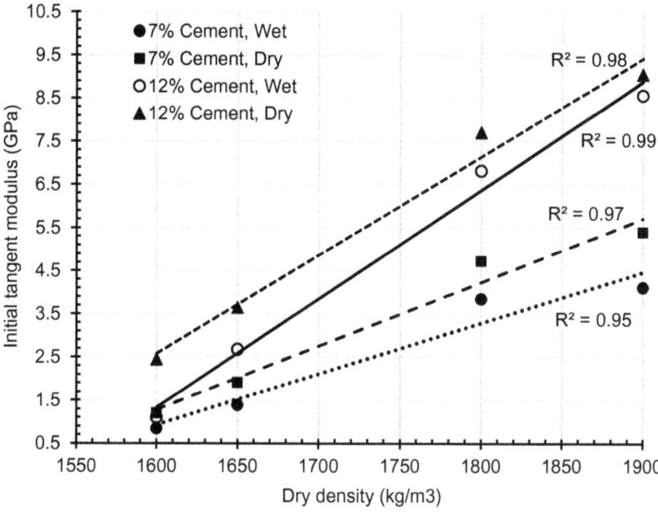

Fig. 13.4 Modulus variation with the density of CSRE (clay: 15.8%)

linearly related. The modulus of CSRE increases with the increase in the density. The modulus increases by over 100% for a 10% increase in the density from 1700 kg/m^3. The contribution of dry density in increasing the modulus of rammed earth can be attributed to the decrease in the porosity of the material with the increase in the density. For example, the void ratio of cylindrical specimens with 1650 kg/m^3 density was 0.546, whereas it was 0.342 for 1900 kg/m^3 specimens. The investigations of Reddy and Latha (2014) show direct correlation between the void ratio and the strength for the cement stabilised compressed earth specimens. As the void ratio decreases, there will be better contact between the soil grains and hence the cementitious products formed in the cement hydration process establish better bonding in the matrix.

13.6 Strength and Modulus Relationships

There are analytical expressions available to assess the modulus of elasticity based on the compressive strength for the materials such as concrete, bricks and masonry, etc. Considering the data available in the literature, an attempt was made to develop an analytical expression relating the modulus and the compressive strength of the CSRE. The modulus and the compressive strength relationships for the CSRE are shown in Fig. 13.5 for both the wet and the dry cases. The compressive strength of the CSRE is dictated by some major parameters such as clay content, cement content and density. The data derived for the relationships shown in Fig. 13.5 only considers the results for the optimum clay content in the range of 11–16%. Figure 13.5 shows linear relationships (passing through origin) between compressive strength and the modulus with coefficient of determination in the range of 0.88–0.93. These plots provide global understanding of the effects of various parameters such as the dry density, the clay content and the cement content on the strength and the modulus of the CSRE. The modulus increases with the increase in the compressive strength of the CSRE in both the dry and the wet states. The following relationships can be derived from the strength and the modulus plots shown in Fig. 13.5.

$$E_{csred} = 0.76 f_{csred} \quad \text{dry state} \tag{13.1}$$

$$E_{csrew} = 1.15 f_{csrew} \quad \text{wet state} \tag{13.2}$$

where E_{csred} = Initial Tangent Modulus of CSRE in GPa in dry state

E_{csrew} = Initial Tangent Modulus of CSRE in GPa in wet state

f_{csred} = CSRE cylinder dry compressive strength in MPa

f_{csrew} = CSRE cylinder wet compressive strength in MPa.

The slope of the linear relationship for the wet state has a higher value than for the one in the dry state. These relationships shall be useful for predicting the CSRE modulus using the soils having clay content in the range of 10–15%.

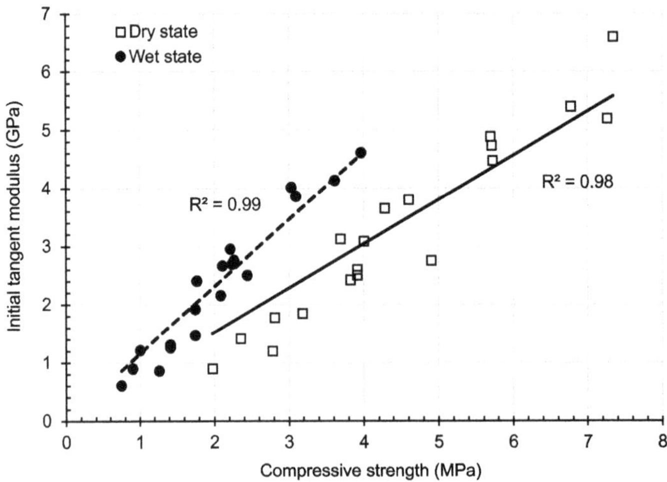

Fig. 13.5 Modulus and strength relationships for CSRE (clay: 11–16%)

13.7 Poisson's Ratio, and Strains at Peak Stress and at Failure

Fig. 13.6 shows the salient points on the stress–strain plot showing strains at the peak stress and at failure. Similar to the strength and the modulus, the strains are also controlled by several factors such as the clay content, the cement content, the strength, the moisture, the density, etc. The ranges for strains at the peak stress and at the failure are given in Table 13.1. The strain values given in the Table are for the mixes with clay content in the range of 11–16%, cement content varying between 7 and 12% and density in the range of 1650–1900 kg/m^3. The strains in the dry condition are much higher than the strains in the wet condition. The strain at failure can reach a value of about 1.5% in the dry condition. The CSRE in the dry condition exhibits larger straining capacity than in the wet condition. The Poisson's ratio for CSRE lies in the range of 0.10–0.16 (CEFIPRA 2016).

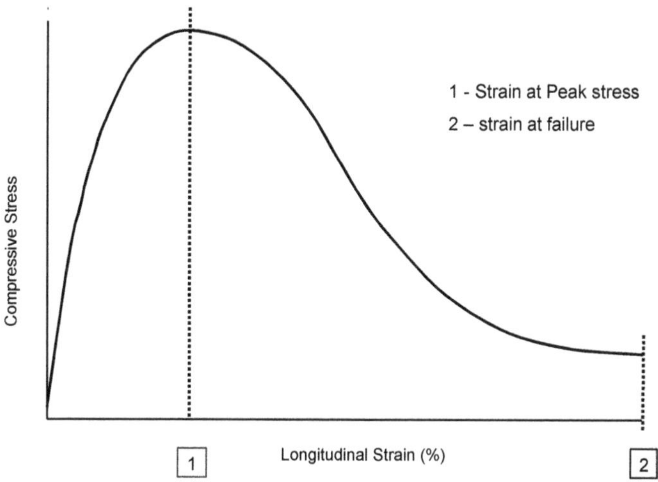

Fig. 13.6 Salient points on the stress–strain curve

Table 13.1 Strains at salient points for CSRE

Sl. No	Longitudinal strain	Strain at	
		peak stress	Failure
1	Strain in dry condition	0.002–0.006	0.004–0.015
2	Strain in wet condition	0.001–0.004	0.004–0.008

References

Bui QB, Morel JC (2014) First exploratory study on the ageing of rammed earth material. Materials 8(1):1–5

Bui QB, Morel JC, Hans S, Meunier N (2009) Compression behaviour of non-industrial materials in civil engineering by three scale experiments: the case of rammed earth. Mater Struct 42(8):1101–1106

Bui TT, Bui QB, Limam A, Maximilien S (2014) Failure of rammed earth walls: from observations to quantifications. Constr Build Mater 51:295–302

CEFIPRA (2016) Project report on "Developing design guidance for rammed earth construction". Indo-French collaborative project report, Indian Institute of Science Bangalore, India

Kumar PP (2009) Stabilised rammed earth for walls: materials, compressive strength and elastic properties. PhD thesis, Department of Civil Engineering, Indian Institute of Science, Bangalore, India

Lepakshi R (2017) Studies on characteristics of cement stabilised rammed earth and flexural behaviour of plain and reinforced rammed earth. PhD thesis, Department of Civil Engineering, Indian Institute of Science, Bangalore, India

Lombillo I, Villegas L, Fodde E, Thomas C (2014) In situ mechanical investigation of rammed earth: calibration of minor destructive testing. Constr Build Mater 51:451–460

Maniatidis V, Walker P (2008) Structural capacity of rammed earth in compression. J Mater Civ Eng 20(3):230–238

Miccoli L, Müller U, Fontana P (2014) Mechanical behaviour of earthen materials: a comparison between earth block masonry, rammed earth and cob. Constr Build Mater 61:327–339

Miccoli L, Oliveira DV, Silva RA, Müller U, Schueremans L (2015) Static behaviour of rammed earth: experimental testing and finite element modelling. Mater Struct 48(10):3443–3456

Reddy BVV, Kumar PP (2009) Compressive strength and elastic properties of stabilised rammed earth and masonry. Mason Int (J Int Mason Soc) 22(2):39–46, ISSN 0950-2289

Reddy BVV, Kumar PP (2011) Cement stabilised rammed earth. Part B: compressive strength and stress–strain characteristics. Mater Struct 44(3):695–707

Reddy BVV, Latha MS (2014) Influence of soil grading on the characteristics of cement stabilised soil compacts. Mater Struct 47(10):1633–1645

Chapter 14
Behaviour of Rammed Earth Under Tension and Shear

14.1 Introduction

The load bearing rammed earth walls in a building transmit loads from the overlying roof/floor system to the underlying support system (foundation). Hence, rammed earth walls are primarily subjected to in-plane compressive loading. However, in addition to the compressive loads, rammed earth walls experience lateral bending (out of plane) and in-plane shear loading during seismic events, wind loading, uneven settlement of foundation, load eccentricity, etc. Figure 14.1 illustrates the walls under out-of-plane bending and in-plane shear, in a typical building under lateral load. Out-of-plane bending can cause lateral bending of the wall with some restraints at the supports. In-plane shear loading is generally associated with vertical pre-compression in load bearing rammed earth wall buildings.

14.2 Flexure Strength of Rammed Earth

14.2.1 Behaviour of Unreinforced Rammed Earth Under Flexure

The rammed earth wall has a layered structure; its capacity to resist flexural tensile stresses varies across the two orthogonal directions of the wall. Figure 14.2 illustrates a rammed earth wall subjected to bending moments in the two orthogonal directions. The flexure strength of the CSRE in the two orthogonal directions can be determined experimentally. Either a beam or wallette specimen can be used. Figure 14.3 shows the schematic representation of the two rammed earth beam specimens for evaluating the flexure strength in the two orthogonal directions. Such beams can be subjected to flexure through four-point bending test as shown in Fig. 14.4. The flexure strength

B. V. V. Reddy, *Compressed Earth Block & Rammed Earth Structures*,
Springer Transactions in Civil and Environmental Engineering,
https://doi.org/10.1007/978-981-16-7877-6_14

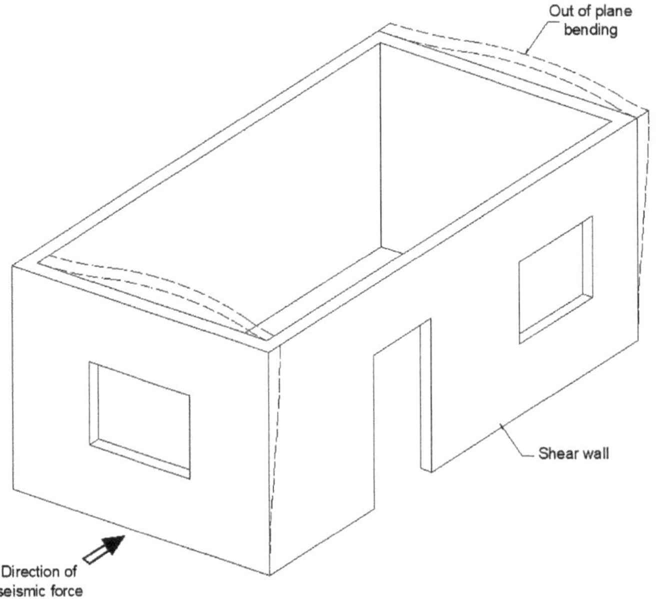

Fig. 14.1 Walls under out-of-plane bending and in-plane shear

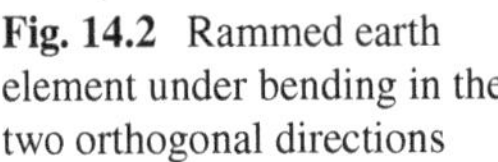
Fig. 14.2 Rammed earth element under bending in the two orthogonal directions

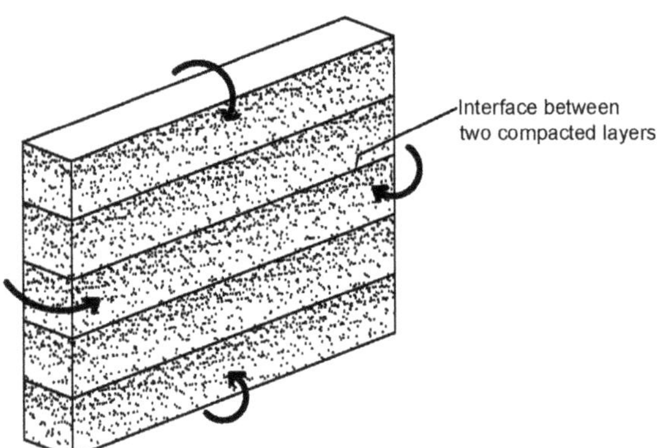

or modulus of rupture can also be evaluated using smaller rammed earth prisms like the ones used for assessing concrete flexure strength. The investigations of Walker (2002), Jayasinghe and Mallawaarchchi (2009), Anita (2009) and Lepakshi (2017) deal with certain aspects of CSRE under flexure. Similar to the compressive strength, the flexure strength of the CSRE is also sensitive to the soil composition, cement content, moisture content and the density.

Table 14.1 gives flexure strength values in the two orthogonal directions for the CSRE. Apart from the flexure strength values, the table gives information on clay content of the mix, the density, the cement content and the moisture content. The

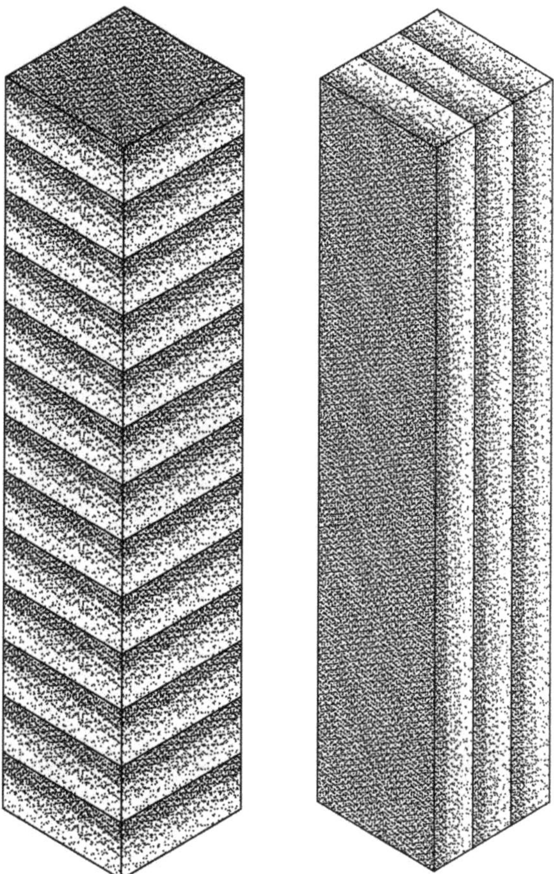

Fig. 14.3 Schematic representation of the CSRE beams with vertically and horizontally compacted layers

flexure strength when the tension developed is parallel to the compacted layers is more than double that of the flexure strength when the tension developed is perpendicular to the compacted layers. The CSRE wall can resist higher tensile stresses when the bending direction is parallel to the compacted layers. In reality, the CSRE walls can experience bending in both the orthogonal directions, depending upon the direction of loading. Figure 14.5 shows the failure patterns for the CSRE beams experiencing flexure in the two orthogonal directions. When the bending direction is perpendicular to the compacted layers, the flexure crack is at the interface of the two compacted layers. The interface is a weak link between the compacted layers and hence lower strength in this direction. The flexure strength of the CSRE in the dry condition will be about 1.0 MPa when the bending direction is parallel to the compacted layers and 0.5 MPa when the bending direction is perpendicular to the compacted layers. A similar behaviour can be noticed in masonry walls using different types of masonry units. The flexure strength of the masonry when the tension developed is parallel to the bed joints is 2 to 3 times higher than the flexure strength when the tension developed is perpendicular to the bed joints (Hendry 1998; Joshi 2015; Gaurav 2015).

Fig. 14.4 Out-of-plane loading of CSRE beams (tension parallel to compacted layers)

14.2.2 Behaviour of Reinforced Rammed Earth Under Flexure

The CSRE walls have low flexure strength especially when the tension developed due to the bending is in the direction perpendicular to the compacted layers as discussed in the previous section. The flexural strength of the CSRE walls can be improved by reinforcing with the steel bars. Plain as well as deformed bars can be used as reinforcement. The flexural behaviour of the reinforced CSRE greatly depends upon the bond strength between the steel bars and the CSRE matrix. Hence, understanding the behaviour of reinforced rammed earth necessitates to examine the bond strength between the steel bar and the CSRE matrix.

14.2.2.1 Bond Strength of Steel Reinforcement Bars in Cement Stabilised Rammed Earth

There are many investigations on the bond development and the bond strength between the steel reinforcement and the concrete (Yeih 1997; Abrishami and Mitchell 1996; Soroushian and Choi 1989; Cairns and Abdullah 1995; Mo and Chan 1996). Also, there are standard test methods to assess the bond strength of the reinforcement bars in the reinforced concrete (RILEM 1994; IS 2770–2002).

The pullout bar test is the method adopted for measuring the reinforcement bar bond strength in the concrete (Naaman and Najm 1991; Soroushian et al. 1991; Abrishami and Mitchell 1996; Yeih et al. 1997; Pecce et al. 2001; Achillides and

Table 14.1 Flexure strength of cement stabilised rammed earth

Rammed earth specimen		Clay (%)	Dry density (kg/m^3)	Cement (%)	Flexure strength (MPa) when tension developed is				Source
Type	size (mm)				Parallel to compacted layers	Moisture content (%)	Perpendicular to compacted layers	Moisture content (%)	
Prism	500 × 200 × 60	5.0	1975	6.0	1.42	Dry	–	–	Walker (2002)
Beam	900 × 600 × 150	Sandy soil	–	10.0	0.92	Air dry	0.46	Air dry	Jayasinghe and Mallawaarchchi (2009)
Prism	570 × 180 × 100	15.8	1800	8.0	1.17	3.8	0.52	3.64	Anita (2009)
Prism	700 × 300 × 250	15.8	1800	8.0	1.27	8.8	0.45	7.1	Anita (2009)
Beam	1300 × 250 × 250	15.8	1900	10.0	1.54	Oven dry	0.73	Oven dry	Lepakshi (2017)

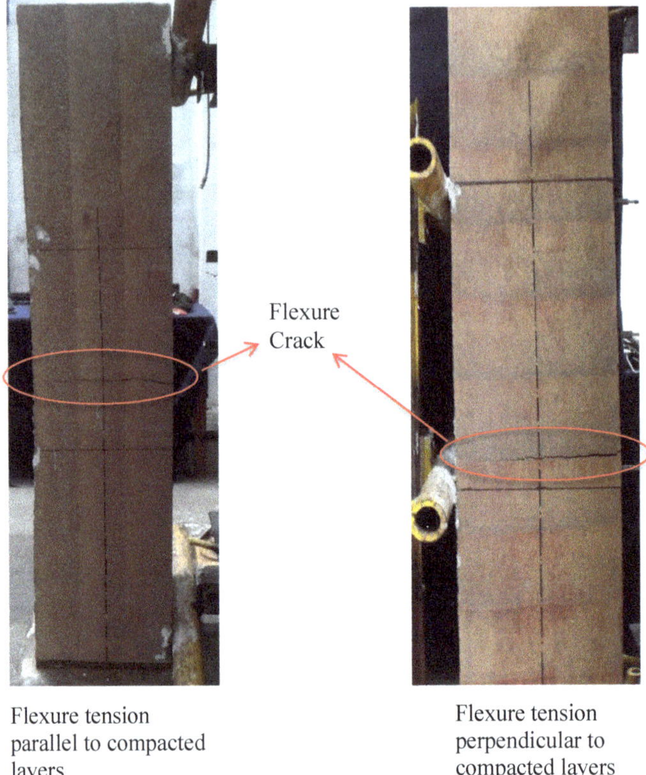

Fig. 14.5 Failure modes of CSRE beams under flexure

Pilakoutas 2004). There are no standard codes prescribing pullout tests for reinforced CSRE. The procedure outlined for the concrete pullout testing can be adopted to assess the pullout bond strength of the rebars in the CSRE. Here, the embedded reinforcement bars are pulled out using a test set-up such as the one shown in Fig. 14.6. A rammed earth cube specimen with embedded reinforcement bar is shown in Fig. 14.7. The test involves pulling out the embedded steel bar and measuring the pullout length/displacement. The limited investigations by Walker and Dobson (2001) and Walker et al. (2002) on the bond strength of the reinforcement bars in the CSRE matrix showed that there is a linear relationship between the bond strength and the compressive strength of the CSRE, and the deformed bars show much higher bond strength than the plain bars.

Several parameters influence the bond strength between the reinforcement bar and the CSRE matrix. Some of the prominent ones are the cement content, density, reinforcement bar type (plain or deformed) and the moisture content (during testing). Detailed investigations on the bond strength of reinforcement bar and the CSRE matrix were examined by Lepakshi (2017). Lepakshi (2017) considered the cement

Fig. 14.6 Experimental set-up for pullout testing

Fig. 14.7 A typical pullout rammed earth test specimen with reinforcement bar

content, density, ribbed steel bars (5 and 8 mm diameter) and the moisture content (during testing) as variables while assessing the bond strength. Tests have been performed in both the dry and the wet conditions. Figure 14.8 shows the bond strength variation with respect to the various parameters considered.

The bond strength increases as the density of the CSRE increases. This is mainly due to the better packing density and the lower porosity at higher densities. Also, the bond strength is sensitive to the cement content of the CSRE. There is a difference in the bond strength between 7 and 10% cement CSRE specimens. The bond strength is higher in the dry condition than in the wet condition. The wet to dry bond strength ratio is in the range of 0.65–0.90. At higher density, the ratio is larger. The difference in the bond strength in the wet and the dry states is mainly attributed to the presence of the residual clay minerals in the CSRE. Upon wetting, the clay minerals soften and loose strength. The differences in the bond strengths between the 5 and the 8 mm

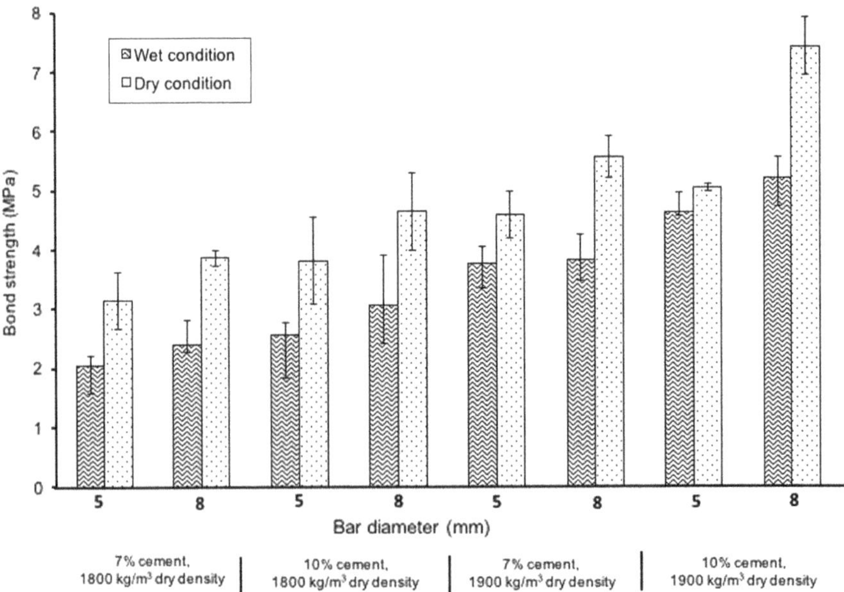

Fig. 14.8 Comparison of bond strength values with bar diameter, dry density and cement content (after Lepakshi 2017)

ribbed bars are mainly due to the differences in the rib volume per unit length of the bar.

The bond strength information will be useful in the design of the reinforced CSRE walls under flexure. Flexural behaviour of the reinforced CSRE is discussed in the subsequent sections. Considering the characteristic compressive strength of CSRE (5–15 MPa), the reinforced CSRE may not be able to accommodate more steel area for it to be under-reinforced. The bond strength for the deformed bars (5–8 mm) in CSRE is in the range of 3–6 MPa and 2–4 MPa for the dry and the wet states respectively when the cement content is about 7% (by mass). These numbers change respectively to 4–7 MPa and 2.5–5 MPa for the dry and the wet states using 10% cement. The bond strength increases as the density of the CSRE increases. Thus, the bond strength of the deformed reinforcement bars in the CSRE matrix has a wide range (2–7 MPa) dictated by the cement content, density and the moisture content. The bond strength of the reinforcement bars in the concrete depends upon the strength of the concrete, type of bar, fresh concrete properties, etc. The allowable bond stress for plain bars in concrete is in the range of 1.2–1.9 MPa (IS 456 - 2000). The code suggests increasing the allowable bond stress by 60% for the deformed bars.

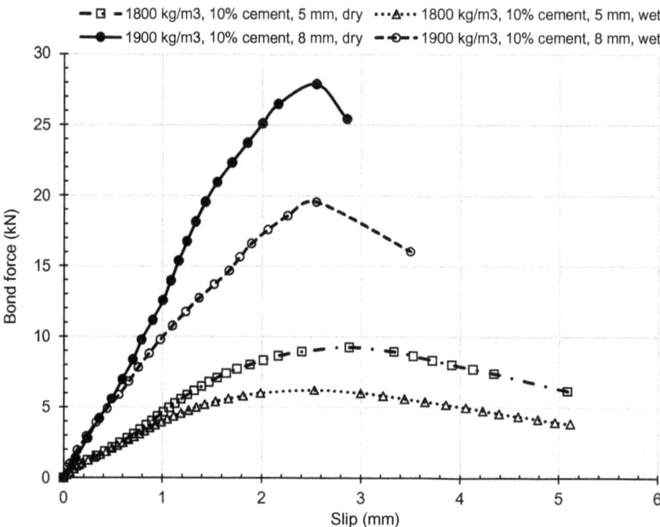

Fig. 14.9 Bond force–slip relationships for reinforced CSRE

14.2.2.2 Bond Force–Slip Relationships for CSRE

The typical bond force and the end slip relationships are shown in Fig. 14.9 for the 5 and the 8 mm deformed bars and for the two different densities using 10% cement in CSRE. These relationships show that the bond force increases with the slippage until the ultimate load is reached and followed by a drop in the bond force with further slippage. For some cases, the end slip can be as high as 5 mm. The slip values for higher density specimen are lower when compared with the slip values of the lower density specimens.

14.2.2.3 Modes of Failure in Pullout Bond Tests

The rammed earth pullout test specimens either fail due to the pullout of the reinforcement bars (Fig. 14.10) or the lateral splitting of the rammed earth specimens (Fig. 14.11). The specimens embedded with plain reinforcement bars fail due to the pulling out of the steel bar. The failure due to the tensile yielding of the reinforcement bars may not occur in CSRE. The bond developed between the plain reinforcement bars and the surrounding matrix is attributed to the adhesion and the friction between the steel bar and the CSRE matrix. The bond strength is dictated by the shear bond strength of the surrounding rammed earth and the surface properties of the steel bar. The pullout failure occurs when the shear strength of the bond between the reinforcement bar and the CSRE was exceeded.

Fig. 14.10 Typical pullout failures for plain and deformed reinforcement bar

Fig. 14.11 Typical splitting failure of deformed reinforcement bar in a pullout test

The failure due to lateral splitting of the specimen is commonly observed in the reinforced concrete specimens embedded with deformed rebar (Cairns and Abdullah 1995). In the case of concrete, the specimens are generally cast with helical reinforcement, which provides confinement, and hence preventing the splitting failure of the concrete and initiating shear pullout failure mode. However, in the case of the CSRE, it is not possible to prepare the test specimens with helical reinforcement. The helical reinforcement hinders the compaction process of the rammed earth specimen.

14.3 Shear Strength of Rammed Earth

The rammed earth is a monolithic material with visible stratified compacted layers. Since the rammed earth is monolithic, it ultimately fails in shear mode even under concentric compressive loads (Walker et al. 2005; Jayasinghe 2007; Jayasinghe and Kamaladasa 2007; Bui et al. 2007; Reddy and Kumar 2009, 2011). Figure 12.15 (Chap. 12) shows typical shear failures of CSRE specimens (cylinder, wallette and wall) when subjected to compression. In rammed earth shear slip can occur due to a shearing action along the interface of rammed earth layers. Also, the diagonal tension failure occurs due to poor shear strength of rammed earth material.

The behaviour of rammed earth under shear needs to be understood with reference to (a) establishing the shear strength parameters and the failure envelopes and (b) global behaviour under raking or in-plane loads causing shear failure of the rammed earth structural elements. For generating the shear strength parameters and the Mohr–Coulomb failure envelopes tri-axial shear test is needed. The shear strength of the rammed earth parallel to the compacted layers can be determined through the triplet shear tests and the raking in-plane shear tests on the wallettes. The diagonal shear test is another technique used for assessing the shear strength indirectly. The ASTM E519-15 code gives diagonal tension (shear) test procedure for the masonry. This procedure can be adopted to assess the diagonal shear strength of the rammed earth. Figures 14.12, 14.13, 14.14 and 14.15 show the schematic representation of the test set-ups for tri-axial shear, triplet shear, diagonal tension shear and raking shear tests, respectively. The investigations of Jaquin et al. (2006), Rao et al. (2011), Cheah et al. (2012), Lepakshi (2017), Arslan et. al. (2017), Pavan et al. (2020a, 2020b) shed some light on the shear strength of the cement stabilised rammed earth.

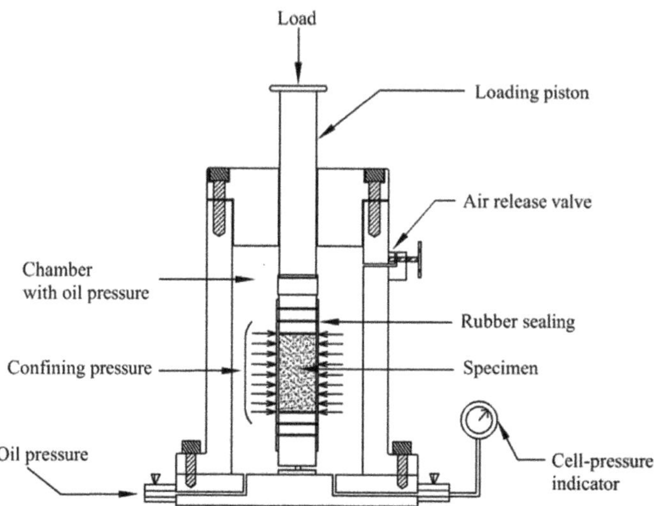

Fig. 14.12 Triaxial shear test set-up

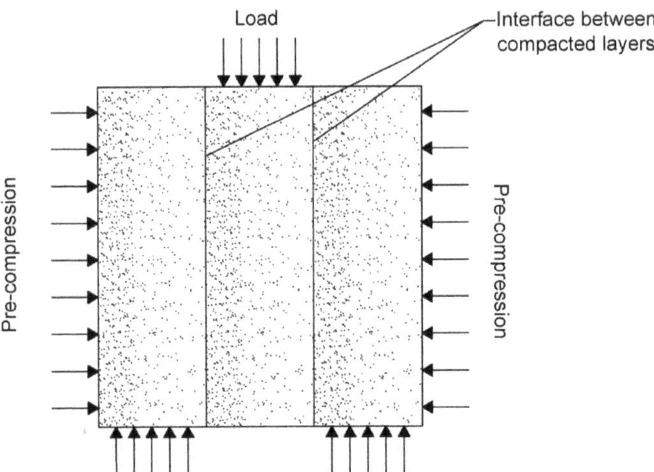

Fig. 14.13 Triplet shear test with pre-compression

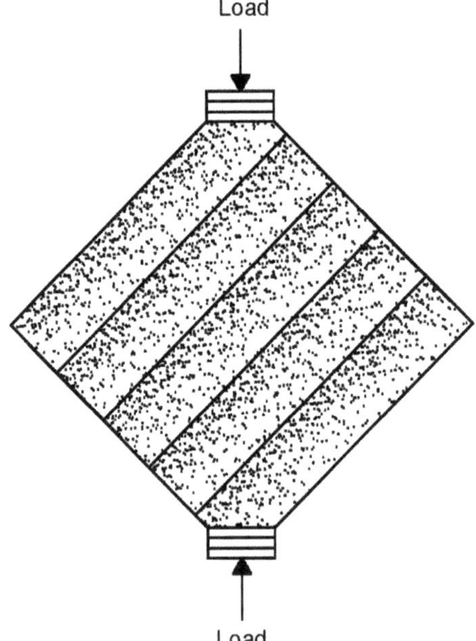

Fig. 14.14 Diagonal tension (shear) test

Fig. 14.15 Raking shear test

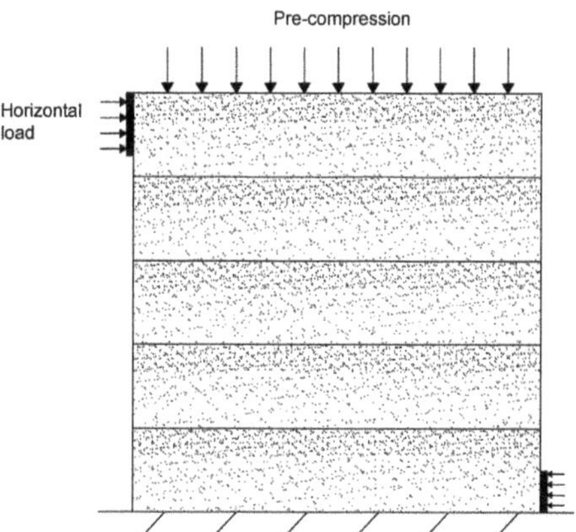

14.3.1 *Shear Strength Parameters and Mohr–Coulomb Failure Envelopes*

The basic shear strength parameters include the cohesion (c) and the angle of internal friction (Φ). The shear strength as well as the shear strength parameters of the stabilised rammed earth depend upon the following parameters.

(a) Soil composition and grading
(b) Stabiliser content
(c) Density of the rammed earth
(d) Moisture content.

Lepakshi (2017) conducted detailed investigations for assessing the shear strength parameters of the cement stabilised rammed earth. Details of the tri-axial shear test procedures and the results are discussed in the following sections.

14.3.2 *Tri-axial Shear Test*

The tri-axial shear test set-up is shown in Fig. 14.12. The cylindrical rammed earth specimen sealed in a rubber membrane (Fig. 14.16) is subjected to the vertical compression and the confined hydrostatic pressure in the tri-axial test cell. For a constant cell pressure, the axial load is gradually increased until failure. The state of stress on the rammed earth specimen is schematically represented in Fig. 14.17 (σ_1: Applied vertical pressure and σ_3: Confining cell pressure). The cylindrical specimen to be used in a tri-axial test depends upon the tri-axial test set-up and cell capacity to apply the confining pressure. The conventional tri-axial cells used in the geotechnical laboratory for assessing the shear strength parameters of soils will have limitations on the maximum cell pressure that can be applied on the rammed earth specimen. The tri-axial cells used for testing the rock specimens will have a metal cell, where

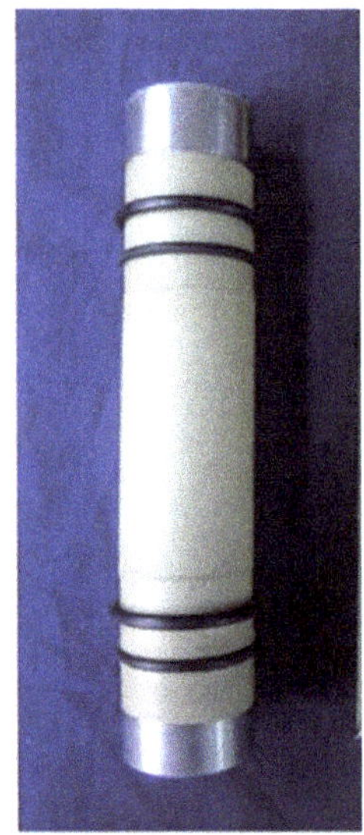

Fig. 14.16 Rammed earth cylindrical specimen and the sealed specimen in rubber membrane

high confining pressures (0.5–1.0 MPa) can be achieved. Such cells are suitable for testing the cement stabilised rammed earth specimen. The size of the cylindrical specimen depends upon the tri-axial cell capacity and the size. The stabilised rammed earth cylindrical specimens of size 35–40 mm diameter and 70–80 mm height can be used. The CSRE cylinder preparation involves compacting the partially saturated (at optimum moisture content) mix in three layers to achieve the specified density.

14.3.3 Shear Strength Parameters and Failure Envelopes for CSRE

The two inherent properties of the soil are its cohesion and the angle of internal friction. The shear strength of the soil or soil compact such as rammed earth can be determined from these two parameters. The shear strength is a function of the cohesion, the angle of internal friction and the normal stress at a given point. The relationship among these parameters and the shear strength is as follows:

$$\tau = c + \sigma \tan \phi \tag{14.1}$$

where τ = shear strength, c = cohesion, σ = normal stress, φ = angle of internal friction.

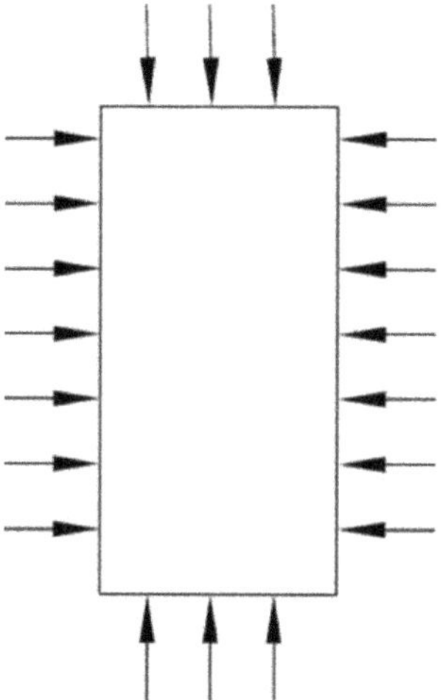

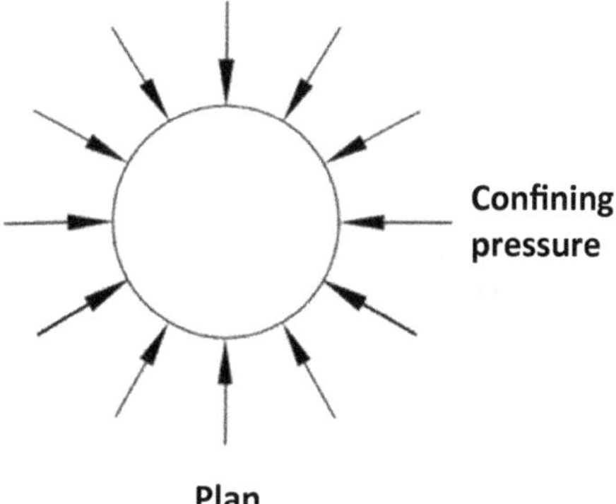

Fig. 14.17 State of stress on the rammed earth specimen under tri-axial testing

The normal stress–shear stress relationships defined here can be graphically represented using the Mohr circle as shown in Fig. 14.18. The Mohr–Coulomb failure criteria in terms of failure stresses is as follows:

$$\sigma_1 = \sigma_3 \tan^2\left(45 + \frac{\phi}{2}\right) + 2c \tan\left(45 + \frac{\phi}{2}\right) \tag{14.2}$$

The Mohr circle provides a graphical representation of the normal stress and the shear stress on any plane in a two-dimensional biaxial state of stress. The principal stresses σ_3 (lateral) and σ_1 (vertical) are plotted on the abscissae that makes the diameter of the Mohr circle.

Considering a soil with clay content of 15.8% and a dry density of 1800 kg/m^3, the shear strength parameters and the failure envelopes generated by varying cement

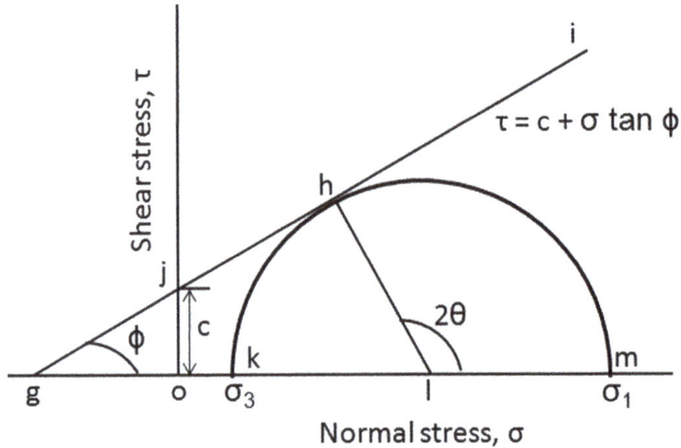

Fig. 14.18 Mohr circle with failure envelope

content (4, 7, 10 and 15%, by mass) are discussed here. The CSRE cylindrical specimens of size: 38 mm diameter and 76 mm height, compacted in three equal layers at standard Proctor OMC (12.36%) were used in generating these experiments. The 28-day moist cured and oven dried at 60 °C to constant weight were tested in tri-axial cell under the dry and the wet conditions (48 h soaking in water). Figure 14.19 shows the mould used in casting three-layered CSRE cylindrical specimen. The typical failure patterns of the CSRE cylindrical specimen in the tri-axial test are shown in Fig. 14.20. The Mohr circles drawn for the tests on four confining pressures

Fig. 14.19 Cylindrical mould used for casting CSRE cylinders of 76 mm height and 38 mm diameter

Fig. 14.20 Typical shear failure of CSRE cylinders after tri-axial test

and the failure envelopes for some cases of CSRE samples are shown in Figs. 14.21 and 14.22. Figure 14.23 shows the typical Mohr–Coulomb failure envelopes for the CSRE. Based on the Mohr–Coulomb failure envelopes, "c", "ϕ", and the shear strength can be estimated.

The confining pressure or the normal stress influences the shear strength of the CSRE. For the two confining pressures of 0 and 0.5 MPa, the relationship between the shear strength and the cement content are displayed in Fig. 14.24. The relationship between the shear strength and the cement content is bi-linear. The slope of the line reduces after 10% cement content. Under zero confining pressure or the normal stress, the shear strength of the CSRE is in the range of 0.22–1.22 MPa and 0.59–2.19 MPa for the wet and the dry states, respectively, as the cement content varies between 4 and 15%. Cheah et al. (2012) report a cohesion value of 0.72 MPa for CSRE using soil having 13% clay and 7.7% cement, based on the tests on cylinders

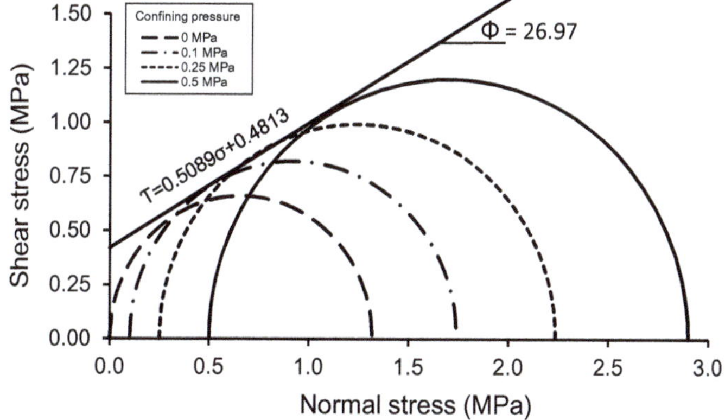

Fig. 14.21 Mohr–Coulomb failure envelope for CSRE (4% cement) in dry state (after Lepakshi 2017)

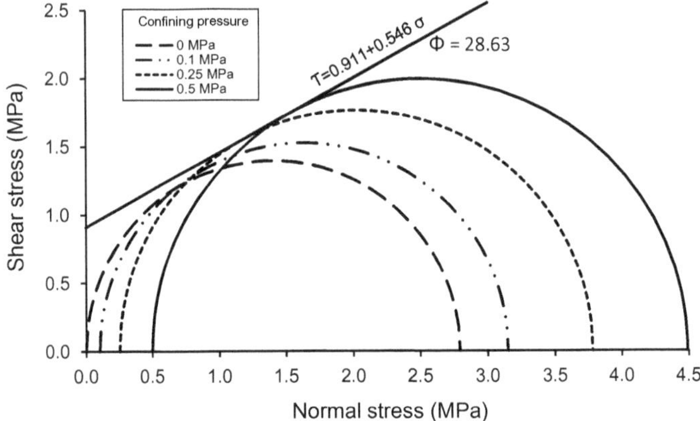

Fig. 14.22 Mohr–Coulomb failure envelope for CSRE (15% cement) in wet state (after Lepakshi 2017)

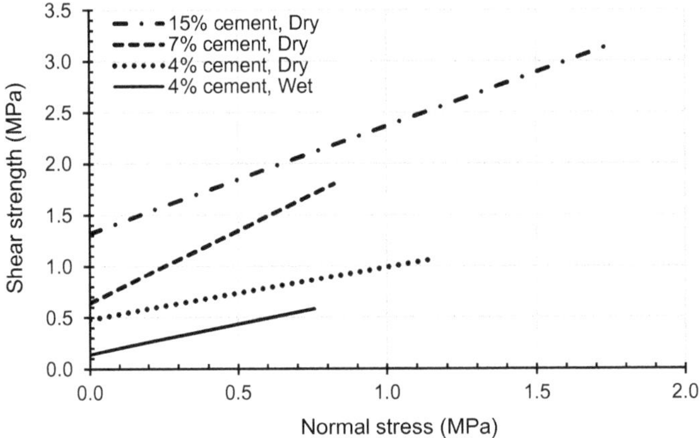

Fig. 14.23 Mohr–Coulomb failure envelopes for cement stabilised rammed earth (after Lepakshi 2017)

(100 mm diameter and 200 mm height) cured for 28 days at 20 °C temperature and 65% relative humidity.

There is a linear relationship between the cohesion and the cement content as displayed in Fig. 14.25 for both the dry and the wet states. The cohesion increases as the cement content increases. The increase is about 1.75 and 5.5 times for the dry and the wet cases, respectively, as the cement content increases from 4 to 15%. Similarly, the value of the angle of internal friction for the CSRE ranges between 26.9°–54.7° and 28.6°–34.3° for the dry and the wet states, respectively. Cheah et al. (2012) report 48° angle of internal friction, for the CSRE using 13% clay and 7.7% cement, based on the tests on cylinders.

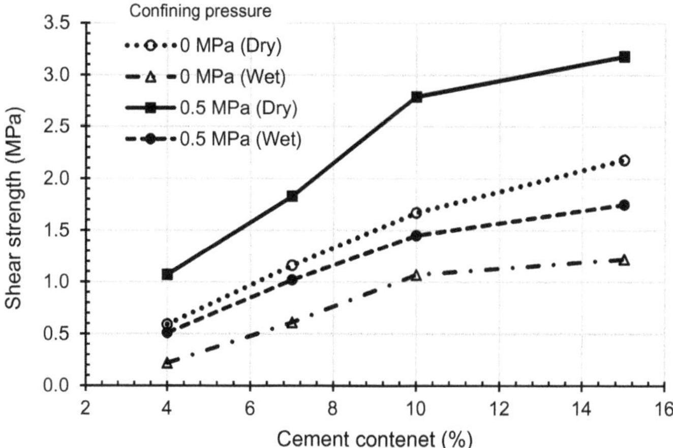

Fig. 14.24 Influence of cement content on the shear strength of CSRE (after Lepakshi 2017)

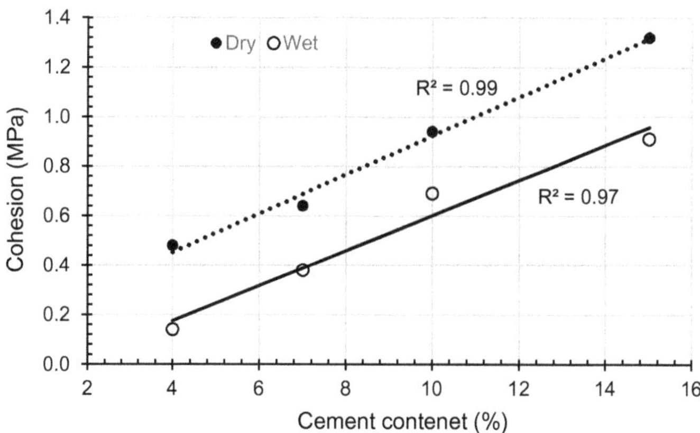

Fig. 14.25 Influence of cement content on the cohesion of CSRE

14.3.4 Triplet Shear Strength

The shear strength parameters determined using triplet shear test are given in Table 14.2, based on the investigations of Cheah et al. (2012) and Pavan et al. (2020a). Using 7–10% cement and soil with 13–15% clay (optimum clay), the results show that the cohesion is in the range 0.3–0.8 MPa. The cohesion value reduces by half in the wet condition. The angle of internal friction varies in between 26 and 45°. Figure 14.26 shows a triplet shear test setup. The test setup ensures that shearing takes place at the interface of the compacted layers.

The CSRE triplets can be subjected to normal stress in the form of pre-compression, while assessing the shear strength of the interface between the

Table 14.2 Shear strength of CSRE from triplet shear tests

Sl. no	Triplet size (mm)	Clay (%)	Dry density (kg/m^3)	Cement (%)	Cohesion (MPa)	Angle of internal friction (degrees)	Moisture content (%)	Source
1	100 × 200 × 200	13.0	2100	7.7	0.33	45	3–4	Cheah et al. (2012)
2	230 × 230 × 75	15.0	1850	10.0	0.78	26	4	Pavan et al. (2020a)
3	230 × 230 × 75	15.0	1850	10.0	0.39	39	12	Pavan et al. (2020a)

Fig. 14.26 Triplet shear test set-up

compacted layers as shown in Fig. 14.13. The investigations of Pavan et al. (2020b) show linear relationships for the normal stress and the shear stress. Figure 14.27 shows typical failure at the interfaces of a CSRE triplet.

14.3.5 Diagonal Tension (Shear) Strength

The diagonal tension test on a square rammed earth panel can be used for assessing the shear strength of the rammed earth indirectly following the test procedure given for the masonry in ASTM E519-15 code. Figure 14.28 shows the diagonal tension test setup, where the displacements and the related strains along both the diagonals can be monitored. These strains can be used to determine shear strains. When the vertical load is applied onto the steel loading shoes along the diagonal, causing compression along the loaded diagonal, and tension along the other horizontal diagonal. A state of pure shear is created in the central region of the diagonal panel.

Fig. 14.27 Failure along the interfaces of triplet specimen

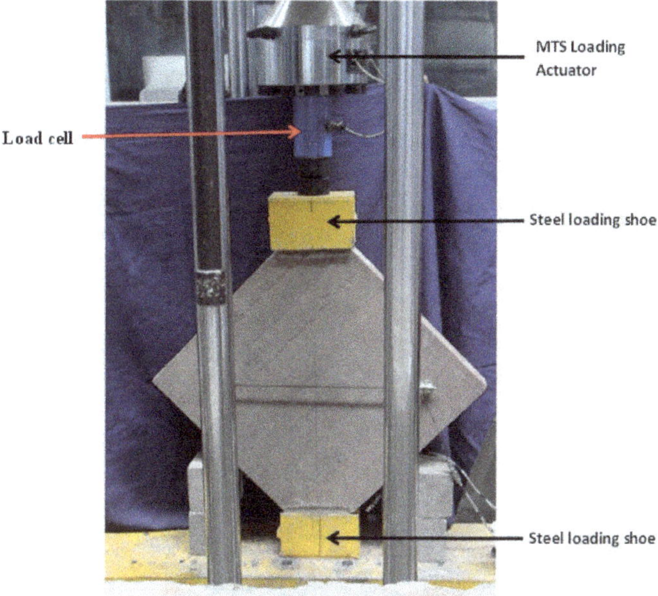

Fig. 14.28 Diagonal tension (shear) test set-up

The shear stress (τ), shear strain (γ) and shear modulus (G) of the rammed earth panels can be calculated as per ASTM standard (ASTM-E519) procedure, as follows.

$$\tau = 0.707\,(P \div A) \tag{14.3}$$

$$\gamma = \varepsilon_h + \varepsilon_v \quad (14.4)$$

$$G = (\tau \div \gamma) \quad (14.5)$$

where

A = [(h + w) t] ÷ 2; h, w, t are height, width and thickness of the rammed earth panel, respectively,

P = Applied load along the vertical diagonal,

ε_h = strain along the horizontal diagonal,

ε_v = strain along the vertical diagonal.

Figure 14.29 shows the typical failure pattern of the CSRE diagonal test panels. The diagonal panels failed due to the development of splitting vertical cracks across the rammed earth layers along the loaded diagonal. The shear slip or sliding shear mode of failure is absent in the diagonal tests. Such failure modes are mainly attributed to higher interfacial shear strength of CSRE than the material shear strength.

The investigations of Pavan et al. (2020b) showed a diagonal shear strength for 10% cement CSRE panels as (1850 kg/m^3 dry density) 1.24 and 0.75 MPa for the dry and wet cases, respectively. The corresponding shear strains at the peak stress were 0.00061 and 0.00043 for the dry and the wet cases, respectively. The secant shear modulus at 50% of peak shear stress was found to be 3700 and 2700 MPa for the dry and the wet cases respectively. The specimen moisture content at the time of the testing, in these investigations was 3 and 12% for the dry and the wet cases, respectively.

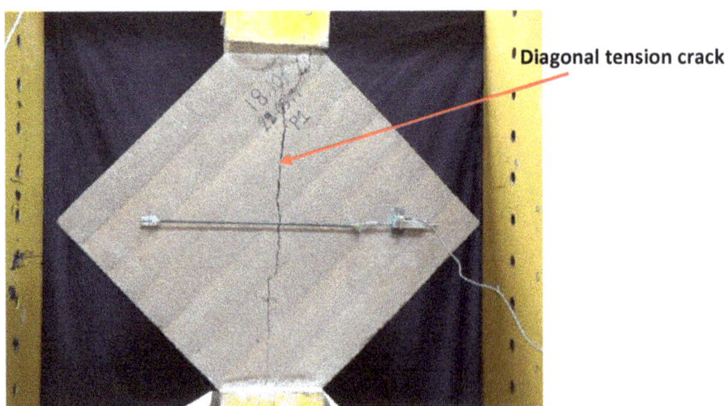

Fig. 14.29 Failure along the loaded diagonal of the diagonal tension test on CSRE panel

References

Abrishami HH, Mitchell D (1996) Analysis of bond stress distributions in pullout specimens. J Struct Eng 122(3):255–261

Achillides Z, Pilakoutas K (2004) Bond behaviour of fiber reinforced polymer bars under direct pullout conditions. J Compos Constr 8(2):173–181

Anita M (2009) Static and dynamic behaviour of cement stabilised rammed earth panels and building models. MSc (Engg) thesis, Department of Civil Engineering, Indian institute of science, Bangalore, India

Arslan ME, Emiro Glu M, Yalama A (2017) Structural behavior of rammed earth walls under lateral cyclic loading: a comparative experimental study. Constr Build Mater 133:433–442

ASTM-E519 (2015) Standard test method for diagonal tension (Shear) in masonry assemblages. ASTM International

Bui QB, Hans S, Morel JC (2007) The compressive strength and pseudo elastic modulus of rammed earth. In: Proceedings of the international symposium on earthen structures, Interline Publishers, Bangalore, India, pp 217–223

Cairns J, Abdullah R (1995) An evaluation of bond pullout tests and their relevance to structural performance. Struct Eng 73(11)

Cheah JSJ, Walker P, Heath A, Morgan TKKB (2012) Evaluating shear test methods for stabilised rammed earth. Proc Inst Civ Eng Constr Mater 165 (6):325–334

Gourav K (2015) Studies on flexural behaviour of Fly-ash lime gypsum brick masonry. PhD thesis, Department of Civil Engineering, Indian Institute of Science, Bangalore, India

Hendry AW (1998) Structural masonry. Macmillan press, London, UK

IS 2770 (Part 1) (2007) Method of testing bond in reinforced concrete—(Part 1): pullout test. Bureau of Indian Standard, New Delhi, India

IS 456 (2000) Plain and reinforced concrete code of practice. Bureau of Indian Standard, New Delhi, India

Jaquin PA, Augarde CE, Gerrard CM (2006) Analysis of historic rammed earth construction. In: Proceedings of the 5th international conference on structural analysis of historical constructions, New Delhi, India, pp 1091–1098

Jayasinghe C (2007) Shrinkage characteristics of cement stabilised rammed earth. In: Proceeding of international symposium on earthen structures, Bangalore, India, pp 212–216. ISBN 81-7296-051-4

Jayasinghe C, Kamaladasa N (2007) Compressive strength characteristics of cement stabilised rammed earth walls. Constr Build Mater 21(11):1971–1976

Jayasinghe C, Mallawaarachchi RS (2009) Flexural strength of compressed stabilised earth masonry materials. Mater Des 30(9):3859–3868

Joshi AA (2015) Static and dynamic behaviour of reinforced masonry: Experimental and analytical investigations. PhD thesis, Department of Civil Engineering, Indian Institute of Science, Bangalore, India

Lepakshi R (2017) Studies on characteristics of cement stabilised rammed earth and flexural behaviour of plain and reinforced rammed earth. PhD thesis, Department of Civil Engineering, Indian Institute of Science, Bangalore, India

Mo YL, Chan J (1996) Bond and slip of plain rebars in concrete. J Mater Civ Eng 8(4):208–211

Naaman AE, Najm H (1991) Bond-slip mechanisms of steel fibres in concrete. Mater J 88(2):135–145

Pavan GS, Ullas SN, Nanjunda Rao KS (2020a) Interfacial behavior of cement stabilized rammed earth: experimental and numerical study. Const Build Mater 257:119327

Pavan GS, Ullas SN, Nanjunda Rao KS (2020b) Shear behavior of cement stabilized rammed earth assemblages. J Build Eng 27:100966

Pecce M, Manfredi G, Realfonzo R, Cosenza E (2001) Experimental and analytical evaluation of bond properties of GFRP bars. J Mater Civ Eng 13(4):282–290

Rao KSN, Venkatarama Reddy BV, Gade MR (2011) Strength and elastic properties of cement stabilized rammed earth. In: National conference on recent developments in civil engineering, VIET, Mysore, Karnataka

Reddy BVV, Kumar PP (2009) Compressive strength and elastic properties of stabilised rammed earth and masonry. Mason Int 22(2):39

Reddy BVV, Kumar PP (2011) Structural behaviour of story-high cement-stabilised rammed-earth walls under compression. J Mater Civ Eng 23(3):240–247

RILEM (1994) Technical recommendations for the testing and the use of construction materials. In: International union of testing and research laboratories for materials and structures, E and FN Spon, London

Soroushian P, Choi K (1989) Local Bond of deformed bars with different diameters in confined concrete. ACI Struct J 86(2):217–222

Soroushian P, Choi KB, Park GH, Aslani F (1991) Bond of deformed bars to concrete: effects of confinement and strength of concrete. Mater J 88(3):227–232

Walker P (2002) The Australian earth building handbook, HB 195–2002. Standards Australia, Sydney, Australia

Walker P, Ayala R, Dobson S (2002) Reinforced composite rammed earth in flexure. In: Proceedings 3rd international conference on non-conventional materials & technologies, 12–13 March, Hanoi, Vietnam, pp 439–451

Walker PJ, Dobson S (2001) Pullout tests on deformed and plain rebars in cement-stabilised rammed earth. J Mater Civil Eng 13(4):291–297

Walker P, Keable R, Martin J, Maniatidis V (2005) Rammed earth: design and construction guidelines. BRE Bookshop, Watford, UK

Yeih W, Huang R, Chang JJ, Yang CC (1997) A pullout test for determining interface properties between rebar and concrete. Adv Cem Based Mater 5(2):57–66

Chapter 15
Structural Design of Rammed Earth Walls

15.1 Introduction

The structural design and behaviour of a rammed earth wall needs understanding on the resistance to external loads causing compression, shear and tension. The design includes assessing the strength or resistance of rammed earth wall under compression, bending and in-plane shear, and also the stability due to slenderness and lateral/eccentric loads. The structural design of the rammed earth wall subjected to gravity loads needs information on the following.

(a) Characteristic compressive strength
(b) Slenderness of the wall element
(c) Effect of load eccentricity and slenderness on the compressive strength
(d) Stress reduction factors
(e) Construction joints and shrinkage behaviour

The following sections delve on these aspects apart from some design exercises.

15.2 Characteristic Compressive Strength of Rammed Earth

The procedure for assessing the compressive strength of the conventional materials such as concrete, bricks, masonry, etc. can be found in the standard codes of practices. For example, the characteristic compressive strength of the brick masonry can be determined using compressive strength of either a prism or the wallette (ASTM C1314 2007; BS EN 1052-1 1999). The characteristic compressive strength of the masonry can be determined using correction factors (available in the codes on masonry) based on the slenderness of the specimen. The standard test procedure for determining the characteristic compressive strength of the stabilised rammed earth

B. V. V. Reddy, *Compressed Earth Block & Rammed Earth Structures*,
Springer Transactions in Civil and Environmental Engineering,
https://doi.org/10.1007/978-981-16-7877-6_15

is still emerging. The specimen size and its slenderness or the aspect ratio can affect the compressive strength of the rammed earth elements. Unlike in other materials, the characteristic compressive strength to be used in the design of the rammed earth walls and other structural elements needs to be understood.

In the case of the cement stabilised rammed earth (CSRE), apart from the soil composition, the cement content, density and moisture content of the CSRE specimen influence the compressive strength. Cube (150 mm), prism (150 × 150 × 300 mm) and the cylindrical specimens (150 mm diameter × 300 mm height) have been used to determine the compressive strength of cement stabilized rammed earth (King 1996; Walker et al. 2005; Tibbets 2001; Walker and Standards Australia 2002; Walker 2002; Hall and Djerbib 2004; Maniatidis and Walker 2008; Reddy and Kumar 2011a, b; Daniela and Joshua 2012; Windstorm and Schmidt 2013). Middleton (1987) recommended a cylinder of size 150 mm diameter and 110 mm height, and gave corrections factors to assess the characteristic compressive strength. Generally, the cylinder or the prism of height-to-width ratio of 2 is used.

The wallette specimens (height-to-width ratio of 4–5) were also used to determine the compressive strength of the CSRE (Reddy and Kumar 2011a, b; Jayasinghe and Kamaladasa 2007). Based on the compressive strength of the wallette, cube, prism or cylinder, the characteristic compressive strength of the rammed earth is sometimes assessed considering the correction and partial safety factors from the masonry codes.

Reddy et. al. (2017) carried out a comprehensive study on the specimen aspect ratio on arriving at the characteristic compressive strength of the cement stabilised rammed earth. The compressive strength of the CSRE wallettes was determined using specimens cast with a soil having optimum clay (14.3%). The CSRE wallettes having different aspect ratio (2.5–6.0), shown in Fig. 15.1. The wallettes compressive strength results were compared with the cylinder strength (aspect ratio 2) and recommendations were made for the assessing the characteristic compressive strength of the CSRE. Figure 15.2 shows compressive strength versus aspect ratio relationships for the CSRE. The relationships show that there is hardly any variation (1–1.5%) in the compressive strength of the specimens as the aspect ratio was increased from 2 to 6.

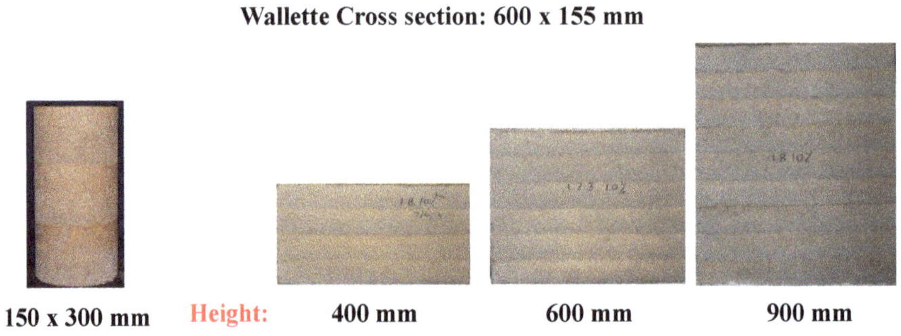

Fig. 15.1 CSRE cylinder and wallette specimens of different aspect ratio

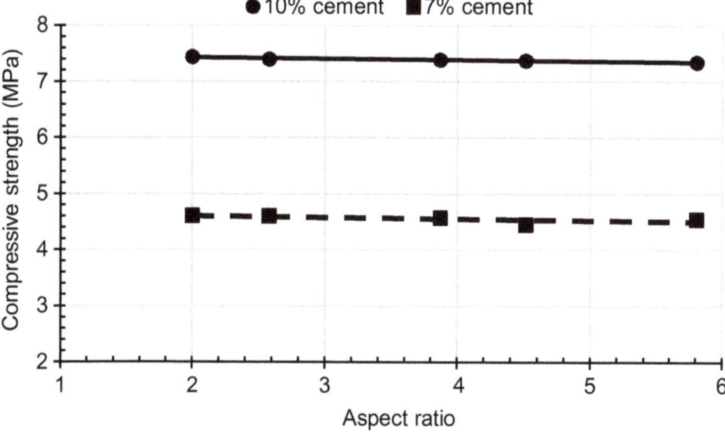

Fig. 15.2 Strength and aspect ratio relationships for CSRE (dry density 1800 kg/m^3)

The masonry codes (ASTM 1314 2007, NBC 2016) give strength-correction factors based on the aspect ratio of the masonry prisms as shown in Fig. 15.3. The plots shown in Fig. 15.3, corresponding to NBC (2016) code is for the blockwork, whereas the plot corresponding to the ASTM 1314 (2007) code is for both the brickwork and the blockwork. Figure also shows a plot of the strength-correction factor and the aspect ratio for the CSRE wallettes.

Considering the strength-correction factor for the masonry prisms as 1.0 for the aspect ratio of 2, the strength-correction factor for the masonry prism aspect ratio of 5 is in the range of 1.22–1.37. For the CSRE, the strength-correction factor for the aspect ratio of 5 is 1.018. This clearly demonstrates that the strength-correction

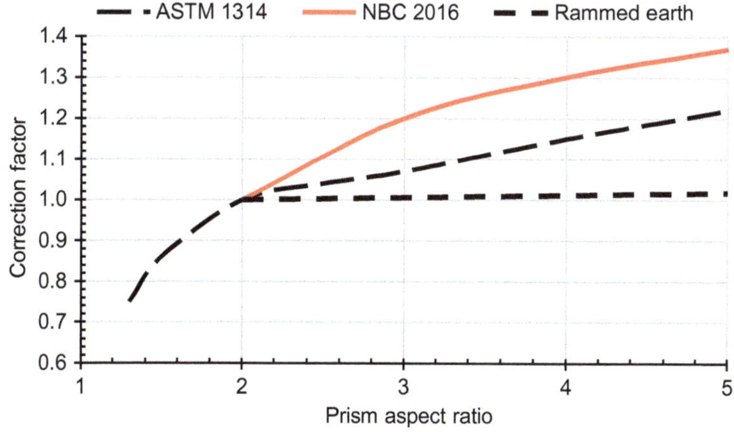

Fig. 15.3 Correction factor–aspect ratio relationships for rammed earth and masonry

factors specified in standard codes of practices for the masonry are not suitable for the monolithic CSRE wallette elements.

There is negligible reduction in the CSRE wallette strength as the aspect ratio is varied between 2 and 6. This analysis suggests that the compressive strength of the CSRE cylinder or the prism specimen with the aspect ratio of 2 can be used for assessing the characteristic compressive strength of rammed earth without any correction factor. The CSRE cylinder or prism should be at least 300 mm height with a minimum of three compacted layers preferably of 100 mm thickness. The density of each compacted layer should be controlled to achieve uniform density for the specimen. The wet condition (48 h soaking in water) is the critical condition for the structural design. Hence, the CSRE cylinders shall be tested in the wet condition to ascertain the mean wet compressive strength and the corresponding characteristic compressive strength for the CSRE. Assuming that the compressive strengths of the CSRE cylinders are normally distributed, the characteristic compressive strength of CSRE can be computed as follows.

$$f_{\text{kcsre}} = f_{\text{mcsre}} - 1.64\sigma_{\text{s}} \tag{15.1}$$

where
f_{kcsre} = Characteristic compressive strength of CSRE
f_{mcsre} = Mean compressive strength of the CSRE cylinder
σ_{s} = Standard deviation.

15.3 Slenderness and Strength

The slenderness ratio or the aspect ratio (height to thickness ratio) in the range of 2–6 does not make much difference to the compressive strength of the CSRE under concentric compression. The slenderness ratio of CSRE walls beyond 6 affects the compressive strength of the wall (Walker et al. 2005, Kumar 2009, Reddy and Kumar 2011c, Tripura and Singh 2014, Reddy et al. 2019). The compression test set-up for a CSRE wall is shown in Fig. 15.4. Figure 15.5 shows the plots for the compressive strength versus slenderness ratio for the CSRE walls. The investigations of Reddy et al. (2019) used a soil with 15.3% clay, 10% cement and dry density of the CSRE walls was 1800 kg/m^3. The moisture content of the CSRE walls during testing was about 1.5%. The studies of Kumar (2009) used a soil with 15.8% clay, 8% cement and dry density of the CSRE walls were 1800 kg/m^3. The moisture content of the CSRE walls in this study was about 6.5% during testing. The compressive strength of the CSRE wall decreases with the increase in the slenderness ratio. For a given slenderness ratio, the difference in the wall compressive strength between the two studies can be attributed to the differences in the cement content and the moisture content of the CSRE during testing. Figure 15.6 shows typical load versus lateral deflection curves for a slender CSRE wall, under concentric compression. Typical failure patterns for the concentrically loaded wall are shown in Fig. 15.7. Even

Fig. 15.4 Compression test set-up for a slender CSRE wall

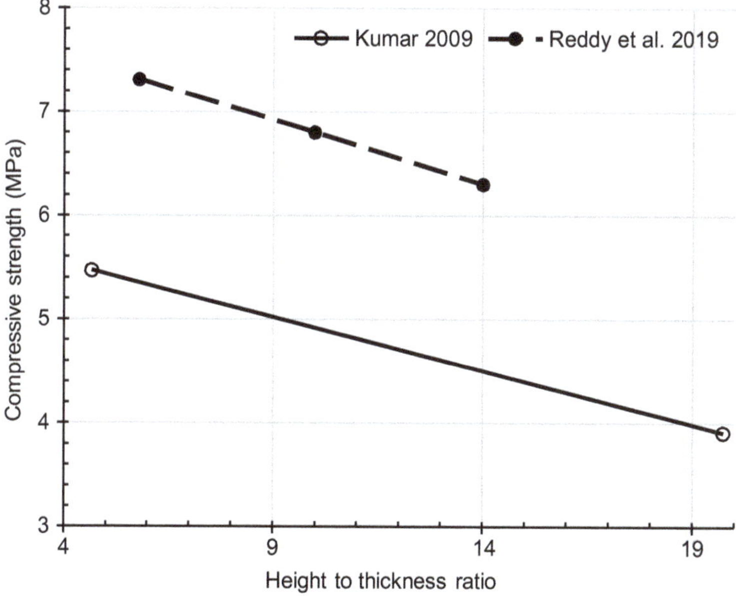

Fig. 15.5 Strength versus slenderness ratio for CSRE walls

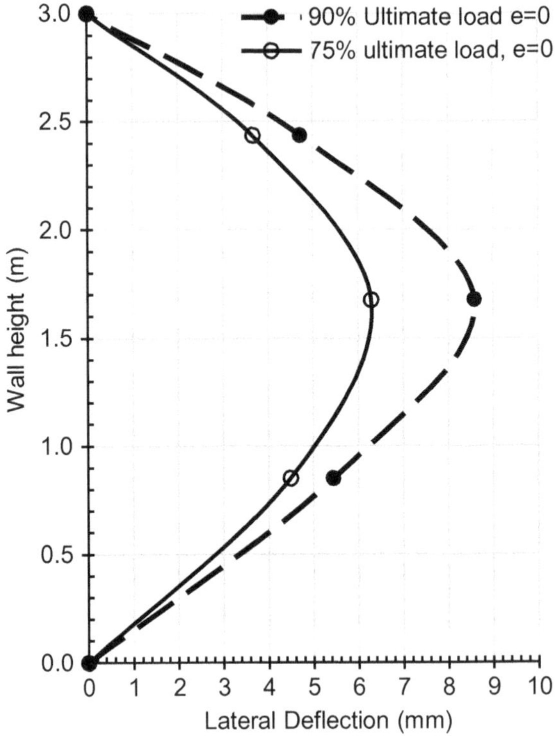

Fig. 15.6 Lateral deflections for concentric loads on a slender CSRE wall (Kumar 2009)

though the wall was under concentric compression, it undergoes considerable lateral deflection before failure. The CSRE walls ultimately show material failure (shear failure) even for slenderness ratio as high as 20.

15.4 Effect of Load Eccentricity and Slenderness on the Compressive Strength

The eccentricity of the load acting on the wall affects the strength of the CSRE wall significantly. Reddy et al. (2019) have examined the effect of eccentricity of loading on the compressive strength of the CSRE walls. Figure 15.8 shows the relationship between eccentric load and the compressive strength of the CSRE wall. The load eccentricity has been varied between 0 and 1/3 for three slenderness ratios (height to thickness ratio) of 6, 10 and 14. The soil used in these walls had 15.3% clay, 10% cement and dry density of the CSRE walls was 1800 kg/m^3. The CSRE walls have been tested in dry condition (moisture content during testing was about 1.5%) with hinged end conditions. The results show that the compressive strength of the CSRE

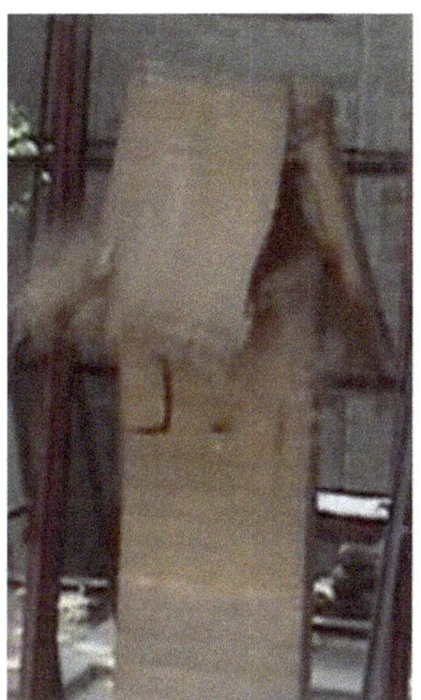

Fig. 15.7 Shear failure of slender CSRE walls

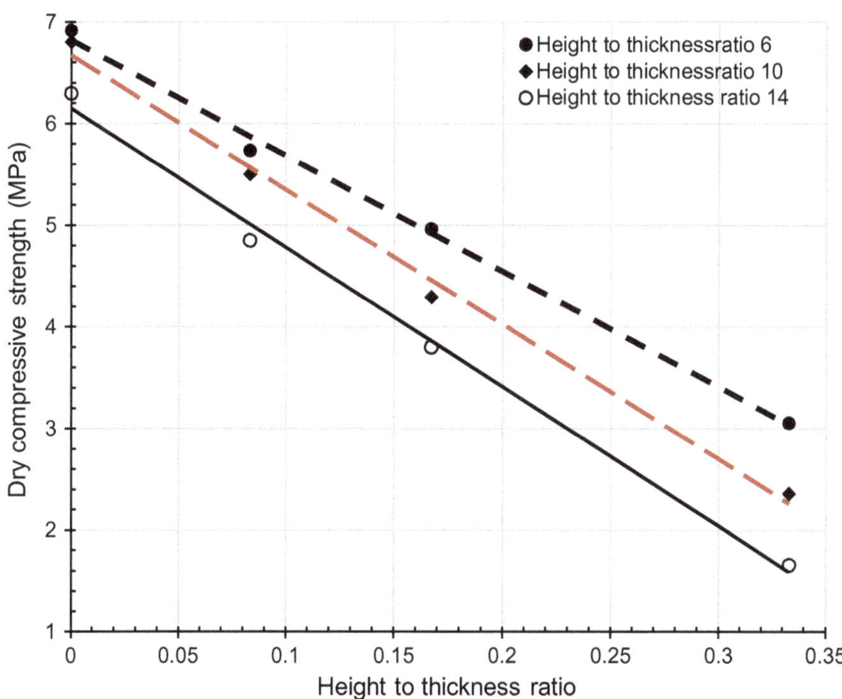

Fig. 15.8 Effect of load eccentricity on the compressive strength of the CSRE walls (Reddy et al. 2019)

wall is sensitive to the load eccentricity. The strength decreases linearly with the increase in the eccentricity ratio. The reduction in the compressive strength is about 55, 65 and 75% as the load eccentricity increases from 0 to 1/3 for the walls with the h/t ratio of 6, 10 and 14, respectively. As the slenderness increases, the load carrying capacity reduces. If the eccentricity ratio crosses 1/3, a portion of the wall will be under tension (Chapman and Slatford 1957).

15.4.1 The Lateral Deflection

Figure 15.9 shows the lateral deflection profiles (for 90% of the vertical ultimate load) for the CSRE wall with h/t ratio of 10. The load eccentricity causes increase in the lateral deflection for the wall. For example, 1.5 m-height wall (h/t = 10) shows a mid-height lateral deflection (at 90% ultimate load) as 2.69, 3.46, 4.29 and 4.77 mm, for the load eccentricities of 0, 1/12, 1/6 and 1/3, respectively. The lateral deflection (at mid height) doubles as the load eccentricity increase from 0 to 1/3.

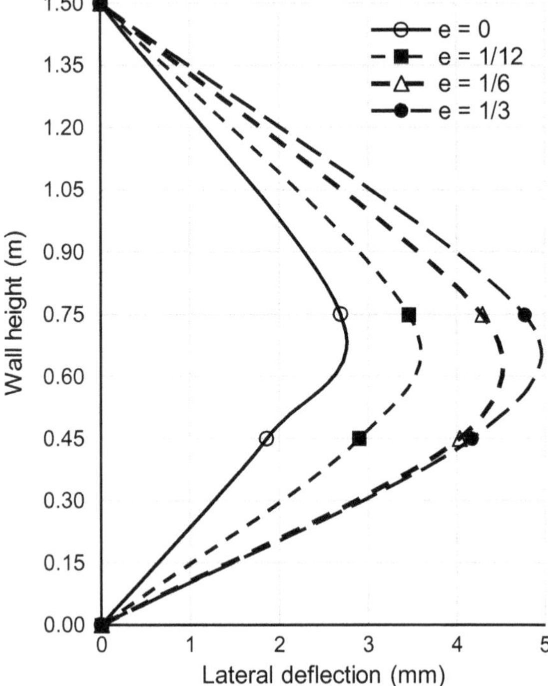

Fig. 15.9 Lateral deflections under concentric and eccentric loads at 90% of the ultimate load for 1.5 m wall (Reddy et al. 2019)

15.4.2 Failure Patterns Under Eccentric Loading

Figure 15.10 shows the typical failure patterns for eccentrically loaded CSRE walls. Due to significant lateral deflection under thc ccccntric loading, flexural tension cracks develop across the width of the wall on one of the wall faces and the opposite side of the wall face there will be crushing of the material due to eccentric compression. The flexure tension crack is initiated at the interface of the compacted layers. Considerable amount of lateral deflection can be noticed before the wall collapses.

15.5 Stress Reduction Factors

The design of rammed earth walls need stress reduction factors to account for the slenderness of the wall and the load eccentricity. Based on the data from Fig. 15.8, the stress reduction factors derived for the CSRE are given in Table 15.1, where t is wall thickness. The stress reduction factors differ slightly with those presented by Walker et al. (2005) for unstabilised rammed earth. The stress reduction factors differ widely from those used for the masonry design (IS 1905; NBC 2016).

Fig. 15.10 Crushing and flexure tension failures for the CSRE walls under eccentric compression (Reddy et al. 2019)

Table 15.1 Slenderness and eccentricity reduction factor (ϕ)

Slenderness ratio (S_r)	Reduction factor (ϕ)			
	Eccentricity ratio (e_{max}/t)			
	0	1/12	1/6	1/3
6	1.00	0.83	0.72	0.43
8	0.97	0.82	0.68	0.39
10	0.93	0.80	0.63	0.35
12	0.90	0.79	0.62	0.31
14	0.86	0.77	0.60	0.26

15.6 Shrinkage and Construction Joints

The clay minerals present in the soils undergo swelling upon moisture absorption and shrink upon drying. The swell–shrink phenomenon is inherent in soils or dense earth materials. The swell–shrink potential in soils greatly depends upon the type and the quantity of the clay minerals present. The soil- or earth-based products such as rammed earth is also prone for swelling and the shrinkage due to the variations in the moisture content during the construction (moulding moisture, curing) as well as the surrounding environment (exposure to rain and the humidity). The cement stabilised rammed earth is also prone for the swelling and the shrinkage. The following factors directly affect the swelling and the shrinkage in CSRE.

(1) Soil and aggregate mixture grading
(2) Cement content and curing duration
(3) Residual clay content (type and quantity)
(4) Moulding moisture content
(5) Density
(6) The surrounding environment (relative humidity)

The consequences of the swelling and the shrinkage in CSRE will be the development of the shrinkage cracks and affecting the durability of the structure. The drying shrinkage cracks are common in monolithic constructions such as CSRE and concrete. In concrete, the shrinkage cracks are controlled using the shrinkage reinforcement. In the case of CSRE, the shrinkage cracks shall be handled through the introduction of the construction joints. Figure 15.11 shows typical shrinkage cracks in CSRE walls. The shrinkage cracks are common at the junctions of the two walls and the field investigations on the shrinkage cracks in CSRE buildings show the width of the vertical shrinkage cracks as much as 5 mm (Kariyawasam and Jayasinghe 2016; Jayasinghe 2007). The unstabilised rammed earth walls are more prone for the shrinkage cracks. The rammed earth walls shrink, vertically, laterally and longitudinally as the material compacted at around 8–10% moisture dries to around 2.0% at the ambient dry conditions. The quantum of the shrinkage depends upon the clay content of the soil mix, the soil grading, the moulding moisture content and the rate of dying.

Fig. 15.11 Drying shrinkage cracks in a CSRE wall in a building

15.6.1 The Swell–Shrink Phenomenon in Cement Stabilised Rammed Earth

In the production process of the stabilised rammed earth, three important stages can be identified as follows.

(a) **Green stage**: The rammed earth is compacted in layers in a rigid formwork. Immediately after casting the rammed earth, the formwork can be stripped but the rammed earth will be fragile for handling. Hence, it is allowed to dry for 24 h before it attains some stiffness for commencing the curing process or the construction of the subsequent layers. Shrinkage caused at this initial stage of casting (basically drying taking place in 24 h, due to the loss of moisture) can be termed as *initial shrinkage.*

(b) **Curing stage**: 24 h after casting the stabilised rammed earth the curing commences. Generally, curing is carried out by covering with the wet burlap and sprinkling water using a low-pressure water spray, such that the entire mass is moist facilitating the formation of the cement hydration products. Here, the rammed earth absorbs water and can undergo swelling. This stage can be termed as *swelling* stage.

(c) **Drying stage**: After the curing period (at least 28 days), the stabilised rammed earth is allowed to dry. This process can last for few weeks to attain dry status. The duration may vary depending upon the surrounding ambient temperature and humidity conditions. The shrinkage during this stage of drying can be termed as *drying shrinkage*.

The shrinkage takes place in the green and the drying stages because there will be loss of moisture from the stabilised rammed earth. During the curing stage, the rammed earth can undergo swelling due to the absorption of the water by the clay minerals present in the rammed earth. Thus, in the production process the stabilised rammed earth undergoes "*initial shrinkage, swelling and drying shrinkage*". Therefore, the volume change phenomenon in all these three stages of the stabilised rammed earth manufacturing process should be examined for understanding the shrinkage phenomenon. In the case of unstabilised rammed earth, the curing process will be absent. It is allowed to undergo drying, and hence, there will be only *drying shrinkage stage* in the unstabilised rammed earth.

Initial shrinkage: Based on the data generated by Latha (2015) and Latha and Reddy (2017), the shrinkage taking place in the initial 24 h after casting the cement stabilised rammed earth is shown in Fig. 15.12. The soil used in the CSRE specimens had 14% clay (liquid limit: 26.9% and plasticity index: 17·5%). The dry density was 1750 kg/m^3 and the moulding moisture content of 12.5%. The relationships show that the initial shrinkage takes place until about 8 h after casting the CSRE specimen irrespective of the cement content. Beyond 8 h, there is marginal increase in the initial shrinkage value. The initial shrinkage is in the range of 0·010 and 0·019%, the highest being for the 4% cement.

Swelling during curing and drying shrinkage: Fig. 15.13 depicts the swelling during curing stage and the drying shrinkage variation for three cement contents (4, 7 and 10%). This is in continuation of the initial shrinkage work explained in the previous paragraph. The CSRE specimens after 28 days curing were oven-dried at 50 °C till constant mass was obtained. During the curing stage, the CSRE undergoes swelling. The maximum swelling at 28 days was in the range of 0.09–0.23%. The maximum swelling value was for the CSRE with 4% cement content. The swelling decreases with the increase in the cement content. The swollen CSRE specimen starts shrinking upon drying. In order to accelerate shrinkage (in the laboratory

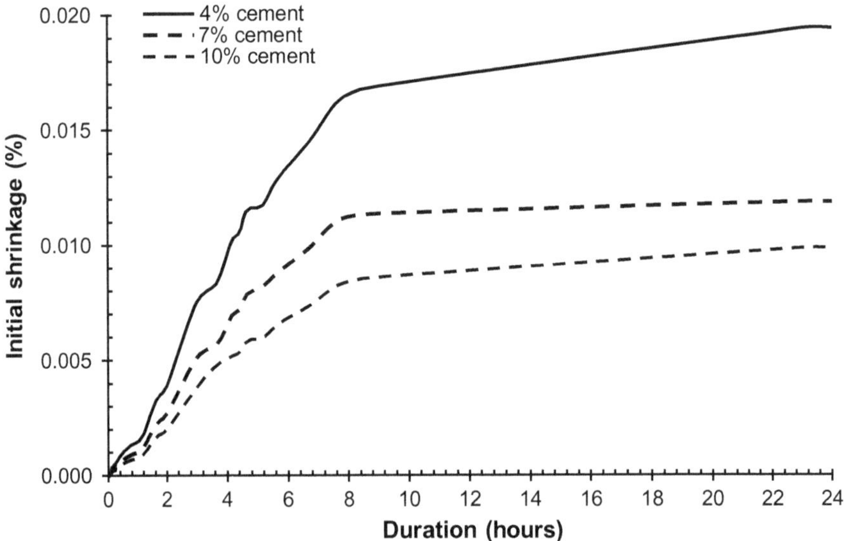

Fig. 15.12 Initial shrinkage up to 24 h after casting the CSRE specimen (Latha 2015)

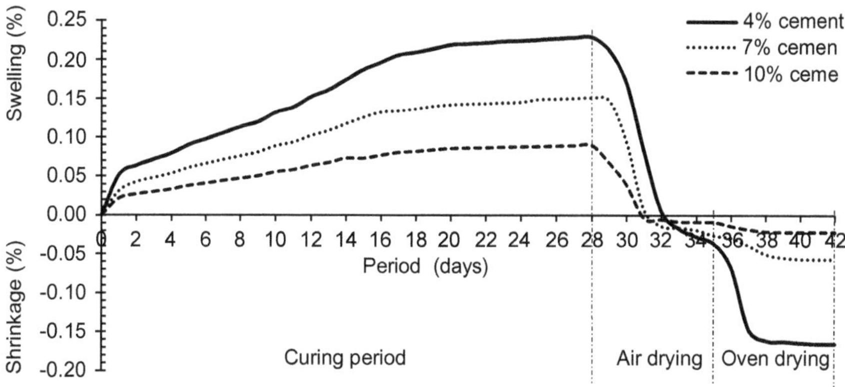

Fig. 15.13 Swelling and drying shrinkage of cement stabilised rammed earth (Latha 2015)

experiments), the specimens have been oven-dried at low temperature. The drying shrinkage variation with the cement content is shown in Fig. 15.14. The drying shrinkage decreases with the increase in the cement content. The drying shrinkage for CSRE was in the range of 0.02–0.17%. Jayasinghe's (2007) investigations on cement stabilized rammed earth using lateritic soil reveal the drying shrinkage strains in the range of 0.17–0.29%. The drying shrinkage in unstabilised rammed earth will be about 0.5% (Walker et al. 2005).

The ambient environmental conditions in the filed affect the drying shrinkage of the rammed earth. Also, it is difficult to monitor the drying shrinkage in the

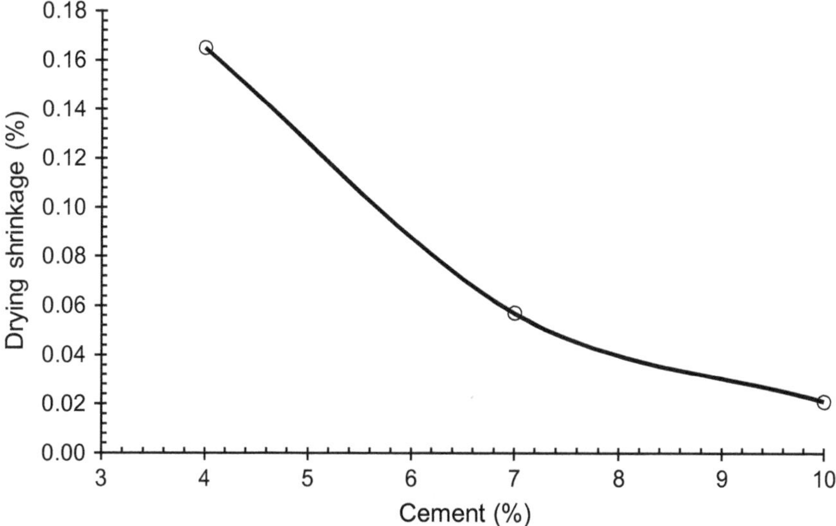

Fig. 15.14 Drying shrinkage versus cement content for cement stabilised rammed earth

field, where measurements need to be monitored over longer time durations. The laboratory results data discussed in the earlier paragraphs more closely resemble the field conditions. Such data can be used to predict the crack widths due to drying shrinkage in the stabilised rammed earth walls.

15.6.2 Movement Construction Joints in Rammed Earth Walls

The movement construction joints (vertical) should be provided in the rammed earth walls, to take care of the drying shrinkage manifesting into irregular shrinkage cracks and the damage. Such joints allow shrinkage to take place without impairing the structural integrity. The crack widths in CSRE can be estimated based on the shrinkage strain values. The vertical shrinkage crack width with a shrinkage strain of 0.15% will be 3.75 mm and 4.5 mm for a 2.5 m and 3.0 m-length CSRE wall, respectively. Kariyawasam and Jayasinghe (2016) observed a crack width of 5 mm in the walls of CSRE buildings. Figure 15.15 shows typical shrinkage cracks in the cement stabilised rammed earth wall in the absence of proper constructions joints. Generally, it is recommended to give a vertical construction joint at regular intervals of 2.5–3.0 m in the rammed earth walls. Figures 15.16 show possible construction joint details in the CSRE walls. The construction joints normally run across the full height of the wall. Such joints are filled or sealed with mastic sealant, or similar, for weather protection.

15.7 Structural Design of Stabilized Rammed Earth Wall

The structural design of a rammed earth building encompasses (a) design of walls for compression, bending and shear (b) stability of the building against horizontal loads and (c) seismic safety. The design concepts for compression and bending, and stability are discussed in the following sections.

15.7.1 Design for Compression and Bending

The compressive stresses in the CSRE walls are mainly due to the vertical gravity loading, whereas the bending or flexure stresses are due to load eccentricity and horizontal loads from wind, seismic effects, etc. The compressive strength of the wall is a function of (a) material compressive strength, (b) slenderness ratio and (c) load eccentricity. The brief design procedure for assessing the compressive strength of the CSRE wall is as follows.

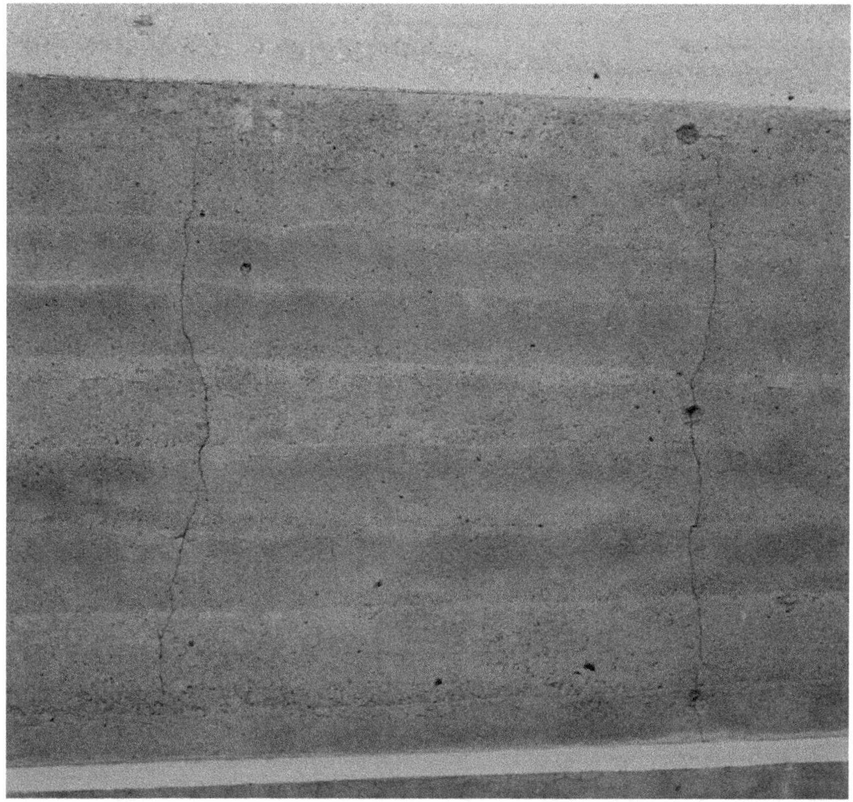

Fig. 15.15 Typical shrinkage cracks in cement stabilised rammed earth wall

1. Assess the loads realised on the wall, considering most unfavourable imposed load actions
2. Determine the slenderness ratio and eccentricity for the wall element under consideration
3. Estimate the stress reduction factor due to slenderness and load eccentricity
4. For sufficient compressive strength, the wall must satisfy the following basic requirement

$$N_{\mathrm{Ed}} \leq (\phi f_{\mathrm{kcsre}} b t_{\mathrm{w}}) \div (\gamma_{\mathrm{m}}) \tag{15.2}$$

where
N_{Ed} = design compressive force
ϕ = capacity reduction factor
b = breadth of the wall
t_{w} = wall thickness
f_{kcsre} = characteristic material compressive strength of CSRE

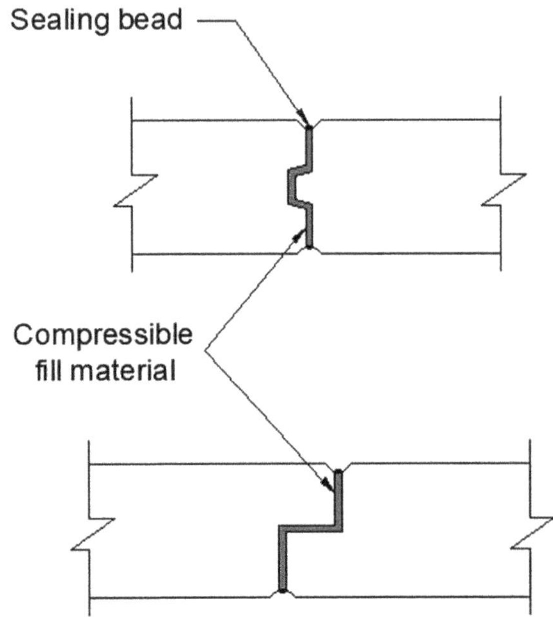

Fig. 15.16 Construction joint in rammed earth wall–tongue and grove

γ_m = material partial safety factor

Characteristic material compressive strength of CSRE was defined in Eq. (15.1).

Imposed loads: The imposed loads and unfavourable loading combination on a specific wall section can be estimated based on the plan, geometry and location of the building following the guidelines specified in the respective country's standard codes on loading actions (such as NBC 2016, BS EN 1991-1-1 2002).

Slenderness ratio: It is the ratio of effective height (h_e) and effective thickness (t_e) of the rammed earth wall. The effective height is a function of the lateral and rotational restraints at the top and bottom of the wall. In the absence of data specific to the rammed earth walls, the information available for the masonry walls for assessing the effective height can be adopted. Some of the typical guidelines in assessing the effective height are as follows.

h = Clear wall height between the restraints

$h_e = 0.75h$ for a wall laterally supported and rotationally restrained at both the ends

$h_e = 0.85h$ for a wall laterally supported both at the top and at the bottom and rotationally restrained at one of the ends

$h_e = 1.0h$ for a wall laterally supported but rotationally free at the top and at the bottom

$h_e = 2.0h$ for a wall laterally supported and rotationally restrained only along its bottom edge.

$$\text{Slenderness ratio} = S_r = (h_e) \div (t_e) \tag{15.3}$$

The effective thickness of rammed earth solid wall is equal to the wall thickness, but for cavity walls it can be taken as $(2 \div 3)(t_1 + t_2)$, as followed for the masonry walls in the absence of data on the structural behaviour of cavity rammed earth walls. Here, t_1 and t_2 are the thicknesses of the two leaves of the rammed earth cavity wall. In the design of unreinforced masonry, there is an imposition on the maximum slenderness value for the structural masonry walls. Similar guidelines can be adopted for the stabilised rammed earth walls, in the absence of specific data for rammed earth walls.

Load eccentricity: Assessing eccentricity of loads acting on the rammed earth wall is required to choose the right strength or the stress reduction factor. The eccentricity of the load may arise due to the eccentric nature of the vertical load, the load reactions realised on a wall due to unequal spans, the type of wall-floor/roof connections, the support restraints (lateral support and rotational restraints) at the wall ends and imperfections in the constructions. The masonry design codes (such as BS EN 1996-1-1:2005+A1 2012, NBC 2016) provide guidelines for assessing the eccentricity. Such guidelines can be adopted for the rammed earth wall design, in the absence of any specific data for the rammed earth walls.

Material partial safety factor (γ_m): A partial safety factor is applied to the material property design values to account for variations in the material quality, the quality of the workmanship and the overall quality of the work. For example, the characteristic compressive strength of the material is assessed using a small cylindrical specimen in the laboratory. The casting of the small specimen will not account for the possible defects arising due to the workmanship and other quality control aspects. The selection of the value for "γ_m" is at the designer's discretion. A partial safety factor value in the range of 3–4 can be used depending upon the workmanship, material quality (grading, optimum mix, etc.) and the quality control exercised during the construction.

Slenderness and eccentricity reduction factor: This aspect was discussed in Sect. 15.4, and the stress reduction factor values are given in Table 15.1. For a given slenderness ratio and the eccentricity ratio, the stress reduction factor can be obtained from this table.

15.8 Design Example for CSRE Using Limit State Method

A design example for cement stabilised CEB masonry was illustrated in Chap. 8, Sect. 8.4 and the Example 1 (Sect. 8.4.1). Figures 8.1 and 8.2 show the plan and the cross section of a four-storey load bearing dormitory building located close to Bangalore city, India. The design of the two critical walls designated as Wall-A and Wall-B has been illustrated. The roof and floor slabs are continuous reinforced concrete in situ construction.

Loadings

Roof

The slab thickness: 150 mm including the weatherproof course and ceiling finishes

The bulk density of the reinforced concrete slab and finishes: 25 kN/m^3

Dead load = 0.15 m × 25 kN/m^3 = 3.75 kN/m^2

Live load or imposed load = 1.5 kN/m^2 (Clause 3.4.1, NBC 2016, Group 2, Part 2, Sect. 1).

Floor

The slab thickness: 150 mm including the floor and ceiling finishes

The bulk density of the reinforced concrete slab and finishes: 25 kN/m^3

Dead load = 0.15 m × 25 kN/m^3 = 3.75 kN/m^2

Live load or imposed load = 2.0 kN/m^2 (Clause 3.3.1, NBC 2016, Group 2, Part 2, Sect. 1).

Walls

CSRE walls of 200 mm thick, without any rendering plaster layer on the surfaces.

Self-weight of the wall = 0.20 m × 1.0 m × 19 kN/m^3 = 3.8 kN/m^2

Loads realised on the walls

Figure 8.3 shows the areas of the roof/floor slab transferring the loads to Wall-A and Wall-B.

Wall-A

The wall receives uniformly distributed loads from Areas (1) and (2) and a concentrated reaction from Beam (1). Due to the concentrated load reaction from Beam (1), the CSRE wall length to be considered as per the code (angle of dispersion 30°) is illustrated in Fig. 8.4.

The CSRE wall length to be considered = 1.73 m

Slab area A1 (triangle) transferring the load on to the wall = $0.5(1.73)^2$ = 1.50 m^2

Slab area A2 (trapezium) transferring the load on to the wall = 0.5(2.25 + 0.315) × (1.73 + 0.16) = 2.42 m^2

Total slab area transferring the load = 1.50 + 2.42 = 3.92 m^2.

Reaction from Beam (1)

The beam is receiving a triangular load as shown in Fig. 8.5.

The reaction (point load) on the edge of Wall-A due to the reaction from the Beam (1) = (1.63 m^2) × load intensity

The calculations of vertical loading on Wall-A are illustrated in Table 15.2.

Wall-B (with an opening shown in Fig. 8.6)

Table 15.2 Calculation of vertical loading on Wall-A

Floor level considered		Load per 1.73 m run (kN)		Cumulative design load to floor (kN)
		Dead	Imposed	
3rd Floor				
Dead weight of roof: 3.92 m^2 × 3.75 kN/m^2	= 14.70 kN	44.08	8.33	52.41
Weight of wall: 3.0 m x (1.73 + 0.2) m × 3.80 kN/m^2	= 22.00 kN			
	= 36.70 kN			
Reaction from Beam (1): (1.63 m^2) × 3.75 kN/m^2	= 6.11 kN			
Self-weight of beam: 0.5(0.20 m × 0.23 m × 2.21 m) 25 kN/m^3	= 1.27 kN			
	44.08 kN			
Imposed load: (3.92 + 1.63) m^2 × 1.5 kN/m^2	= 8.33 kN			
2nd Floor				
Dead weight of floor: 3.92 m^2 × 3.75 kN/m^2	= 14.70 kN	88.16	17.48	105.64
Weight of wall: 3.0 m x (1.73 + 0.2) m × 3.80 kN/m^2	= 22.00 kN			
	= 36.70 kN			
Reaction from Beam (1) + self-weight: (6.11 + 1.27)	= 7.38 kN			
Dead weight from above:	= 44.08 kN			
	88.16 kN			
90% Imposed load: 0.90(3.92 + 1.63) m^2 x (2 + 1.5) kN/m^2	= 17.48 kN			
1st Floor				
Dead weight (as in 2nd floor): (36.70 + 7.38) kN	= 44.08 kN	132.24	24.42	156.66
Dead weight from above:	= 88.16 kN			
	132.24 kN			
80% Imposed load: 0.80(3.92 + 1.63) m^2 x (2 + 2 + 1.5) kN/m^2	= 24.42 kN			
Ground Floor				
Dead weight (as in 2nd floor): (40.64 + 7.38) kN	= 44.08 kN	176.32	29.14	205.46
Dead weight from above:	= 132.24 kN			

(continued)

Table 15.2 (continued)

Floor level considered		Load per 1.73 m run (kN)		Cumulative design load to floor (kN)
		Dead	Imposed	
	176.32 kN			
70% Imposed load: 0.70(3.92 + 1.63) m^2 x (2 + 2 + 2 + 1.5) kN/m^2	29.14 kN			

The wall receives uniformly distributed loads from the roof and the floor slab portions (4) and (5). Area (4) is one way slab. It is assumed that the reaction from the Beam (1) is not realised on the wall-B. Area (5) is part of the two way slab. The CSRE wall has a door opening, and hence, the entire wall length can be considered for the design calculations.

The CSRE wall length considered = 4.03 m

Slab area A4 (one way slab) transferring the load to the wall = 0.5(2.21) × (4.03) = 4.45 m^2

Slab area A5 (two-way slab) transferring the load to the wall = 0.5(4.03 × 2.015) = 4.06 m^2

Total slab area transferring the load = 4.45 + 4.06 = 8.51 m^2

The calculations of vertical loading on Wall-B are illustrated in Table 15.3.

Wind loading

General stability is met as per Sect. 4.2.2.2 of NBC (2016) (Part 6, Group 2, Sect. 4). Height-to-width ratio of the building is about 0.60 $<$ 2.0, and cross walls spacings are within the limits specified in the code. The solid reinforced concrete slab rests on the cross walls. Hence, the general stability need not be checked.

In the limit state design, the design load (actions) shall be matched with design resistance as detailed in Eqs. (8.2) and (8.3). The design exercise involves assessing the values for the following.

(a) Partial safety factors for the ultimate limit state of strength
(b) Partial safety factors for the materials
(c) Capacity reduction factor (Φ) accounting for slenderness and load eccentricity

The partial safety factors for the permanent load actions (γ_G) and variable load actions (γ_Q) can be taken conservatively as 1.35 and 1.5, respectively, as per the Eurocode (BS EN 1990-2002) guidelines. Assessing the material partial safety factors (from the Tables given in the relevant EC codes) is difficult, because of the absence of stabilised rammed earth in those tables. The selection of the partial safety factor is at the discretion of the design engineer. The recommended values for the partial safety factor vary between 1.0 and 3.0 for the masonry materials (EN 1996-1-1:2005+A1:2012). For rammed earth, Walker et al. (2005) suggest partial safety factor in the range of 3.0–6.0. Assuming the rammed earth building under consideration has been executed by the experienced contractor exercising good quality control

Table 15.3 Calculation of vertical loading on Wall-B

Floor level considered		Load per 4.03 m run (kN)		Cumulative design load to floor (kN)
		Dead	Imposed	
3rd Floor				
Dead weight of roof: 8.51 m^2 × 3.75 kN/m^2	= 31.91 kN	77.83	12.77	90.60
Weight of wall: 3.0 m × 4.03 m × 3.80 kN/m^2	= 45.92 kN			
	77.83 kN			
Imposed load: 8.51 m^2 × 1.5 kN/m^2	= 12.77 kN			
2nd Floor				
Dead weight of floor: 8.51 m^2 × 3.75 kN/m^2	= 31.91 kN	155.56	26.81	182.37
Weight of wall: 3.0 m × 4.03 m × 3.80 kN/m^2	= 45.82 kN			
	77.73 kN			
Dead weight from above:	= 77.83 kN			
	155.56 kN			
90% Imposed load: 0.90 × 8.51 m^2 x (2 + 1.5) kN/m^2	= 26.81 kN			
1st Floor				
Dead weight (as in 2nd floor):	= 77.73 kN	233.29	37.44	270.73
Dead weight from above:	= 155.56 kN			
	233.29 kN			
80% Imposed load: 0.80 × 8.51 m^2 x (2 + 2 + 1.5) kN/m^2	= 37.44 kN			
Ground Floor				
Dead weight (as in 2nd floor):	= 77.73 kN	311.02	44.68	355.70
Dead weight from above:	= 233.29 kN			
	311.02 kN			
70% Imposed load: 0.7 × 8.51 m^2 x (2 + 2 + 2 + 1.5) kN/m^2	= 44.68 kN			

measures, a material partial safety factor of $\gamma_M = 3.0$ has been considered in this design example.

Capacity reduction factor (*Φ*) for Wall-A and Wall-B

Effective height of the wall = 0.75 × (3.15) m = 2.36 m (wall laterally supported and rotationally restrained both at the top and the bottom)

Effective thickness = 0.20 m

Slenderness ratio = (2.36) ÷ (0.20) = 11.80

Consider a load eccentricity = 1/12, arising out of construction difficulties in keeping the wall to plumb, though the floor slab is continuous over the wall and the slab spans on either side of walls does not vary much.

Consider capacity reduction factor (Φ) = 0.791 for Wall-A and Wall-B (from Table 15.1).

Characteristic and mean compressive strength for the CSRE in Wall-A and Wall-B

$$N_{\text{Ed}} \leq N_{\text{Rd}}$$

$$N_{\text{Ed}} = \text{Design load}$$

$$N_{\text{Rd}} = \text{Design resistance}$$

$$\text{For } N_{\text{Ed}} = N_{\text{Rd}}$$

$$N_{\text{Ed}} = (\phi f_{\text{kcsre}} b t_{\text{w}}) \div (\gamma_{\text{m}}) = (0.791 \times f_{\text{kcsre}} \times 1.0 \times 200) \div (3.0) = (52.73) f_{\text{kcsre}}$$

$$f_{\text{kcsre}} = (N_{\text{Ed}}) \div 52.73$$

where

N_{Ed} = design compressive force in N/mm

ϕ = capacity reduction factor

b = breadth of the wall in mm

t_{w} = wall thickness in mm

f_{kcsre} = characteristic material compressive strength of CSRE in MPa

γ_{m} = material partial safety factor

Table 15.4 gives design details for Wall-A and Wall-B.

The mean wet cylinder compressive strength of CSRE required for Wall-A in the ground floor is 4.32 MPa, and it reduces to 2.02 MPa in the top floor. Similarly, strength requirements can be seen for Wall-B. The CSRE strength requirement in each of the floors is given in Table 15.4.

Table 15.4 Design details for Wall-A and Wall-B

Wall-A			
Ultimate design load (N_{Ed}) for 1.73 m wall length	N_{Ed} (N/mm)	f_{kcsre} = (N_{Ed}) ÷ (52.73) (MPa)	*Compressive strength of *CSRE* (MPa)
3rd Floor 1.35(44.08) + 1.5(8.33) = 72.00 kN	41.62	0.79	2.02
2nd Floor 1.35(88.16) + 1.5(17.48) = 145.24 kN	83.95	1.59	2.82
1st Floor 1.35(132.24) + 1.5(24.42) = 215.15 kN	124.37	2.36	3.59
Ground Floor 1.35(176.32) + 1.5(29.14) = 281.74 kN	162.86	3.09	4.32
Wall-B			
Ultimate design load (N_{Ed}) for 4.03 m wall length. The wall has 1.0 m opening. Actual wall length = (4.03 – 1.0) = 3.03 m	N_{Ed} (N/mm)	f_{kcsre} = (N_{Ed}) ÷ (52.73) (MPa)	*Compressive strength of *CSRE* (MPa)
3rd Floor 1.35(77.83) + 1.5(12.77) = 124.23 kN	41.00	0.78	2.01
2nd Floor 1.35(155.56) + 1.5(26.81) = 250.22 kN	82.58	1.57	2.80
1st Floor 1.35(233.29) + 1.5(37.44) = 371.10 kN	122.48	2.32	3.55
Ground Floor 1.35(311.02) + 1.5(44.68) = 486.90 kN	160.69	3.05	4.28

* Wet compressive strength of CSRE cylinder 150 mm diameter and 300 mm height assuming a standard deviation of 0.75 MPa

References

ASTM C1314-07 (2007) Standard test method for compressive strength of masonry prisms. ASTM, West Conshohocken, PA, USA

BS EN 1052-1 (1999) Methods of test for masonry—determination of compressive strength. British Standards Institution, London, UK

BS EN 1990:2002+A1:2005 (2005) Eurocode. Basis of structural design. The British Standards Institution, UK

BS EN 1991-1-1:2002 (2002) Eurocode 1, Actions on structures, General actions, Densities, self-weight, imposed loads for buildings. British Standards Institution, London, UK
BS EN 1996-1-1:2005+A1:2012 (2012) Eurocode 6. Design of masonry structures. General rules for reinforced and unreinforced masonry structures. British Standards Institution, London, UK
Chapman JC, Slatford J (1957) The elastic bucking of brittle columns. Proc Inst Civ Eng 6:107–125
Daniela C, Joshua G (2012) Experimental investigation on the compressive strength of cored and molded cement-stabilized rammed earth samples. Constr Build Mater 28(1):294–304
Hall M, Djerbib Y (2004) Rammed earth sample production: context, recommendations and consistency. Constr Build Mater 18(4):281–286
IS 1905 (1987) Code of practice for structural use of unreinforced masonry. Bureau of Indian Standards New Delhi, India
Jayasinghe C, Kamaladasa N (2007) Compressive strength characteristics of cement stabilized rammed earth walls. Constr Build Mater 21(11):1971–1976
Jayasinghe C (2007) Shrinkage characteristics of cement stabilised rammed earth. In: Proc. international symposium on earthen structures (ISES-2007), 22–24 August. Indian Institute of Science, Bangalore, India, pp 212–216
Kariyawasam KKGKD, Jayasinghe C (2016) Cement stabilized rammed earth as a sustainable construction material. Constr Build Mater 105:519–527
King BPE (1996) Buildings of earth and straw—structural design for rammed earth and straw bale architecture. Ecological Design Press, Sausalito, CA, USA
Kumar PP (2009) Stabilised rammed earth for walls: materials, compressive strength and elastic properties. PhD thesis, Department of Civil Engineering, Indian Institute of Science, Bangalore, India
Latha MS, Reddy BVV (2017) Swell-shrink properties of stabilised earth products. Constr Mater Proc ICE (London) 170(CM1):3–15. https://doi.org/10.1680/jcoma.15.00032
Latha MS (2015) Studies on characteristics of stabilised soil compacts for structural applications. PhD.thesis, Department of Civil Engineering, Indian Institute of Science, Bangalore, India
Maniatidis V, Walker P (2008) Structural capacity of rammed earth in compression. J Mater Civ Eng 20:3(230):230–238
Middleton GF (Revised by Schneider) (1987) Earth wall construction, Bulletin 5, 4th edn. Commonwealth Scientific and Industrial Research Organisation (Division of Building Construction and Engineering), North Ryde, NSW, Australia
NBC (2016) National building code of India 2016, group 2, sections 1 and 4. Bureau of Indian Standards, New Delhi, India
Reddy BVV, Kumar PP (2011b) Cement stabilised rammed earth. Part B: compressive strength and stress–strain characteristics. Mater struct 44(3):695–707
Reddy BVV, Suresh V, Nanjunda Rao KS (2017) Characteristic compressive strength of cement stabilized rammed earth. J Mater Civ Eng 29(2):04016203-1–04016203-7
Reddy BVV, Suresh V, Nanjunda Rao KS (2019) Behaviour of cement stabilised rammed earth walls under concentric and eccentric gravity loading. In: Reddy BVV, Mani M, Walker P (eds) Earthen dwellings and structures. Springer Transactions in Civil and Environmental Engineering. Springer, Singapore, pp 293–303
Reddy BVV, Kumar PP (2011a) Cement stabilised rammed earth. Part A: compaction characteristics and physical properties of compacted cement stabilised soils. Mater Struct 44(3):681–693
Reddy BVV, Kumar PP (2011c) Structural behaviour of story-high cement-stabilised rammed-earth walls under compression. J Mater Civ Eng 23(3):240–247
Tibbets JM (2001) Emphasis on rammed earth—the rational. Interaméricas Adobe Build 9:4–33
Tripura DD, Singh KD (2014) Behavior of cement-stabilized rammed earth circular column under axial loading. Mater Struct 49(1–2):371–382. https://doi.org/10.1617/s11527-014-0503-4
Walker P, Standards Australia (2002) The Australian earth building handbook. Standards Australia International, Sydney, Australia
Walker P, Keable R, Martin J, Maniatidis V (2005) Rammed earth design and construction guidelines. BRE Bookshop, Watford, UK

Walker PJ (2002) Reinforced composite rammed earth in flexure. In: Proc., Int. Conf. on non-conventional materials and technologies (NOCMAT/3), Hanoi, Vietnam, pp 439–445

Windstorm B, Schmidt A (2013) A report of contemporary rammed earth construction and research in North America. Sustainability 5(2):400–416

Part IV
Energy, Carbon Emissions and Sustainability

Chapter 16
Status of Clay Minerals in the Stabilised Earth Materials

16.1 Introduction

Soil formation is mainly due to the weathering process of the rocks. The soil formation process takes millions of years. The soil characteristics (especially strength, plasticity and shrinkage) are controlled by the quantity and types of clay minerals present in the soil. Clay minerals have a definite structure and are essentially alumino-silicates. The soil forms the basic raw material for the manufacture of unbaked earth products and burnt clay bricks/blocks. The stabilised compressed earth bricks (CEBs) and the stabilised rammed earth represent the products with inorganic binders such as the cement and or the lime. The cement/lime stabilised CEBs and the rammed earth have emerged as alternative building materials since the last 6–7 decades. Such earth-based products have low embodied carbon when compared to the embodied carbon in the conventional earth-based materials such as burnt clay bricks. The burnt clay bricks are manufactured by firing at 800–1000 °C. The firing process causes irreversible structural changes to the clay minerals present in the soil and the soil gets nearly transformed to rock form. It is difficult to retrieve the natural clay minerals from the burnt clay products, unless allowed to undergo weathering for millions of years (Monto and Reddy 2012). But in the case of cement stabilised earth products (such as stabilised CEB and stabilised rammed earth), there is a possibility for retrieving the natural clay minerals.

The status of clay minerals in the Portland cement (PC) and the lime stabilised compressed earth specimens can be assessed. This will mainly address the question of recyclability of soil-based materials as soil at the end of life of the stabilised earth products such as CEB and rammed earth.

B. V. V. Reddy, *Compressed Earth Block & Rammed Earth Structures*,
Springer Transactions in Civil and Environmental Engineering,
https://doi.org/10.1007/978-981-16-7877-6_16

16.2 Soil Composition and Clay Minerals

The soils used for earthen products such as CEB and rammed earth are mainly composed of two types of particles: (a) inert particles such as gravel, sand and silt and (b) clay minerals. In addition, non-clay minerals and minor quantity of organic matter may be present in some soils. The inert particles are basically crystalline silica. Kaolinite, montmorillonite and illite are the commonly found clay minerals in the soils used for the cement and the lime stabilised CEBs and the rammed earth. More discussions and details on the soils and the clay minerals are provided in Chap. 2.

16.3 Cement and Lime Stabilisation

The soils with expansive clay minerals such as montmorillonite need lime stabilisation, while the soils with non-expansive clay minerals (kaolinite, illite, etc.) can be efficiently stabilised with the Portland cement. Selection of a suitable stabilising additive (cement or lime) greatly depends upon the types of the clay minerals present in the soils. Ordinary Portland cement (OPC) and lime (calcium hydroxide) are the most commonly used inorganic binders for the manufacture of the stabilised earth products. The mechanism of strength development in the cement and the lime stabilised soils is distinctly different. The lime and the cement stabilisation mechanisms are discussed in detail in Chap. 3.

16.3.1 Cement Stabilisation

In a mixture of soil, cement and water, cement hydrates resulting in the formation of cementitious products such as hydrates of silicate, aluminate, silicate-aluminates, etc. and the release of a small quantity of calcium hydroxide (CH). The cement hydration products are responsible for the development of the bond between the gravel, sand and the silt particles in the soil. The cementitious products (hydrates of silica and alumina) do not directly react with the clay minerals, whereas the calcium hydroxide released in the cement hydration process can react with the clay minerals forming additional cementitious products. The strength of the cement stabilised compressed earth increases with the increase in the cement content.

16.3.2 Lime Stabilisation

Generally, hydrated lime (calcium hydroxide) is used for the manufacture of the lime stabilised earth products. The clay minerals are pozzolanic in nature. When lime is mixed with the soil, the clay minerals in the soil react with the lime in the presence of moisture. Hence, the lime-clay reactions lead to the formation of the hydrates of silicate and the silicate-aluminates, which will coat the inert gravel, sand and silt particles and establish bonds between them. At ambient temperature and curing conditions, the strength gain rate due to the lime-clay reactions is very slow. The lime-clay reactions can continue over longer periods of time due to the humidity present in the air. There are arguments favouring the lime stabilisation of soils instead of the cement stabilisation with the assumption that there will be re-carbonation of the lime in the lime stabilised soil leading to the mitigation of the carbon emissions. A comparison between the mechanisms of the strength development in the cement and the lime stabilised soils reveals that in the lime stabilised soils, both the clay minerals and the calcium hydroxide get consumed, whereas major quantity of the clay minerals is nearly intact in the cement stabilised soils.

16.4 Assessing Residual Clay Minerals in the Stabilised Compressed Earth Products

The presence of residual clay minerals in the stabilised earth products can be indirectly inferred through some physical tests on the cured stabilised earth samples. The cured, dried and powdered samples can be tested and analysed for (1) particle size distribution (PSD), (2) Atterberg's limits, (3) X-ray diffraction (XRD) patterns and (4) SEM and EDS analysis. The PSD analysis will reveal the clay size fraction, the Atterberg's limits indirectly indicate the presence of the clay minerals, qualitatively, XRD patterns show the clay minerals present in the powdered samples, and the SEM/EDS analysis identifies the cementitious products such as calcium silicate hydrate (CSH), calcium aluminate hydrate (CAH) and calcium hydroxide (CH) in the stabilised earth products. Reddy and Latha (2014), and Latha (2015) performed controlled tests on the cement and the lime stabilised compressed earth specimens to decipher the presence of the residual clay content in the cement and the lime stabilised compressed earth specimens. Some of the results from these studies are discussed here.

The particle size distribution curves for the natural and the ground (for 12 min) soil samples are shown in Fig. 16.1. Table 16.1 gives the characteristics of the two soil samples. The 15.8% clay size fraction in the natural soil enhances to 19.6% after 12 min of grinding. This is mainly attributed to the breaking of sand/silt size particles into smaller sizes. As the fines in the soil increase, the water-holding capacity increases resulting in the increased liquid limit. The BET-specific surface area increased from 5.4 to 8.1 m^2/g after grinding.

The stabilised compressed earth products are cemented materials and when exposed to normal weathering process can take a very long time (may be several decades) to disintegrate into smaller size particles. In order to examine the status or presence of residual clay minerals in the stabilised compressed earth specimens, the samples have to be crushed or ground, and examined.

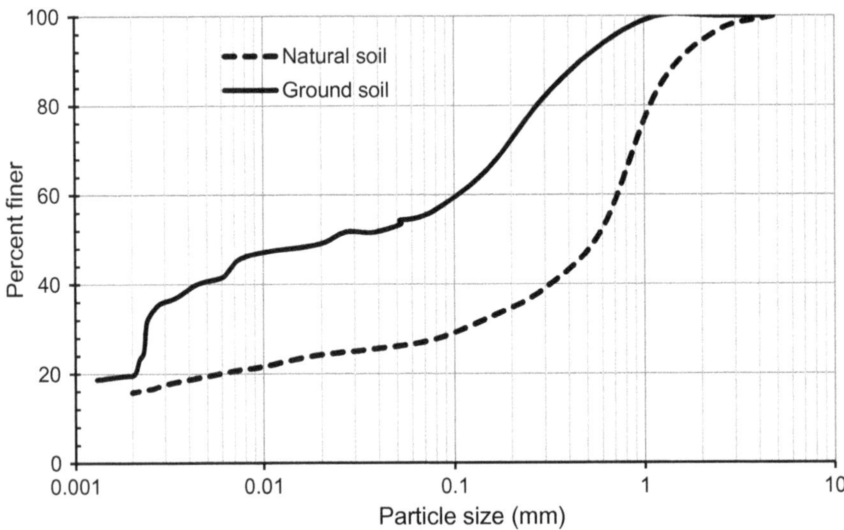

Fig. 16.1 Particle size distribution curves for the natural soil and the ground soil (Reddy and Latha 2014, Copyright Elsevier)

Table 16.1 Characteristics of soils

Soil characteristics	Natural soil	Ground soil
1. Textural composition (% by weight)		
Sand (4.75–0.075 mm)	72.6	44.0
Silt (0.075–0.002 mm)	11.6	36.4
Clay (<0.002 mm)	15.8	19.6
2. Atterberg's limits		
Liquid limit(%)	26.9	33.9
Plastic limit (%)	9.4	8.8
Plasticity index	17.5	25.1
3. Unified soil classification	SC	CL
4. Clay mineral type	Kaolinite	Kaolinite
5. pH	8.06	7.84
6. Specific surface area (m^2/g)	5.4	8.1

16.5 X-Ray Diffraction (XRD) Patterns, SEM Examination and EDS Analysis

The XRD patterns of the particles reveal the presence of the clay minerals in the powdered samples. Figure 16.2 shows the XRD patterns generated from less than 2 μ size particles obtained from the ground (12 min) stabilised (cement and lime) compressed earth samples and the natural soil. The XRD pattern of the natural soil

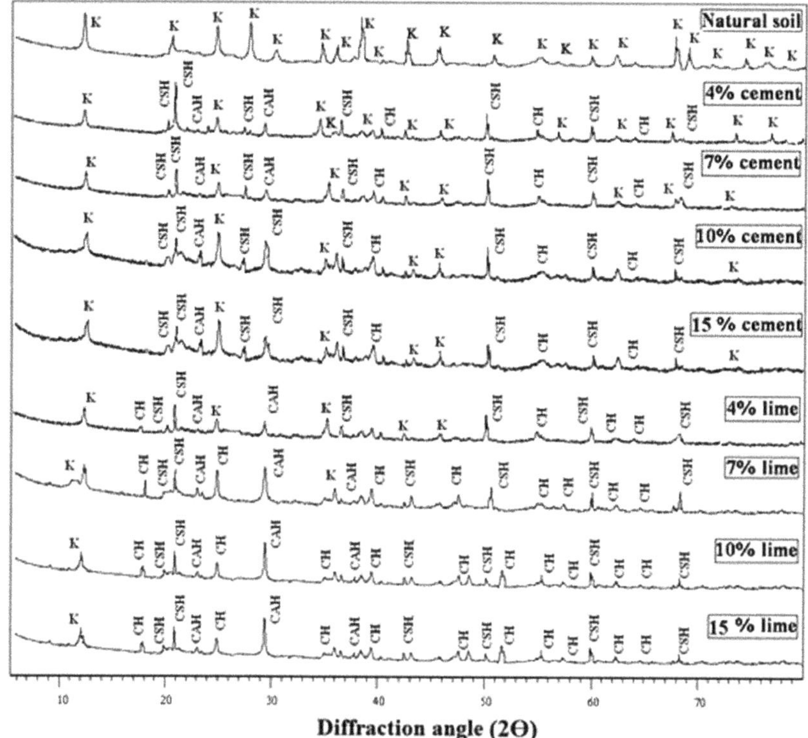

CH - Calcium hydroxide, CSH - Calcium silicate hydrate, CAH - Calcium aluminate hydrate, K - Kaolinite

Fig. 16.2 XRD patterns for ground natural soil and stabilised compressed earth samples (Reddy and Latha 2014, Copyright Elsevier)

clay size fraction shows the presence of only kaolinite clay mineral. The XRD patterns for the cement and the lime stabilised samples show the presence of kaolinite as well as the cementitious products such as CSH and CAH. Also, calcium hydroxide (CH) peaks can be seen in both the lime and the cement stabilised samples. The CH peaks in cement stabilised ground samples arise from the lime released during the cement hydration process. In the case of lime stabilised samples, the CH peaks indicate the un-reacted lime. There are only a couple peaks showing the kaolinite mineral in the XRD of 7% lime, whereas there is only one peak showing kaolinite mineral in the 10% and the 15% lime samples. The quartz peaks are missing in the XRD patterns shown in Fig. 16.2. This is mainly because of the samples (<2 μ) for XRD have been collected through settling process where the heavier silt/sand particles settle quickly and the decanted materials from the suspension in the jar are finer particles of clay and cement hydration products (CSH, CAH and CH). The particle size of the aggregated CSH particles in the hydrated cement paste will be in the range of 0.8–3 μ (Mehta and Monteiro 2017). The cement stabilised samples show more kaolinite peaks, whereas the lime stabilised samples show a single kaolinite peak.

The XRD patterns for the silt and sand size fractions collected from the ground (12 min) stabilised compressed earth samples and the ground natural soil are shown in Fig. 16.3. The XRD patterns clearly show the presence of kaolinite clay mineral in the sand and the silt size fractions of the cement stabilised samples. The kaolinite clay mineral peaks are missing in the sand and the silt size fractions of the lime stabilised samples and the natural soil. It is interesting to note that kaolinite clay mineral is present in the sand and the silt size fractions of the cement stabilised compressed earth material. This can be attributed to the fact that the clay mineral particles or the lumps of clay mineral particles get entrapped in the interstices of coarser cemented soil lumps (i.e. silt and sand size particles), and if such coarser soil particles are allowed for further grinding or natural weathering, the clay minerals entrapped in their interstices can get released.

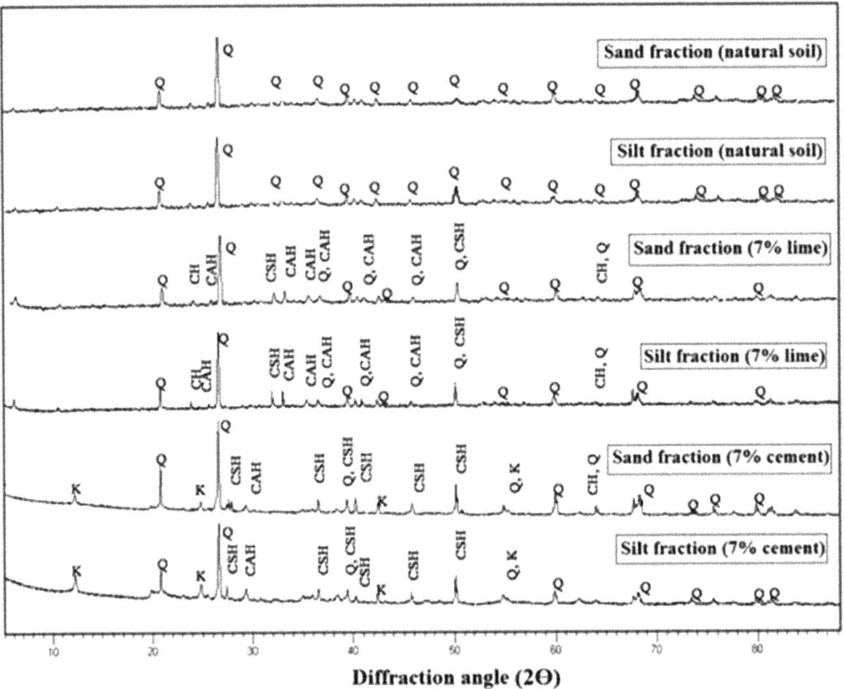

CH - Calcium hydroxide, CSH - Calcium silicate hydrate, CAH - Calcium aluminate hydrate, K - Kaolinite

Fig. 16.3 XRD patterns for sand and silt size fractions collected from ground natural soil and stabilised soil compacts (Reddy and Latha 2014, Copyright Elsevier)

The CAH and CSH products are formed during the cement hydration process in the cement stabilised samples. The lime stabilised samples also show the formation of CSH and CAH products due to the lime-clay reactions. The natural soil does not show the presence of cementitious products (such as CSH and CAH).

The SEM images and EDS analysis for the lime and the cement stabilised soil samples indicate the formation of the cementitious products (CSH, CAH) either due to cement hydration or due to the reaction between the clay mineral and the lime. The observations of Millogo and Morel (2012) on the SEM and EDS analysis of the cement stabilised adobe brick samples showed the presence of cementitious compounds as well as the kaolinite clay mineral. The lime and the kaolinite react to form CSH and CAH (Bell 1996). It is not possible to quantify the clay minerals from the XRD patterns and the SEM images. The particle size analysis of the powdered samples can give an idea of the clay size fractions in the powdered samples.

16.6 Particle Size Analysis and Clay Size Fraction

The particle size analysis of the samples reveals the clay size fraction. Kaolinite clay particle size is <2 μ (Ian et al. 1993).

16.6.1 Effect of Grinding Duration on Particle Size Fractions

The grinding duration affects the fineness of the ground soil. Figure 16.4 shows the variation in the clay size fraction (<2 μm) with the grinding duration for the natural soil, the cement (7% cement) and the lime (4% lime) stabilised compressed earth specimens. In the case of ground samples, the percentage of sand size fraction reduces while the percentage of the silt and the clay size fractions increase. The plots in Fig. 16.4 show that the clay size fractions increase with the grinding duration irrespective of the type of stabilisation. For the case of natural soil, the clay size fraction increased from 17.5 to 22% as the grinding duration was increased from 8 to 16 min. In the case of cement (7%) and the lime (4%), the clay size fraction increased by 40%. This is obviously the result of breaking of the larger particles (manly cemented lumps) in the grinding process.

16.6.2 Particle Size Analysis and Clay Size Fractions in Cement and Lime Stabilised Compressed Earth Specimens

Fig. 16.5 shows typical particle size distribution curves for the ground (for 12 min) natural soil and the stabilised compressed earth specimens with the cement (7%) and the lime (7%). Figure 16.6 shows the plot of the clay size fraction versus the

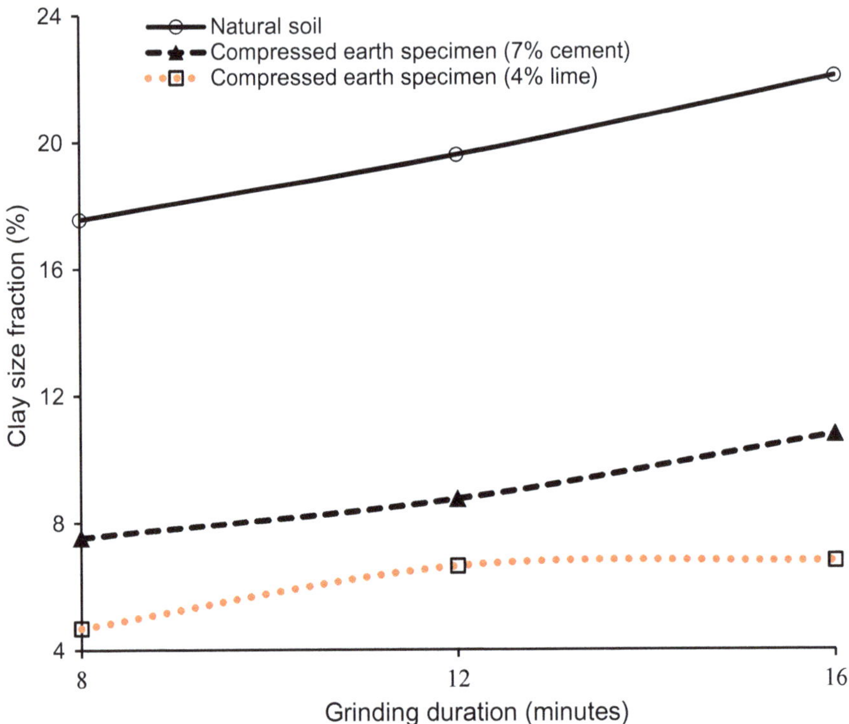

Fig. 16.4 Clay size fraction versus grinding duration (Latha 2015)

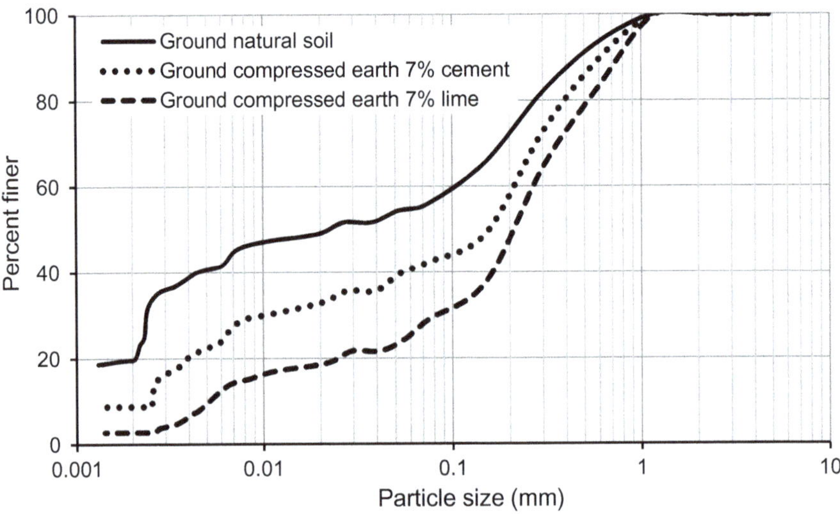

Fig. 16.5 Particle size curves for the natural soil and the stabilised compressed earth specimens ground for 12 min

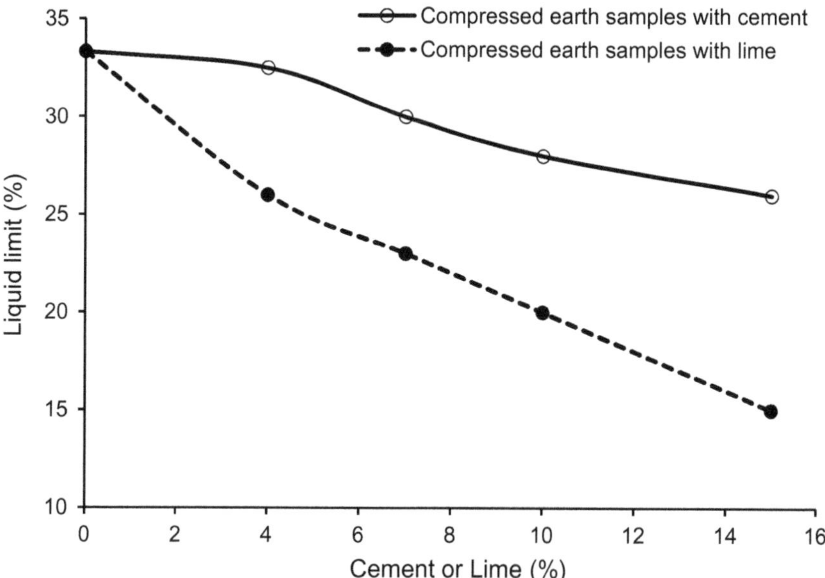

Fig. 16.6 Clay size fraction in the stabilised compressed earth specimens ground for 12 min (Reddy and Latha 2014, Copyright Elsevier)

stabiliser content (cement or lime) in the ground (12 min) stabilised compressed earth specimens. The clay size fraction in the ground (12 min) stabilised compressed earth specimens steadily decreases with the increase in the stabiliser content. It reduces from 11.9 to 6.3% as the cement content increases from 4 to 10%. In the case of lime stabilisation, the clay size fraction reduces from 6.6 to 2.0% as the lime content goes up from 4 to 10%. The clay size fraction of the ground (12 min) natural soil sample was 19.6%. There is a considerable difference in the clay size fractions among the cement and the lime stabilised compressed earth specimens. The lower percentage of clay size fraction in the lime stabilised specimens can be attributed to the consumption of the clay minerals in the lime-clay reactions responsible for the strength development. In the case of lime stabilised soils, both the clay minerals and the lime get transformed into new products such as CSH, CAH, etc. Therefore, there is not much free lime (calcium hydroxide) left for the re-carbonation.

The stabilised compressed earth samples ground for 12 min show specific surface in the range 6–7 m^2/g. The clay size fraction of the ground samples contains a mixture of the clay mineral particles as well as the ground sand and the silt particles. The clay size fraction increases with the grinding duration. In the case of lime stabilised samples, the increase in the clay size fraction with the grinding duration is mainly attributed to the increased percentage of the broken clay sized, sand and the silt particles, as the clay mineral is absent in the coarser particles (silt and sand) range (Fig. 16.3). Whereas in the case of the cement stabilised samples apart from the broken clay-sized sand and the silt particles, some entrapped clay particles in the coarser particle lumps (sand and silt size lumps) will also get released and adds to the total clay content of the ground sample.

16.6.3 Atterberg's Limits

The Atterberg's limits (liquid limit and plasticity index) of a soil are greatly influenced by the quantity and type of the clay minerals present. For the powdered samples from the stabilised compressed earth specimen, Fig. 16.7 shows the relationship between the liquid limit (LL) and the stabiliser content (lime and cement) and Fig. 16.8 shows the relationship between the plasticity index (PI) and the stabiliser content. The stabilised compressed earth samples show LL and PI values and hence are plastic. The Atterberg's limit values of the natural soil reduce when it is stabilised with the cement and the lime. There is a drastic reduction in the LL (56%) and PI (88%) values for the lime stabilised soil sample as the lime content goes from 0 to 15%. In the case of the cement stabilised samples, this reduction is 18% and 36% for LL and PI, respectively. In the case of lime stabilisation, the clay fraction in the stabilised soil mix greatly reduces due to the lime-clay reactions. This reduction in the clay content affects the Atterberg's limits of the resultant stabilised soil mixture. Much less reduction in the Atterberg's limits for the cement stabilised compressed earth mixes, when compared to the lime stabilised compressed earth samples, can be mainly attributed to the presence of the higher percentage of the clay minerals in the cement stabilised compressed earth samples. In the case of the cement stabilisation, a small percentage of lime is released during the cement hydration, and it can react with the clay minerals. Less reduction in the LL and the PI values for the cement

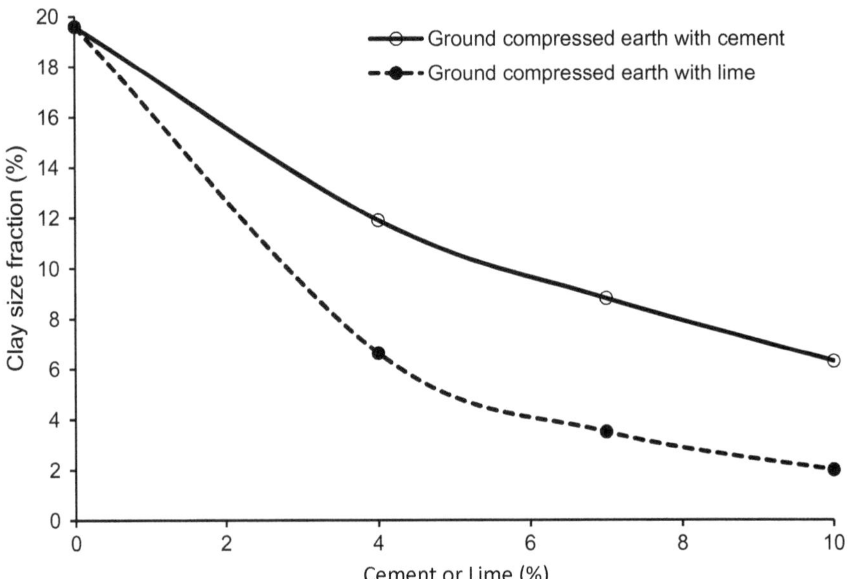

Fig. 16.7 Clay size fraction versus stabiliser content in compressed earth specimens (Reddy and Latha 2014, Copyright Elsevier)

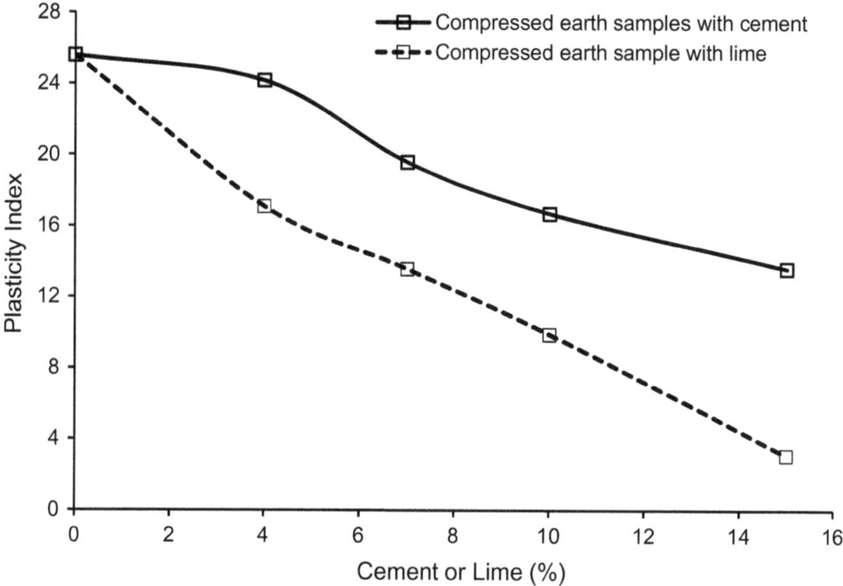

Fig. 16.8 Plasticity Index versus stabiliser content in compressed earth specimens (Reddy and Latha 2014, Copyright Elsevier)

stabilised compressed earth again substantiates the fact that considerable percentage of the clay minerals present in the original soil are still intact even after the cement stabilisation process in producing the CEB.

16.7 Summary

It is possible to retrieve a large percentage of the natural clay minerals in the cement stabilised compressed earth products. Whereas in the lime stabilised earth products, only a small fraction of the clay minerals which was present in the original soil can be retrieved, as the natural clay gets consumed in the lime-clay reactions. The quantity of unreacted clay mineral left in the lime stabilised earth product depends upon the quantity of lime used. At very high lime content (15%), practically no clay mineral is left in the stabilised earth product. At higher cement contents (15%), the retrievable residual clay content reduces drastically. The cement stabilisation appears superior from the point of view of retrieving back the precious clay minerals after the end of life of the building built with cement stabilised earth products such as CEB or rammed earth or stabilised adobe.

In the case of the lime stabilised earth, both the lime and the clay get transformed into products such as hydrates of silicates and aluminates. Hence, not much lime left for re-carbonation. The calcium hydroxide used in the lime stabilisation process also gets consumed in the lime-clay reactions. Therefore, the quantity of residual free lime (calcium hydroxide) left in the stabilised earth product is small, and hence, only a small fraction of the lime added is available for the re-carbonation process.

References

Bell FG (1996) Lime stabilization of clay minerals and soils. Eng Geol 42:223–237

Ian DR, Mackinnon PJR, Uwins AY, Page D (1993) Kaolinite particle sizes in the <2 micron range using laser scattering. Clay & Clay Miner 41(5):613–623

Latha MS (2015) Studies on characteristics of stabilised soil compacts for structural applications. PhD thesis, Department of Civil Engineering, Indian Institute of Science, Bangalore, India

Mehta PK, Monteiro PJM (2017) Concrete: microstructure, properties, and materials, 4th edn. McGraw-Hill Education

Monto M, Reddy BVV (2012) Sustainability in human settlements: imminent material and energy challenges for buildings in India. J Indian Inst Sci 92(1):145–162

Millogo Y, Morel JC (2012) Microstructural characterisation and mechanical properties of cement stabilised adobes. Mater Struct 45:1311–1318

Reddy BVV, Latha MS (2014) Retrieving clay minerals from stabilised soil compacts. Appl Clay Sci 101:362–368

Chapter 17
Energy, Carbon Emissions and Sustainability of Construction Materials and Buildings

17.1 Energy and Carbon Emissions

The production of construction materials requires two main resources: (1) raw materials and (2) energy. Bulk of the raw materials used in the production of the building materials are mined, and hence, the building products have heavy environmental costs. Energy is expended in the manufacturing processes and the transportation of building materials. The energy generation as well as consumption results in emissions and pollution.

The lifecycle energy (LCE) of a building comprises (a) embodied energy (EE), (b) maintenance or operational energy (OE) and (c) energy consumed in the demolition and the disposal. The energy in the demolition and the disposal is a very minor part of the LCE. Figure 17.1 shows the EE and OE share in the building systems. The major component of EE in a building is in the initial construction phase and has a smaller component during the buildings life cycle. After the initial EE investment, there will be a minor quantity of EE in the form of repairs and any additions to the building system during its life cycle. The OE in a building system is a recurring one and cumulative over its life cycle. The share between EE and OE in a building varies and considerably dependent upon the type of space conditioning adopted in the building and the local climate. In the life cycle of a building, the EE will be a small fraction of the OE in highly conditioned buildings located in extreme climatic conditions. Whereas in the passively conditioned buildings, the OE during the life cycle can be a small fraction of EE. In such situations, OE is mainly for lighting. Figure 17.2 illustrates few examples of the two situations examined by Praseeda (2014). These buildings are located in India across two different climatic zones (composite and moderate). The EE share in the LCE ranges between 8 and 80% in these cases.

There are different methods for the EE assessment of building materials: (a) process analysis, (b) input–output analysis and (c) hybrid analysis. The process analysis involves assessing the energy consumption based on the energy consumed in the production process. Accounting the energy consumed in both upstream activities and

B. V. V. Reddy, *Compressed Earth Block & Rammed Earth Structures*,
Springer Transactions in Civil and Environmental Engineering,
https://doi.org/10.1007/978-981-16-7877-6_17

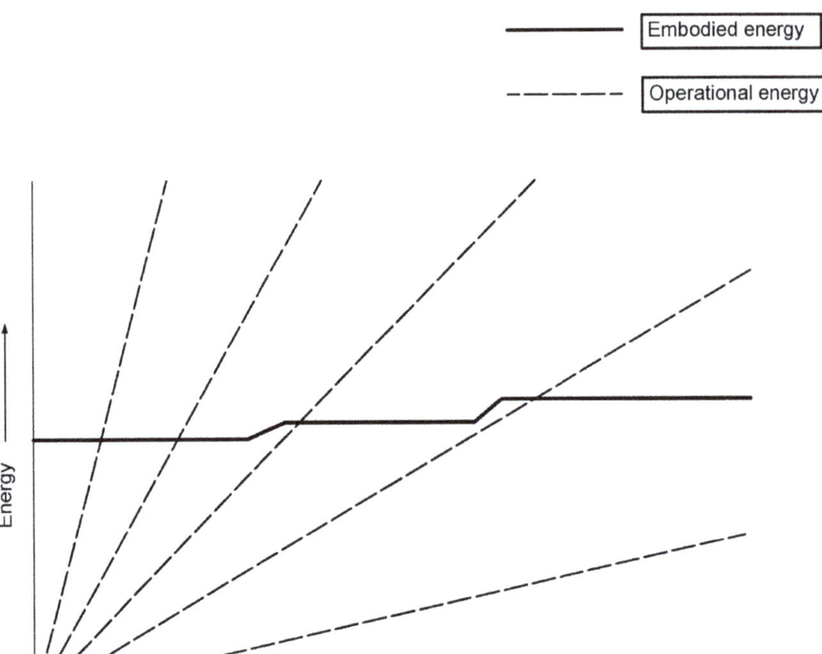

Fig. 17.1 EE and life cycle OE for buildings in India (Praseeda 2014)

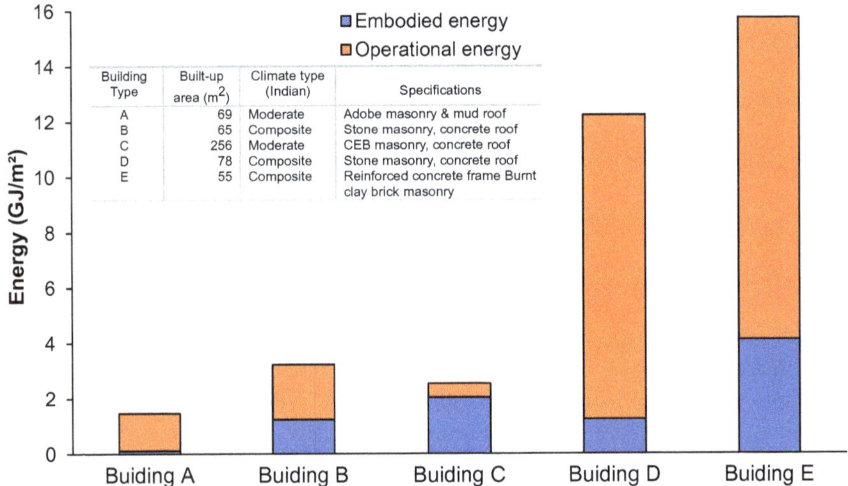

Fig. 17.2 Share of EE and OE in a building life cycle

the actual production processes, and considering a proper system boundary, the EE value of a material can be assessed. The input–output analysis is an indirect method of evaluating the EE of a material. This method demands input–output data Tables on energy consumed by each industrial sector at a national level and further bifurcating the data to a specific material produced in a sector. The assessment heavily hinges upon the accuracy of the data available in the input–output data Tables. The hybrid analysis attempts to combine both these methods to determine the EE value. For example, assessing the upstream processes energy consumption using input–output analysis and combining with the production process energy data from the process analysis. The EE value from the process analysis is sensitive to the system boundary considered. Majority of the EE analysis studies lack consensus on the assessment method and a comprehensive assessment framework. Another discrepancy will be in the energy assessment in terms of primary and end use energy. The primary energy accounts for the transmission losses and the fuel source for the energy generation. The EE of a product considering the primary energy will be higher when the electricity is used in the material production process (Praseeda 2014).

A framework for the EE assessment based on the process analysis is shown in Fig. 17.3. The embodied energy of a building consists of the energy consumed in: (a) the material production process (EE of material), (b) transportation of the product, (c) the construction process and (d) demolition, recycling and disposal. The conventional building materials such as cement, steel, burnt clay bricks, concrete, reinforced concrete, composites, etc. are energy intensive and responsible for the higher carbon emissions. There is a quest for developing low embodied carbon building materials.

17.2 Embodied Energy and Carbon Emissions in Stabilised Earth Products

The stabilised earth products such as compressed earth block (CEB) and rammed earth are produced using soil/earth, sand/aggregates and the stabiliser (cement/lime). The CEBs are used for the masonry, and hence, CEB masonry has two different materials: the CEB's and the mortar, whereas the rammed earth is a monolithic material. The EE and embodied carbon (EC) will differ in these two cases. Plots of EE and EC versus cement content (used in the mix used for CEB and cement stabilised rammed earth) for the CEB masonry and cement stabilised rammed earth (CSRE) are shown in Fig. 17.4. The EE and EC shown in the figure are based on the decentralised production systems for the CEB and the CSRE. The difference between the EE and the EC for the CEB masonry and the CSRE reduces towards higher cement contents. This is attributed to the constant value of EE and EC of the mortar, in the CEB masonry. The EE and EC of the CEB masonry and the CSRE are sensitive to the cement content.

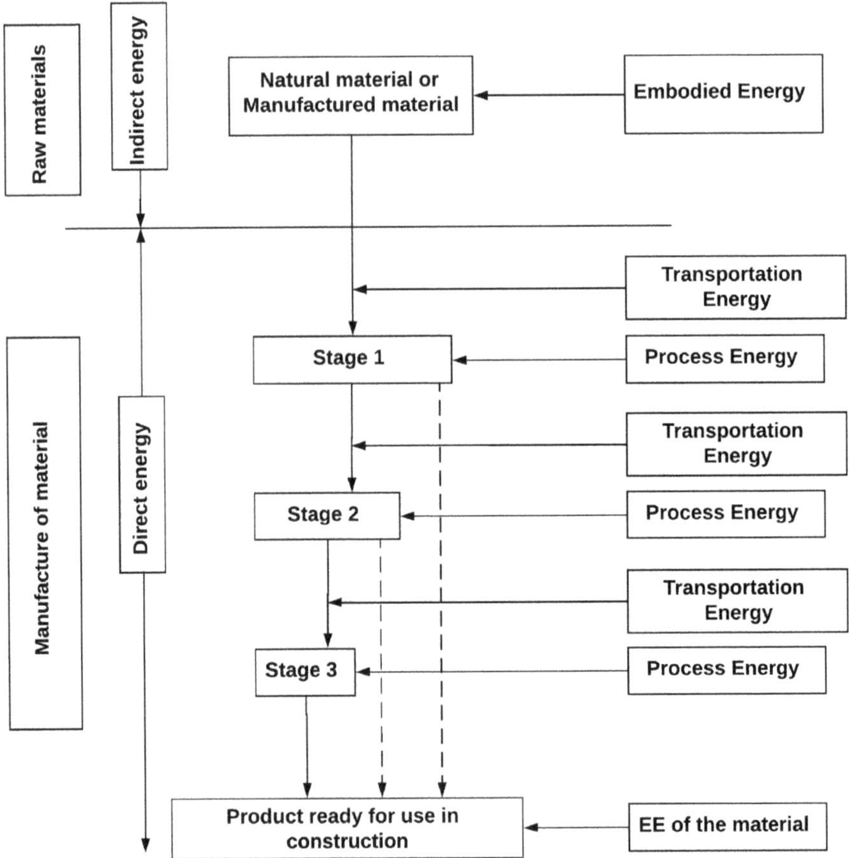

Fig. 17.3 Framework for EE assessment based on the process analysis (Praseeda 2014)

17.3 Embodied Energy and Carbon Emissions in Geopolymer and Lime-Pozzolana Stabilised CEB and Rammed Earth

The geopolymer and the lime-pozzolana are the alternative binders for the Portland cement and can be exploited for the production of stabilised CEB and rammed earth. The geopolymer basically consists of an alkali such as NaOH or KOH and a source for silica/alumina. The fly ash and the ground granulated blast furnace slag (GGBS) are the good sources for the alumina and the silica for the geopolymer binders. The characteristics of the geopolymer stabilised, lime-fly ash and lime-GGBS stabilised compressed earth products are discussed in Chaps. 9 and 10. The EE and the EC of a masonry include the EE and the EC of masonry units (bricks or blocks), and the EE and the EC of the mortar. Considering a composite mortar such as cement-soil mortar of proportion 1:2:5 (cement:soil:sand, by volume), the range of EE and EC

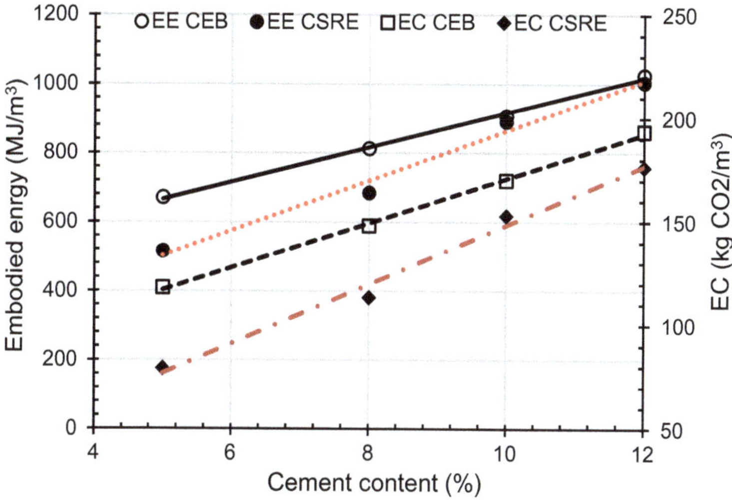

Fig. 17.4 Embodied energy and embodied carbon in CEB and CSRE

values for the masonry using the following alternative masonry units is compared in Figs. 17.5 and 17.6, respectively. The figures also show the range for the EE and the EC values, for the cement stabilised rammed earth (CSRE). The details of the masonry designations shown in these figures are as follows.

(a) Geo-FA brick masonry: Fly ash-based geopolymer brick masonry
(b) Geo-GGBS brick masonry: GGBS-based geopolymer brick masonry
(c) FAL-G brick masonry: Fly ash—lime—gypsum brick masonry
(d) CEB—masonry: Cement stabilised compressed earth brick masonry
(e) CSRE: Cement stabilised rammed earth

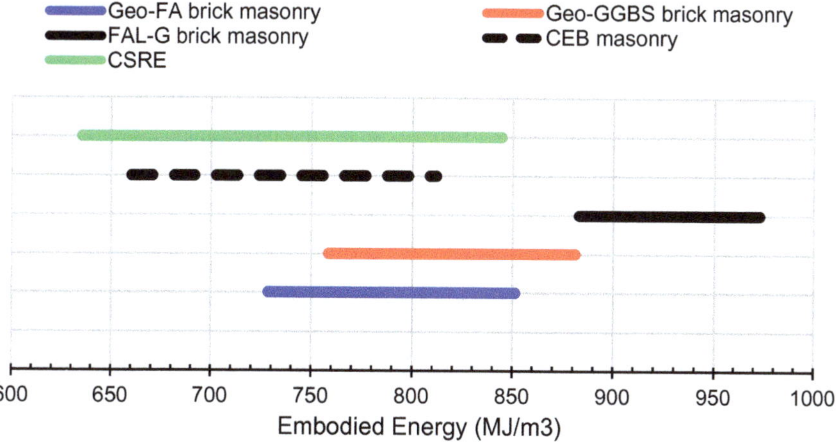

Fig. 17.5 Comparison of embodied energy (EE) in different types of masonry

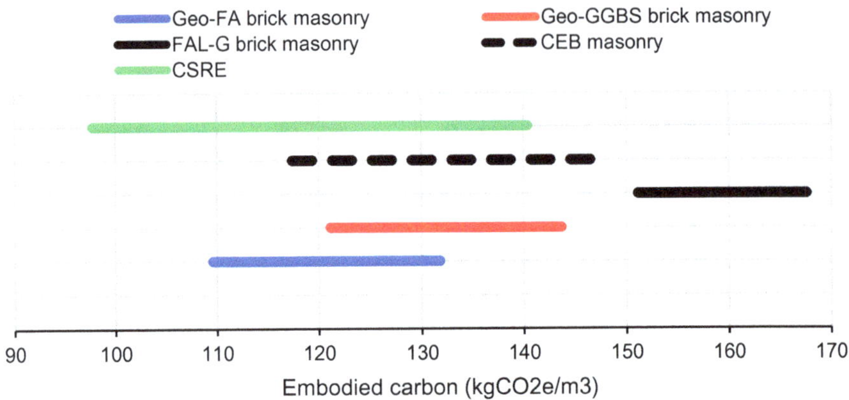

Fig. 17.6 Comparison of embodied carbon (EC) in different types of masonry

Figures 17.5 and 17.6 show the range of values for the EE and the EC of the masonry. The EE and EC for geopolymer bricks using fly ash, GGBS and alkali (NaOH) depend upon the mix propositions. Considering 15% fly ash and 15% GGBS, the molarity of NaOH has been varied between 6 and 12 N. Similarly, for the FAL-G brick, the lime content varies between 9 and 12%. In the case of CEB and CSRE, the cement content ranges between 5 and 8%.

The range for EE is maximum for FAL-G brick masonry (900–1000 MJ/m^3). The geopolymer brick masonry has EE in the range of 700–900 MJ/m^3. The cement stabilised CEB masonry and CSRE possess EE in the range of 600–800 MJ/m^3. Similar trends follow for the EC in the masonry using these alternative construction methods (100–1750 kgCO_2e/m^3).

17.4 Embodied Energy and Carbon Emissions in Buildings

The EE and EC in buildings greatly depend upon the types of materials used in the envelope, floors, openings, interiors, exterior finishes and other components of the structure. The range of EE and EC values for the buildings using conventional and alternative materials is given in Table 17.1. The values given in the Table were extracted from the literature (Fernando et al. 2018; Victoria et al. 2015; Praseeda et al. 2016; Reddy et al. 2020) and computed from the actual constructions. The conventional reinforced concrete framed structures show high EE and EC values, followed by conventional burnt clay brick masonry buildings and the buildings using alternative construction materials.

Table 17.1 EE and EC values for buildings

Sl. No	Type of building	EE (MJ/m^2)	EC ($kgCO_2e/m^2$)
1	Residential buildings with RC frame and infill masonry, and monolithic RC construction	4.5–12.0	400–600
2	Burnt clay brick masonry buildings	2.0–3.0	250–350
3	Cement stabilised CEB, geopolymer and lime-pozzolana binder bricks, and stabilised rammed earth buildings	1.5–2.0	150–250

17.5 Sustainability

The environment, the economy and the society represent the three dimensions of sustainability (Mani et al. 2005). The economy and the society dimensions of sustainability ultimately lead to the consumption of material and energy resources. The society's development needs economic resources in order to meet the energy and material resources. Extraction of materials and transforming into products/distribution necessitates energy expenditure. All these processes ultimately end up in emission, pollution and depleted resources, where the environment gets into picture. Thus, addressing any dimension of sustainability ultimately end up in material and energy resource consumption. In the context of sustainable construction, sources for the materials and the energy can be either from a renewable or non-renewable source.

17.6 Construction Materials

The construction practices have evolved through ages. The earliest building construction practices were mainly dependent upon the natural materials such as stones, soil and biomass. There was hardly any thermal energy expenditure in the production of natural building materials. The natural building materials have zero embodied carbon and at the end of the life, can go back to their native state without causing much irreversible environmental damage. The problems associated with the durability of majority of the natural materials resulted in devising methods to produce durable building materials. The processes used in producing durable building materials result in irreversible changes to the natural materials and such changes cause irreversible damage to the natural materials and alteration/disruption to the prevalent ecosystem services. In such processes, both energy and environmental costs are involved.

Table 17.2 highlights the types of building materials used from the historical period to the present day (Reddy 2004). Prior to 4400 BC, the building construction was mainly dominated by the use of the natural materials. The earliest example of thermal energy usage in the production of building materials can be seen in the manufacture of durable materials such as the burnt clay brick (4400 BC). The use of lime and

Table 17.2 Developments in building materials

Type of material	Period of development	Energy	
Natural materials • Stone • Soil • Adobe brick • Biomass	Prior 4400 BC	Zero embodied energy materials	Zero Embodied Carbon
Burnt clay brick	4400 BC	Medium embodied energy materials	↓
Lime	4000 BC		
Glass	3500 BC		
Use of iron	3000 BC		
Glassware	1500 BC		
Pozzolana	500 BC		
Lime-pozzolana	300 BC		
Modern materials • Bakelite • Aluminium and other metals • Portland cement • Plastics, polymers • Smart/nanomaterials	After 1500 AD	High embodied energy materials	High Embodied Carbon

iron can be seen during 4000–3000 BC. The first glassware products appeared in 1500 BC. The period of 500–300 BC represents the use of lime-pozzolana. The R&D on cementitious binders paved the way for the invention of Portland cement in 1824. The Portland cement and the steel brought revolutionary changes in the construction practices since the early part of the twentieth century. Later on, the plastics, nanomaterials, special alloys, smart materials, etc. entered the construction industry.

Hardly any thermal energy was expended in the use of natural building materials pre-4400 BC period. Such natural materials can be designated as zero energy or zero embodied carbon materials. The later period till the industrial revolution (4400 BC–1800 AD) saw the use of thermal energy for the manufacture and use of the building materials. Some of these materials include the metals, burnt clay bricks, lime and lime-based products, etc. These materials possess some amount of embodied carbon and can be classified under the medium-energy-consuming materials. The modern era (after 1800 AD) saw proliferation of the manufacture and use of energy-intensive building materials (Portland cement, glass, steel, aluminium, plastics, nano materials, smart materials, etc.) with the industrialisation and the standardisation resulting in loss of diversity, skills, etc. Thus, we moved away from the zero embodied energy materials to more modern, energy-intensive materials for the construction activities. It became imminent to spend more energy, emitting more carbon and consuming more natural resources by the construction sector. The global carbon emissions increased phenomenally from about 9 million tonnes to 36 billion tonnes during 1750–2019 (www.ourworldindata.org).

The modern materials (listed in the Table 17.2) are energy intensive and are hauled over long distances before being used for the construction. In the context of global warming and the carbon emission reduction, there is a need to pay attention to (i) energy intensity/embodied carbon of the materials, (ii) consumption of non-renewable natural resources and raw materials, (iii) recycling and safe disposal of materials at the end of life and (iv) impact on the environment. Use of indiscriminately mined natural resources and the energy-intensive processes in the manufacture of the construction materials will not lead to the sustainable construction options. Certain issues pertaining to the mining of material resources, energy, carbon emissions, rating systems and sustainability of the construction practices are discussed in the following sections. Also, the potential of non-organic solid wastes and the by-products in the manufacture of the construction materials, and in mitigating/augmenting the demand for mined raw material resources has been demonstrated.

17.7 Sustainability and Greenness

The sustainability is associated with renewability or regeneration. With reference to the sustainability of the construction sector, it refers to the use of material resources in a renewable fashion. There are several definitions for sustainable development. An oldest definition for sustainable development can be found in Kumarappa's (1945) book entitled "Economy of Permanence". The Brundtland report (1987) provides the most quoted definition on sustainable development. The two definitions are highlighted in the Table 17.3.

The sustainable society is the one which manages its economic growth while keeping in focus the environment and the needs of the future generations (Kumarappa 1945). A similar definition on sustainable development has been stated in the Brundtland report (1987). The prime focus in both these definitions is on sustainable extraction of resources from the planet, without causing irreparable damage to the environment. The construction industry is greatly dependent upon the use of two important

Table 17.3 Definition of sustainable development

Kumarappa (1945)	Brundtland (1987)
Sustainable society: • Manages its economic growth without causing irreparable damage to the environment • Satisfies people's needs without jeopardizing the prospects of future generations	Sustainable development: • Meeting the needs of the present without compromising the ability of future generations to meet their own needs

resources: the materials and the energy. The raw materials and the energy become essential for manufacturing the construction products, the transportation of the materials/products and the construction processes. The raw materials extracted from the earth are processed into construction materials and then used in the construction processes. The energy is expended in the transportation of materials (raw and manufactured), manufacturing of construction products, the construction processes and the disposal at the end of life. The building life cycle encompasses consumption of the energy throughout its active life and the end of life disposal.

The greenness is about decarbonisation. It refers to the changing relative amounts of carbon and hydrogen in the fuels burnt to generate energy (Bradford 2006). There are different types of fuels. The carbon-to-hydrogen ratio in some of the fuels is given in Table 17.4. The firewood has the highest carbon content as compared to other fuels. When these fuels combust, the carbon and hydrogen in the presence of oxygen are responsible for releasing the energy. Only the carbon part of the fuel is responsible for undesirable emission (carbon dioxide, carbon monoxide, etc.) problems.

Typical examples for moving towards greenness can be seen in the manufacture of burnt clay and ceramic products (bricks, floor tiles, etc.), and in the space heating in residences. Initially, the burnt clay/ceramic products were manufactured using solid fuels (biomass, coal) and slowly moved towards liquid and gaseous fuels, electricity (derived from renewables like PV, wind, etc.) and hybrid systems, where carbon emissions are minimised (decarbonisation). Apart from moving towards hydrogen-rich fuels, there were marked improvements both in the efficiency and emission control associated with the burning processes. Similarly, the space heating in the buildings (in colder climates) saw revolutionary changes, starting from biomass-based heating systems to gaseous fuel, electricity and solar-based systems. Also, there are considerable improvements in the insulation systems used in the building envelopes.

The greenness is all about emission reduction. There is an attempt to move towards the low carbon emission or zero carbon emission while using fuels for energy generation. Improving the energy efficiency of the production/transportation systems and the heating/cooling systems in the built environment does not address the issues pertaining to the mining of the material resources and the material resources depletion. The sustainable construction should address both the greenness and the material resource consumption.

Table 17.4 Carbon-to-hydrogen ratio in fuels (Bradford 2006)

Type of fuel	Carbon	Hydrogen
Firewood	10	1
Coal	2	1
Oil	1	2
Natural gas	1	4
Hydrogen	0	1

17.8 Material and Energy Resources Consumption

17.8.1 Material Resources

The fundamental sustainability challenge for the construction sector and in particular for the buildings lies in managing the material/energy resources and the wastes generated. This implies conservation of non-renewable/mined resources and efficiently utilising renewable resources. Particularly, in the context of future generations, consideration should be given to the usage of three different types of planetary assets (UNCHS 1990).

(a) Finite stock of non-renewable material resources (fossil fuels, minerals, soil, stone, ore, coal, etc.).
(b) Renewable resources (solar energy, wind energy, forest produce, biomass, etc.).
(c) Capacity to absorb pollutants/by-products of human development (e.g. toxic chemicals, plastics, chlorofluorocarbons).

The planet is a host for several living organisms, of which humans form a fraction. The human societies occupy ~2% of the planet's surface area but consumes 75% of the planet's resources (O'Meara 1999).

The manufacture of the construction materials requires two essential resources: the raw materials and the energy. The raw materials include soil, stone, sand, varieties of minerals and chemicals, and biomass apart from the water. Except the biomass and the water, all the other raw materials are limited in quantity and are mined. Hence, these materials are exhaustible. The global consumption of material resources has been highlighted in Table 17.5. The annual global consumption of construction materials is in excess of 60 billion tonnes, which amounts to per capita annual consumption of about 8 tonnes. The bulk of these materials are mined from the planet earth. It is alarming to note that aggregates used in the construction sector account for 80% of the total construction materials consumed. The aggregates are mainly derived from mining the river beds and crushing of the rocks. The aggregate

Table 17.5 Global material resources consumption

Type of product	Annual production/consumption		Source
	Total (billion t)	per capita (t)	
Crude steel production (2017)	1.69	0.225	WSA (2018)
Food grains production (2017)	2.62	0.349	www.statista.com
Cement production (2017)	5.50	0.733	www.worldcement.com
Aggregates (2015)	48.30	6.541	www.concreteconstruction.net
Plastics (2015)	0.32	0.044	https://theconversation.com
Burnt clay bricks	4.50	0.592	https://cdn.cseindia.org
Ceramic tiles (2016)	0.16	0.022	http://www.materialicasa.com

extraction has severe burden on the environment. The exploitation of natural stone for the aggregates can wipe out the rocky outcrops. This will further stress already dwindling natural resources and biodiversity. The cement consumption is double that of food grain consumption. The per capita annual burnt clay brick consumption touches 0.6 tonnes. The burnt clay brick production consumes precious fertile soil, which supports agriculture and forestry.

The major problems with the consumption of the raw materials for the manufacture of construction products are permanent changes occurring to the mined raw materials. The basic raw materials undergo permanent structural or physical changes during the processing and the manufacture of the construction materials. One typical example is the case of the burnt clay products. The burnt clay bricks/blocks, terracotta and ceramic products represent some of the burnt clay building materials. Generally, soils with high clay content (fertile soils) are used to produce such building products. In such products, clay minerals present in the soil undergo structural changes during the firing process (800–1000 °C) resulting in the formation of water insoluble bonds. The transformed clay minerals are the binding materials in such products. The soil and the clay minerals are formed due to the weathering of the rocks over a period of millions of years. Considering the geological time scales, the soil reserves on the planet earth are limited. The soils support the plant life, and in turn, the entire ecosystem on the earth is dependent on the plant life. The fertile soil when used for the manufacture of the burnt clay products, gets transformed into a nearly a rock form. Recovering clay minerals from the burnt clay product again needs millions of years of natural weathering. Therefore, depletion of soil resources can threaten the plant and the animal life on the earth.

17.9 Rating Systems and the Material Resources Consumption

The buildings consume material and energy resources, and generate wastes, causing pollution. There are many green building rating systems, mainly addressing the following aspects.

- Site planning, location and the linkages—sustainable site development
- Design, materials and the construction aspects
- Water and waste management
- Energy consumption and generation—energy conservation
- Indoor environment quality and space conditioning
- Healthy living conditions
- Awareness and education.

The rating systems aim to quantify the energy and the material resources consumption in the buildings, ultimately encouraging pollution mitigation and waste management. The procedures adopted in rating the buildings are biased towards quantification of energy consumption and energy generation in a building system. The weightage assigned to different aspects of buildings in the rating process using few selected green rating systems is highlighted in the Table 17.6. The weightage assigned to the materials and the construction methods is in the range of 6–13.5%, whereas the weightage for energy related issues is in the range 39–57%.

Based on the data presented in Table 17.6 and analysis of the rating systems, the limitations of the rating systems can be listed as follows.

- Rating systems are efficiency based—not effectiveness based. Advocate the use of energy efficient devices, but its effective utilization is not questioned.
- Appreciate the use of the local materials, but their significance on rating is inadequate. They are unquantified and no scientific basis for evaluation.
- Irrational comparison between building materials and design, i.e. evaluation compared based on the performance of the material (e.g. insulations or finishes, viz. bamboo & aluminium/glass placed in the same category).
- Zero energy can be high energy. The embodied/process energy and delayed (energy) payback are not accounted properly. High O&M implications (e.g. Solar PV) not highlighted.
- The energy in buildings encompasses embodied energy, transportation energy, energy for construction, energy for operation and maintenance, and finally energy for the demolition and recycling/disposal (Sengupta 2008; Praseeda et al. 2015). These different aspects of energy expenditure lack clarity in the rating systems.

The current rating systems lay too much emphasis on the energy conservation and the pollution reduction. The rating systems attempt to link the concept of green buildings to sustainable construction. There is little or less emphasis on (a) conservation of dwindling basic material resources and (b) environmental damage due to indiscriminate mining of materials resources from the planet earth. The construction sector is surviving on the mined material resources, anything mined is un-sustainable; therefore, sustainable construction in the current scenario is a mirage.

Table 17.6 Weightage assigned in different rating systems

Parameter	Weightage		
	LEED-USA (%)	BREEAM-UK (%)	GRIHA-India (%)
Materials and construction methods	6–9	13.5	10
Energy (consumption/generation, Indoor environment quality, space conditioning, etc.)	57	39	50

17.10 Raw Materials Extraction/Management

The utilisation of mined raw materials from the planet earth for the construction can never lead to the construction sector becoming sustainable. The over exploitation of the raw material resources and the widespread use of energy-intensive materials can drain the energy and material resources and can adversely affect the environment. Some of the possible options for addressing the issues on depleting materials wealth due to the manufacture of the construction materials are as follows.

(a) Use the materials judiciously and completely exploit the potential of the material.
(b) Effect minimum changes to the natural materials during the production processes, such that the discarded materials can easily go back to their native state with minimum environmental costs. A simple example is the use of earth-based materials where the soil is not fired/burnt, thereby conserving the natural clay minerals in the soil. Such products when discarded after the end of the life of the product can easily be recycled as natural soil.
(c) Recycle the solid wastes and the by-products into the construction products. The industrial and mining activities generate huge quantities of non-organic solid wastes or by-products.
(d) Encourage the use of the construction products from the renewable materials. The biomass is a renewable and carbon neutral resource. It is available in woody and non-woody forms. Traditionally, biomass in various forms has been successfully used for the construction of buildings. Processed timber, glue laminated timber/bamboo, plywood panels, wood-based composite panels, bamboo mat, bamboo and wooden poles, etc. represent different forms of woody biomass used in the construction sector. There are limited applications (such as straw bale, fibre reinforced products) or products of non-woody biomass in the construction. There is a great potential for developing structural materials using non-woody biomass as well as agro-residues. The biomass-based construction materials are the real renewable and green materials. Such materials can be grown without threatening the food security. There is a need for utilising the non-woody biomass for the manufacture of the construction materials for both the structural and the non-structural applications.

17.11 Recycling Solid Wastes and By-Products into Building Products

The construction industry heavily depends upon the mined raw materials, and hence, the depletion of the material resources is imminent. The construction sector is growing at an alarming rate in China, India and many other countries. Meeting the demand for the construction products encourages mining of the raw material resources. Therefore, it becomes essential to explore alternative resources to satisfy the demand for the raw materials. In this context, recovery and recycling of non-organic solid wastes (NOSWs) become essential. Large quantities of non-organic solid wastes are being generated by the industrial and mining activities. The annual global production of the solid wastes is in excess of 12 billion tonnes (Asokan et al. 2007, 2009). These include pulverized fuel ash, mine tailings, coal mine wastes, slag, marble dust, kiln dust, red mud, construction and demolition wastes, etc. There are huge quantities of accumulated NOSWs over several decades (wastes from thermal power plants, coal mine wastes, metal ore tailings of several mines, etc.). The recycling of NOSWs into construction products can mitigate the pressure on the depleting raw materials resources due to the mining. The non-organic solid wastes can become the resources in future for the manufacture of the construction materials across the globe.

17.12 Sustainability of Green Buildings

The buildings need material and energy resources. The mass of the urban dwellings in India is in the range of 3–4 t/m^2 (Praseeda 2014). The embodied energy is in the range 1.2–11 GJ/m^2 (Praseeda et al. 2016). The operational energy (OE) in the buildings differs widely in the conditioned and the unconditioned buildings. The range for OE is 0.50–59 GJ/m^2 per annum (Praseeda et al. 2016). The sustainability issues of the buildings are mainly influenced by the energy and material resources consumption and can be pictorially visualized as shown in Fig. 17.7. The energy expenditure in the buildings is associated with embodied and operational energy.

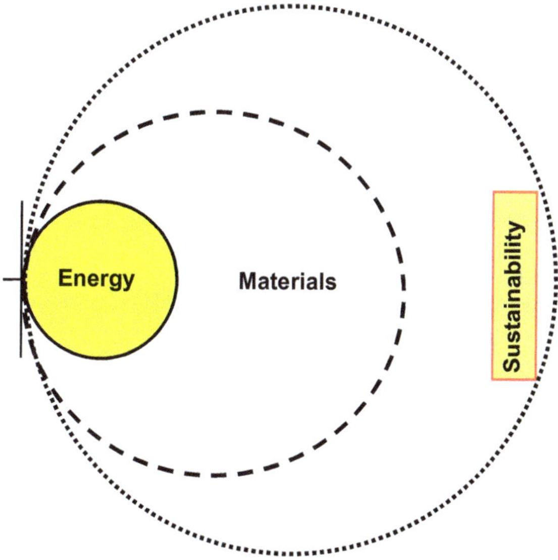

Fig. 17.7 Share of sustainability parameters in buildings

The former one is a one-time investment, and the latter one is a recurring one. Both these forms of energy in the buildings constitute life cycle energy (LCE). The green building concepts mainly focus on the LCE aspects. The conservation of energy and minimising energy consumption in the buildings addresses only a small component of the sustainable building. The major issue needing attention in sustainable buildings or sustainable constructions is the use of mined and non-renewable material resources. Therefore, all green buildings need not be sustainable options. The real sustainable building/construction is the one which uses renewable materials/renewable energy or passively conditioned systems, having least impact on the environment.

17.13 Summary

There are successful attempts in evolving alternative construction materials having low embodied carbon. The sustainability of construction sector should encompass two major issues: the energy and the material resources. The annual per capita consumption of construction material is alarming at 8 tonnes. The lion's share of it is from the aggregate's consumption, and the construction materials hailed from the mined resources from the planet earth. The green building rating systems address only a part of the sustainability aspect, with very little emphasis on the consumption of dwindling mined material resources. In the business as usual scenario, the construction sector is surviving on the mined materials, which is unsustainable. Due attention should be paid in evolving the renewable construction materials and the recycling of the solid wastes.

References

Asokan P, Saxena M, Asolekar SR (2007) Solid wastes generation in India and their recycling potential in building materials. Build Environ 42:2311–2320

Asokan P, Saxena M, Shyam R, Asolekar S, Anusha S (2009) Cross sector recycling opportunities. In: Proceedings of the international seminar on waste to wealth, 12–13 Nov, New Delhi, India

Bradford T (2006) Solar revolution: the economic transformation of the global energy industry. The MIT Press Cambridge, Massachusetts

BREEAM—UK. bream@bre.co.uk

Brundtland Report (1987) Report of the World Commission on Environment and Development—our common future, United Nations

Fernando NG, Victoria MF, Ekundayo DO (2018) Embodied carbon emissions of buildings: a case study of an apartment building in the UK. In: The 7th world construction symposium 2018: 29 June–01 July 2018, Colombo, Sri Lanka

GRIHA India, The national rating system for green buildings. Association for Development and Research of Sustainable Habitats (ADaRSH), The Energy and Resource Institute (TERI), New Delhi, India

http://www.materialicasa.com/file/Home/materialicasa/pdf/tile-international/2017/3/042_049%20Statistic%20PROD%20CONS%20Mondiale.pdf

https://cdn.cseindia.org/docs/photogallery/slideshows/TP-cleaner-brick-production-20171211-15-Overview-of-Brick-Kiln-Sector-Environmental-issues-Nivit-Kuma.pdf
https://theconversation.com/the-world-of-plastics-in-numbers-100291
https://www.concreteconstruction.net
https://www.ourworldindata.org
https://www.statista.com
https://www.worldcement.com
Kumarappa JC (1945) Economy of permanence. Sarva-seva-sangh Prakashan, Varanasi, India
LEED-USA, U.S. Green Building Council, Washington DC, USA. https://new.usgbc.org/leed
Mani M, Ganesh LS, Varghese K (2005) Sustainability and human settlements. Sage Publications, New Delhi, Thousand Oaks, London
O'Meara (1999) Reinventing cities for people and the planet. Worldwatch Institute, Washington, D.C., p 147
Praseeda KI (2014) Studies into embodied and operational energy in traditional and conventional residential buildings in India. PhD thesis, Department of Civil Engineering, Indian Institute of Science, Bangalore, India
Praseeda KI, Venkatarama Reddy BV, Mani M (2015) Embodied energy assessment of building materials in India using process and input–output analysis. Energy Build 86:677–686
Praseeda KI, Venkatarama Reddy BV, Mani M (2016) Embodied and operational energy of urban residential buildings in India. Energy Build 110:211–219
Reddy BVV (2004) Sustainable building technologies. Curr Sci 87(7):899–907
Reddy BVV, Ullas SN, Sugantha M, Mani M (2020) Report on "Low-C buildings under the project Centre for Bio-Energy and Low carbon Technologies (C-BELT)". Centre for Sustainable Technologies, Indian Institute of Science, Bangalore, India
Sengupta N (2008) Use of cost-effective construction technologies in India to mitigate climate change. Curr Sci 94(1):38–43
UNCHS (1990) People, settlements, environment and development: improving the living environment for a sustainable future. Intergovernmental meeting on Human Settlements and Sustainable Development, The Hague
Victoria MF, Perera S, Davies A (2015) Developing an early design stage embodied carbon prediction model: a case study. In: Raidén AB, Aboagye-Nimo E (eds) Procs 31st annual ARCOM conference, 7–9 September 2015. Lincoln, UK, pp 267–276
WSA (2018) World steel association. World steel in figures, Brussels, Belgium

Index

B. V. V. Reddy, *Compressed Earth Block & Rammed Earth Structures*, Springer Transactions in Civil and Environmental Engineering, https://doi.org/10.1007/978-981-16-7877-6

D

X

Z

www.ingramcontent.com/pod-product-compliance
Ingram Content Group UK Ltd.
Pitfield, Milton Keynes, MK11 3LW, UK
UKHW021008290726
14059UKWH00001BA/19